Development and Management of Eco-Conscious IoT Medical Devices

Chich-Jen Shieh
Huaqiao University, China

Vamsi Krishna Reddy Bandaru
Cisco, USA

Prashant Upadhyay
Sharda University, India

Pranav Dass
University of Delhi, India

Monica Bhutani
Bharati Vidyapeeth College of Engineering, New Delhi, India

IGI Global
Scientific Publishing
Publishing Tomorrow's Research Today

Vice President of Editorial	Melissa Wagner
Director of Acquisitions	Mikaela Felty
Director of Book Development	Jocelynn Hessler
Production Manager	Mike Brehm
Cover Design	Jose Rosado

Published in the United States of America by
IGI Global Scientific Publishing
701 East Chocolate Avenue
Hershey, PA, 17033, USA
Tel: 717-533-8845
Fax: 717-533-7115
Website: https://www.igi-global.com E-mail: cust@igi-global.com

Library of Congress Cataloging-in-Publication Data

Names: Sheih, Chich-Jen, 1956- editor | Upadhyay, Prashant, 1991- editor | Bhutani, Monica, 1983- editor | Bandaru, Vamsi Krishna Reddy, 1998- editor | Dass, Pranav, 1983- editor
Title: Development and management of eco-conscious IoT medical devices / edited by Chich-Jen Sheih, Prashant Upadhyay, Monica Bhutani, Vamsi Krishna Reddy Bandaru, Pranav Dass.
Description: Hershey, PA : IGI Global Scientific Publishing, [2026] | Includes bibliographical references and index. | Summary: "This book aims to analyze the role of IoT in advancing sustainable healthcare by examining energy-efficient designs, e-waste reduction, and carbon-neutral solutions while addressing ethical and regulatory challenges"-- Provided by publisher.
Identifiers: LCCN 2025022196 (print) | LCCN 2025022197 (ebook) | ISBN 9798337341347 hardcover | ISBN 9798337341354 paperback | ISBN 9798337341361 ebook
Subjects: LCSH: Medical instruments and apparatus--Technological innovations | Internet in medicine | Internet of things | Artificial intelligence--Medical applications
Classification: LCC R856 .D478 2026 (print) | LCC R856 (ebook)
LC record available at https://lccn.loc.gov/2025022196
LC ebook record available at https://lccn.loc.gov/2025022197

British Cataloguing in Publication Data
A Cataloguing in Publication record for this book is available from the British Library.

Table of Contents

Detailed Table of Contents

Sustainable Internet of Things (IoT) technologies integrated in healthcare systems create a new way to deliver medical services by resolving significant environmental problems. The development of sustainable IoT technology for healthcare matters is analyzed in this chapter through medical sector trends and relevant innovations which deeply affect healthcare delivery systems. This section presents an extensive investigation into medical IoT devices starting from their historical development while focusing on eco-friendly solution adoption and the fusion between IoT and AI as well as edge computing infrastructure and green energy platforms. The book presents real-world applications with their use cases and examines environmental and operational difficulties which affect the industry. Researchers explore future directions and innovative approaches which will drive sustainable healthcare systems based on resiliency and intelligence in the chapter's final section.

This chapter proposes a strategic definition of Smart Healthcare IoT solutions development and deployment strategies for the year 2030. It provides a roadmap which lays out policy recommendations, technological innovations, investment priority, collaborative frameworks. Global trends like climate change, aging populations, healthcare digitalization, drive the need for green and intelligent systems that are the subject matter of the chapter. It combines analysis of current gaps, as well as projections of future demands to provide an overall roadmap to stakeholders looking to create resilient, green, fair digital health ecosystems in line with South Africa's sustainable development goals.

Environmental impact, depleting resources, and the production of wastes pose significant challenges to the healthcare industry. This chapter explores how a healthier, more sustainable healthcare system can be constructed through the integration of the Internet of Things (IoT) and the principles of the circular economy. Enhanced patient outcomes, minimization of wastes, and the optimization of resources are all achievable for healthcare professionals who employ IoT technology for data collection, monitoring, and analysis. A closed-loop system minimizing environmental impact is developed by the circular economy model, which promotes the development of resilient, reusable, and recyclable medical devices and gadgets. How IoT and circular economy strategies have the potential to transform healthcare, minimize its ecological footprint, and enhance its sustainability in the long term is analyzed in this chapter.

In this chapter, the idea of the circular economy is proposed as a transformative strategy on how IoT based healthcare devices can be designed for sustainable use and disposal. However, as connected medical technology exponentially grows, CycleTimber's environmental and economic implications of devices all the way through to their lifecycles are emerging as an important issue. Models in the chapter strive towards resource efficiency, reuse, refurbishment and recycling to decrease its electronic waste while maintaining performance and safety standards. It provides case studies and emerging policy supports the ways a circular economy can help to aid in both ecological responsibility and long term cost savings in the healthcare sector.

The surge in the number of medical IoT devices has created big sustainability concerns esp. in terms of energy consumption, resource utilization, and electronic waste. Despite plenty of attention received by hardware optimization, the role of firmware and software in making affordable devices eco-friendly still lacks proper exploration. This chapter delivers an exhaustive study of the sustainable medical IoT applications of green software engineering and low power firmware design. It discusses such principles as energy-aware coding, modular architecture, and efficient memory control. Discussion on techniques such as real-time operating systems, interrupt-driven programming and lightweight communication protocols is included. The chapter also discusses the trade-off between the energy efficiency and data privacy and the need for consolidation of edge-cloud infrastructures in a way that will reduce computational burden. By pointing out best practices and future research trends, this chapter urges an end to a software-centric approach to creating sustainable healthcare technologies.

The integration of renewable energy sources with that of healthcare IoT systems provides a very powerful solution for serving and minimizing carbon emissions, enhancing energy resilience, and enhancing the feasibility of sustainable medical practice. The purpose of this chapter is to design and implement the models of IoT healthcare with renewable resources, such as the solar, wind and bioenergy. It researches in what way decentralized energy systems can power wearable sensors, diagnostic tools, remote monitoring devices and hospital infrastructure. Real world deployments, energy management techniques and the impact of energy harvesting technology in maintaining uninterrupted service are presented in the chapter. Through these models, it aligns healthcare innovation with clean energy principles to help create a clean energy and environmentally conscious healthcare product delivery.

The combination of Edge Artificial Intelligence (Edge-AI) with Internet of Things (IoT) medical devices is revolutionizing real-time health monitoring as it is made possible with low-latency energy-efficient, and privacy-preserving real-time data processing at the edge of the network. In this chapter we will outline the architecture and implementation of Edge-AI systems that are tasked with the work of continuous intelligent health monitoring. It dissects the key challenges (energy bottlenecks, model optimization, and secure data management), and offers state-of-the-art hardware and software tools. With use cases such as wearable ECG monitoring and fall detection, the chapter reveals how Edge-AI improves healthcare delivery and outcomes. It also calls out the future trends – like federated learning and explainable AI – that would bring a new era of proactive and personalized medicine.

There is a whole range of technical, regulatory, and ethical challenges facing recycling of medical IoT devices. This chapter identifies the barriers to the effective recycling of plastics posing a data security threat, materials for a complexity reason, and risks from contamination and a lack of standard process. It examines current practices of recycling and evaluates their viability in the healthcare application where patient safety and regulatory compliance take precedence. In addition, the chapter considers eco-design, modular architecture, and de/module/safe ways of dismantling that may help device management become more sustainable. Alignment of recycling efforts with the overall green healthcare and circular economy routes is proposed.

The prevalent trend of internet of things (IoT) devices adoption in the healthcare sector has created a massive concern about its environmental effect. Although medical IoT devices have huge benefits for patients and overall operational efficiency, they come at a cost to the environment, which could be missed in the supply chain, usage, and disposal stages. This chapter examines the environmental impact of medical IoT devices on the carbon footprint, energy consumption, e waste generation, and also discusses the impact of extraction of raw material. It points out the hidden environmental costs in the development of medical IoT technologies and tries to provide a framework to guide how to design, produce, and dispose more sustainably. The chapter further focuses on whether eco-friendly innovations can be emphasized in order to maximize data usage while minimizing energy and raw material consumption and eventually finding possible measures to reduce the environmental footprint of medical IoT devices to meet the world sustainability goals.

The remarkable rise of Medical Internet of Things (IoT) equipment has significantly changed healthcare services, providing real-time monitoring, diagnostics, as well as data-driven treatments. But, as with any mass-produced machine, the universally outsourced nature of the production of these devices has brought up huge ethical and transparency issues on where the raw materials are sourced from, the way in which labour is conducted, counterfeit components, and compliance with local regulations. In this paper, the authors are presenting the usage of the blockchain technology to set up an ethical and tamperproof transparent supply chain for the Medical IoT manufacturing. With based on the decentralised and immutable environment provided by blockchain, the framework guarantees full traceability of each part, from raw material to device assembly. Smart contracts facilitate verification of compliance and supplier audit Sn, and the consensus mechanisms grant data integrity requiring no central control.

Preface

The rapid evolution of technology has profoundly reshaped the healthcare landscape, offering unprecedented opportunities to improve patient care, enhance operational efficiency, and reduce environmental impacts. Among these technological innovations, the integration of the Internet of Things (IoT) in healthcare has emerged as a transformative force, enabling real-time monitoring, predictive diagnostics, and data-driven decision-making. However, the proliferation of IoT medical devices also presents critical challenges, particularly in terms of energy consumption, resource depletion, e-waste generation, and the ethical management of global supply chains.

This edited volume, *Development and Management of Eco-Conscious IoT Medical Devices*, brings together a diverse group of researchers, engineers, and industry experts to explore the intersection of sustainable development, IoT technology, and healthcare. The book emphasizes not only technological innovation but also environmental responsibility, circular economy practices, and ethical governance in the development and deployment of medical IoT devices. The chapters collectively offer a roadmap toward building resilient, energy-efficient, and eco-conscious healthcare systems that are capable of meeting both current and future demands.

CHAPTER OVERVIEW

Chapter 1: The Rise of Sustainable IoT in Modern Healthcare Through Trends, Technologies and Future Directions

Bandaru and Bandaru provide an in-depth analysis of sustainable IoT technologies in healthcare, tracing historical developments, examining eco-friendly innovations, and exploring AI and edge computing integrations. Real-world applications and environmental challenges are discussed alongside emerging trends that will shape future healthcare systems.

Chapter 2: Creating a Roadmap to 2030 for Achieving Sustainable Healthcare IoT

Pokkalla presents a strategic roadmap for Smart Healthcare IoT development through 2030. The chapter highlights policy recommendations, technological innovations, and investment priorities, emphasizing climate-conscious, equitable, and resilient digital health ecosystems.

Chapter 3: Sustainable Healthcare through IoT and Circular Economy

Tyagi et al. explore the synergy between IoT and circular economy principles, showing how data-driven monitoring and resource optimization can minimize waste and enhance sustainability in healthcare operations.

Chapter 4: Adopting a Circular Economy Approach to IoT Healthcare Devices

Narravula Reddy and Manikonda discuss designing IoT healthcare devices for sustainability, emphasizing resource efficiency, reuse, refurbishment, and recycling. Case studies illustrate ecological and economic benefits within the healthcare sector.

Chapter 5: Sustainable Health Tech Managing Eco-Conscious IoT Devices for Medical Innovation

Chaudhary et al. examine strategies to balance innovation with ecological responsibility, including energy-efficient design, lifecycle management, and circular economy practices, supported by practical case studies in hospital energy management and device design.

Chapter 6: Designing Resilient and Scalable Systems for Green Healthcare IoT

Gondi focuses on building IoT systems that are both environmentally sustainable and resilient to disruptions, detailing modular architectures, energy-efficient communication protocols, and renewable power integration.

Chapter 7: Green Firmware and Software Architectures for Sustainable Medical IoT Devices

Nettem et al. investigate software and firmware strategies for eco-friendly IoT devices, including energy-aware coding, modular architectures, and lightweight communication protocols, highlighting the trade-offs between efficiency, privacy, and performance.

Chapter 8: Developing Renewable Energy-Powered IoT Models for Sustainable Healthcare

Yellela and Narravula Reddy examine how decentralized renewable energy systems can power IoT healthcare devices, reducing carbon emissions and increasing operational reliability in hospitals, wearables, and monitoring systems.

Chapter 9: Edge-AI for Energy-Efficient Real-Time Health Monitoring in IoT Medical Devices

Busi et al. analyze the integration of Edge-AI with IoT devices to achieve low-latency, energy-efficient, and privacy-preserving health monitoring, with real-world applications such as wearable ECG and fall detection systems.

Chapter 10: Overcoming Challenges in Recycling Medical IoT Devices for a Sustainable Future

Manikonda and Yellela address technical, regulatory, and ethical barriers to recycling medical IoT devices, proposing eco-design, modularity, and safe dismantling practices aligned with circular economy principles.

Chapter 11: Unveiling the Hidden Footprint While Assessing the Environmental Cost of Medical IoT Devices

Bandaru et al. examine the hidden environmental costs of IoT devices across production, usage, and disposal stages, offering frameworks for reducing carbon footprints, energy consumption, and e-waste in sustainable medical technology design.

Chapter 12: Blockchain for Ethical Supply Chains – Transparency in Medical IoT Manufacturing

Parmar et al. present a blockchain-based framework to ensure ethical and transparent supply chains in IoT device manufacturing, enabling traceability, compliance verification, and data integrity without centralized control.

Chapter 13: AI and IoT in Global Health – Ethical Lessons from Pandemic Response

Pokhrel et al. reflect on the deployment of AI and IoT during the COVID-19 pandemic, assessing both successes and ethical challenges in global health monitoring, diagnostics, and service delivery, with insights into privacy, algorithmic fairness, and international governance.

By presenting interdisciplinary perspectives on sustainable, ethical, and technologically advanced healthcare IoT systems, this volume serves as a comprehensive guide for researchers, practitioners, and policymakers striving to develop environmentally conscious and socially responsible digital health solutions.

Chich-Jen Shieh
Huaqiao University, China

Prashant Upadhyay
Sharda University, India

Monica Bhutani
Bharati Vidyapeeth College of Engineering, New Delhi, India

Vamsi Krishna Reddy Bandaru
Cisco, USA

Pranav Dass
University of Delhi, India

Chapter 1
The Rise of Sustainable IoT in Modern Healthcare Through Trends, Technologies, and Future Directions

Vamsi Krishna Reddy Bandaru
https://orcid.org/0009-0001-9396-8250
Cisco, USA

Srinivas Reddy Bandaru
AIIMS, India

ABSTRACT

Sustainable Internet of Things (IoT) technologies integrated in healthcare systems create a new way to deliver medical services by resolving significant environmental problems. The development of sustainable IoT technology for healthcare matters is analyzed in this chapter through medical sector trends and relevant innovations which deeply affect healthcare delivery systems. This section presents an extensive investigation into medical IoT devices starting from their historical development while focusing on eco-friendly solution adoption and the fusion between IoT and AI as well as edge computing infrastructure and green energy platforms. The book presents real-world applications with their use cases and examines environmental and operational difficulties which affect the industry. Researchers explore future directions and innovative approaches which will drive sustainable healthcare systems based on resiliency and intelligence in the chapter's final section.

DOI: 10.4018/979-8-3373-4134-7.ch001

1. INTRODUCTION

Digital transformation has played a notable role in the rapid evolution of healthcare technologies of the 21st century. One of the innovations that stand out amongst the rest is the Internet of Things (IoT). Using IoT technologies, patient's data can be collected, transmitted and analysed in real time through interconnected sensors, devices and platforms, and this fosters personalized and proactive medical care. Yet, the concern about the environmental footprint of the IoT devices in healthcare remains as more and more such devices hit the market. Such demand has further compounded the complexity of sustaining healthcare development by pushing the demand of electronic components, the operational of data centers, and the energy needs of certain systems beyond what was even imagined just a few years ago (Krishnamoorthy, Dua, & Gupta, 2023).

Digital technologies increasingly have an environmental footprint, something that is becoming more and more known around the world in the last few years. Currently healthcare is seen as a highly resource intensive industry and is a large contributor to energy consumption and waste generation. But, when not used in a responsible way, the deployment of IoT in healthcare will worsen the existing sustainability problems. The energy efficiency, e-waste and lifecycle management of manufacturing and disposing of IoT devices is a concern. All these concerns have thus given rise to the idea of sustainable IoT (Nti, Adekoya, Weyori, & Keyeremeh, 2023).

Achieving sustainability is not just an environmental need but also a moral, economic and social need in healthcare. With increased pressure placed on health systems to improve service quality, increase accessibility, reduce costs, and decrease ecological footprint, there is a need for any possible particular point to ambulances to be reduced. Environmental degradation and climate change directly and indirectly affect public health, and the operation of healthcare infrastructure should be in synch with the environment.

In the context at hand, sustainable healthcare implies more than green buildings or sustainable operations—green issues. It covers the use of technologies and processes that cause or engage less harm to the environment and use fewer resources and which promote the health of the current and future generations. Stated simply, these goals can be achieved through sustainable IoT, which utilizes smart and interconnected technologies, conceived on an eco-friendly basis. Using energy efficiently, reducing waste, and improving the overall operational efficiency can provide remarkable contribution towards a more resilient and environment friendly healthcare ecosystem that is the sustainable IoT solutions (Kumar et al., 2023).

The dependency on IoT devices has increased exponentially because of the development of personalized medicine alongside continuous patient monitoring systems. The recent industry trend creates a favorable situation to reengineer these technologies toward long-term sustainability frameworks. Medical IoT devices attain sustainability through implementation of green materials and energy harvesting mechanisms and recyclable components that extend from design through to deployment phases.

The Internet of Medical Things known as IoT in healthcare revolutionizes medical health data collection through analysis and utilization processes. Medical IoT provides a transformative patient care model by enabling various devices spanning implantables to wearable trackers and smart systems for hospitals and remote monitoring platforms. Health tracking through real-time technology results in early medical diagnosis which produces better treatment results for patients. IoT technology enables operational system optimization and enhances error reduction as it optimizes resource utilization for hospitals (Mishra & Singh, 2023).

The Internet of Things enables healthcare delivery expansion toward rural and underserved areas through its vital role. Medical platforms built by IoT technology let doctors provide services from anywhere without needing their patients to physically move or build additional healthcare facilities. The recent COVID-19 pandemic and other public health emergencies demonstrated that IoT systems proved vital by enabling population tracking and patient remote checks as well as evidence-based healthcare decisions.

The undeniable advantages of IoT in healthcare lead to environmental trade-offs in healthcare applications. The extensive roll-out of sensors combined with wearables and connected devices needs a complete infrastructure which includes data centers as well as communication networks with power supply systems. The installed systems require significant energy usage and emit notable amounts of emissions during operation. Electric devices with short lifespans increase the fatal electronic waste problem which presents dangers to human health and environmental stability.

The research and development of sustainable principles for IoT design deployment and management emerged as an urgent priority to resolve these concerns (Bhutani, Lall, & Agrawal, 2022). The goal of sustainable IoT for healthcare is to achieve digital health service transformation without sacrificing environmental health. Energy-efficient IoT implementation combined with circular economy methods and environmentally progressive solutions enables healthcare to benefit from IoT technology while sustaining sustainability targets worldwide.

The chapter outlines detailed information about the role of sustainability concepts in the development of Internet-of-Anything advancements for healthcare systems. This paper examines the intersection between technological advancement and environmental awareness by studying recent patterns and enabling frameworks within possible sustainable development methods.

The main goals of this chapter consist of the following:

- This chapter establishes fundamental principles of sustainable IoT healthcare along with its definition as well as its application in contemporary medical environments.
- This section explores the history of Internet of Things technology adoption in healthcare systems while examining how sustainability matters became essential.
- The evaluation of new technological advancements enables recognition of applicable solutions which produce eco-friendly healthcare IoT systems.
- A discussion will be provided to illustrate real-world healthcare service transformations through sustained Internet of Things applications by showing their practical implementations and notable achievement examples.
- This research assesses both technological and economical and regulatory obstacles that hinder mass acceptance of sustainable IoT practices.
- The authors will provide research opportunities and future directions that help stakeholders establish sustainable IoT healthcare systems.

The chapter supports existing knowledge that helps developers and other stakeholders build IoT systems which are both environmentally sustainable and innovative and efficient. Child et al. (2018) deliver valuable knowledge that helps stakeholders make intelligent choices and create eco-friendly technologies enabling better healthcare delivery for patients.

2. UNDERSTANDING SUSTAINABLE IoT IN HEALTHCARE

Medical organizations face dual challenges in digital advancement and environmental accountability as sustainable implementation of Internet of Things deserves deep research. The deployment of smart clinical devices forms only one aspect of sustainable IoT because it requires complete redesign of connected technology's development process and lifecycle durations to match ecological with economic and societal requirements (Settia, Bhutani, & Saini, 2025). The understanding of this emerging trend requires analysis of sustainable IoT fundamentals and healthcare system characteristics for environmental stewardship together with main factors enabling IoT integration into contemporary medical applications.

2.1 Definition and Principles of Sustainable IoT

Sustainable IoT in healthcare combines connected medical devices and healthcare systems which minimize environmental impact to deliver better healthcare services to patients. The combination between IoT technology and sustainability goals along with social responsibility makes up this concept. Healthcare delivery experiences a constant transformation through IoT which demonstrates why it is now crucial to ensure this innovation follows green practices.

The principles of sustainable IoT encompass energy efficiency, resource optimization, lifecycle responsibility, and ecological compatibility. IoT development follows these principles to produce energy-efficient systems that use eco-friendly materials and recycling or refurbishment possibilities. Sustainable IoT advocates decentralized computing architectures using edge and fog computing methods to minimize power dependence on cloud servers while accomplishing data processing at its source (Osama et al., 2023). The combination of these systems minimizes energy usage and proves better data protection while simultaneously lowering healthcare application response times.

Sustainable IoT measurements need to match the operational framework of the circular economy model. The model presents a system where resources are maintained continually while it encourages recycling and remanufacturing activities combined with reuse operations. Healthcare domain designers should create medical IoT devices with separable components to preserve device functionality beyond original parameters and reduce environmental impact through reduced waste. Patient information management needs to follow ethical practices under sustainable IoT frameworks by conducting data protection with security and transparency.

Healthcare needs sustainable IoT development which demands collaborative multi-disciplinary efforts because this change affects both system dynamics and technical requirements. The successful deployment of IoT requires engineers to unite with medical professionals and policymakers as well as environmental scientists to achieve health benefits together with environmental goals. Sustainable IoT functions as an essential tool to enable a healthcare system which maintains its smartness together with environmental sustainability and social responsibility.

2.2 Key Characteristics of Sustainable Healthcare Systems

A healthcare system becomes sustainable when it improves population health during long periods of time through optimal usage of environmental and economic resources. The digital transformation era requires healthcare practitioners to place sustainability at the core when they develop and deploy IoT healthcare technologies (Bhushan et al., 2023). These systems blend modern technology with ecological mindfulness together with inclusive social practices.

A sustainable healthcare system requires a necessary element which is energy efficiency. The implementation of sustainable healthcare technology includes power-efficient IoT sensors together with smart power control algorithms and renewable energy system infrastructure. Voiced through IoT-enabled energy monitoring systems hospitals and clinics achieve real-time energy optimization which decreases operating expenses and environmental effects. A medical facility benefits from better uptime capabilities alongside reliability when using energy-efficient systems.

The third essential trait involves waste reduction focusing on electronic waste disposal. The growing number of smart medical devices in use brings about a direct increase in discarded electronic waste including devices having dangerous substances. The sustainability of healthcare systems establishes procedures for managing electronic waste and supports recycling initiatives and chooses manufacturers based on their adherence to green product standards. The market now favors biodegradable sensor technology along with disposable eco-friendly devices instead of traditional components.

Interoperable systems which integrate seamlessly work to define sustainable healthcare ecosystems. Sustainable systems work to establish uninterrupted data interchange among medical equipment and digital platforms in contrast to independent solution utilization. Patient outcomes benefit from complete data analysis and duplication of resources can be prevented when this system is implemented. Patient monitoring systems that combine functions into one platform enable medical facilities to buy fewer devices thereby controlling electricity usage and equipment expenditures per patient.

Healthcare sustainability strongly relies on both accessible healthcare for all and fair distribution of resources among populations. Any healthcare technology that serves only city residents or high-income people does not fulfill the objective of achieving health care accessibility for everyone. Sustainable Internet of Things systems include purposes-built scalability and inclusivity features to achieve effective smart health solutions for all communities (Chataut, Phoummalayvane, & Akl, 2023). Lightweight IoT devices through mobile health (mHealth) applications contribute significantly to care expansion outside of regular facilities by minimising physical infrastructure requirements for efficient resource-based service delivery.

2.3 Drivers of IoT Adoption in Healthcare

Localization of forces enables the understanding of why some companies have already adopted the IoT in healthcare, while others are at different stages of their advancement. And among the foremost is the need to improve the quality of care with a reduced cost. By continuously monitoring patients, detecting abnormalities early, and creating tailored treatment plans, healthcare providers have the opportunity to provide better outcomes and decreased hospital readmissions through IoT. These capabilities are quite appealing in an era where value-based care has emerged as the choice over volume based models of service delivery (Alsharif, Jahid, Kelechi, & Kannadasan, 2023).

The global turn toward preventive and remote care are also another major driver. Patient centric models of healthcare that focus on wellness, early detection and home based monitoring has replaced the traditional ways of healthcare that was founded around hospitals and clinics. Through smart wearables, biosensors, mobiles etc. IoT technologies along with collection of vital health data in real time and sending it to the healthcare providers, this transition is being established. This diminishes the burden on healthcare facilities, lowers operational cost, and provides more equable healthcare particularly in remote or underserved regions.

Advances in technology have also made it possible as well as more cost effective to utilize the IoT. The reason for the miniaturization of sensors, the enhancing of the battery life, the appearance of 5G networks, and the designing of cloud and edge computing have all made IoT in medical applications bloom. With these developments, it has been possible to deploy high performance devices which consume low energy and consume very minimal to zero maintenance, as aligned to sustainability goals. Additionally, AI integration into IoT systems has contributed to better diagnostics, predictive analytics with automated decision-making, boosting the value of connected health technologies.

Now, policy frameworks, government initiatives and other factors are facilitating the rat race of adoption rate in the healthcare IoT. More and more national and international bodies are beginning to realise the potential of IoT for the resolution of public health issues, as well as healthcare inequalities, and to meet sustainability targets (Bhutani, Lall, & Agrawal, 2023). Smart health technologies become attractive through incentive programs, regulatory support and funding for research and development processes that allow public and private sector investments. Additionally, global sustainability goals, such as the United Nations Sustainable Development Goals (SDG) also serve as a central unifying operational frame through which healthcare innovation can be reconciled with environmental and social outcomes.

Figure 1. Integrated framework for sustainable IoT in healthcare

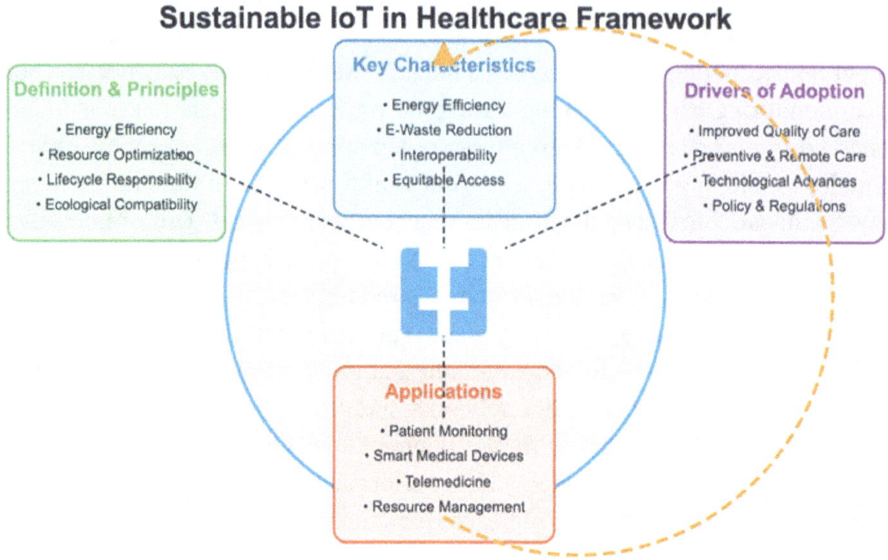

Figure 1 depicts such integrated framework of principles, characteristics, drivers and applications in healthcare that collectively drive environmentally responsible digital dynamic in such medical settings. This conceptual model has its electronics at the heart of the sustainably delivered healthcare delivery, surrounded by the foundational principles that guide sustainable IoT development, the necessary traits of the sustainably developed healthcare delivery systems, and the critical enablers facilitating the adoption. While circular economy principles follow the circular flow to demonstrate the continuous cycle of innovation, implementation, evaluation and refinement, the circular flow connecting these elements illustrates such a flow. This framework combines the multitude of disciplines needed for developing IoT healthcare solutions that achieve level set improvements in patient outcomes and level set reductions in their environmental footprint by improving energy efficiency, reducing waste, and optimising resources end to end throughout the technology life cycle.

3. EVOLUTION OF IOT IN THE HEALTHCARE ECOSYSTEM

Throughout digital transformation reshaping global healthcare systems, the integration of IoT technologies become one of the most important milestones of the former. Starting with the simplest of telemetry systems, the path of IoT in healthcare has been of rapid innovation and growing complexity from the most basic of telemetry systems to what today is affordable wearable health monitors and smart hospitals (Bhutani, Lall, & Agrawal, 2023). However, tracing the roots, understanding the main innovations that have brought sustainable IoT in the day to day world, and the paradigm shift from normal to smart healthcare are essential to fully understand the consequences of sustainable IoT in today's healthcare world. This section traces this evolution and lays context for how IoT has become the key component of the patient center, environmentally conscientious, and intelligent care of today.

3.1 Historical Development and Milestones

IoT in Healthcare is the result of the early use of communication technologies in the 1960s and 70s, where they were used to help support remote consultations and patient monitoring. These primitive systems, based on analog signals and rudimentary data transmission, helped in establishing more advanced developments in the coming years. In the late 1990s the idea for interconnected medical devices emerged further in the flesh as the idea of internet and wireless communication began to develop.

This is an important time of transition, which began in the early 2000s, when wireless sensor networks (WSN) and Bluetooth enabled devices became available for much more dynamic, continuous and real time monitoring. Several examples of medical IoT applications right from the start included sensors with wireless telemetry, glucose monitoring systems, and remote cardiac care systems. Especially important in advancing these developments was the control of chronic disease where patients could get service delivered in the home and maintained connection to their healthcare providers.

The proliferation of mobile phones and mobile applications in the late 2000s greatly spurred growth of IoT within healthcare. Mobile health (mHealth) applications allowed the users to monitor their vital signs, medication schedules, and fitness activities whereas the healthcare professional could access the patient data more easily as well as in real time. However, with the advent of cloud computing, the IoT infrastructure was further strengthened by providing storage and processing on demand, cloud integration with analytics tools, among others (Alahi et al., 2023).

As the 2010s unfolded, the healthcare IoT landscape matured along with the bringing together of new data analytics, machine learning algorithms and, by means of the cloud, the exchange of constantly occurring data. Fitness trackers and smartwatches became mainstream wearable devices and allowed users to measure their heart rate, sleep patterns, oxygen saturation etc. Also smart initiatives for the hospital were also taken root, with the IoT to include the systems facilitating asset tracking and inventory management, and patient flow optimization.

Another critical milestone for the COVID-19 pandemic was in 2020. Remote patient monitoring, virtual consultations, and real time health data analytics fueled the need of the hour – to adopt IoT. In pandemic response strategies, remote temperature sensors, oxygen monitors and contactless diagnostics became essential tools. It was understood by governments and health organizations worldwide that IoT could not only aid in better care delivery, but also be used to improve system resilience in the face of health crises.

3.2 Key Innovations and Disruptions

IoT in healthcare has seen the evolution of series of exciting breakthrough in the technologies and innovations that transformed the manner of designed and delivered health services. The most important innovation has been the development of miniaturized, low power sensors that can capture physiological data at a high accuracy without compromising user convenience. In the end, these sensors are now planted in everything from wearable fitness trackers to implantable medical devices, making it possible to glean continuous health status information about a patient.

Edge and fog computing is another great transformative innovation for Healthcare IoT systems. One challenge of traditional cloud-based architectures is their lack of a low latency and data privacy in time critical medical applications. Edge computing solves these problems by processing the data as close as possible to the source of data generation at the device level or near it. In the process, it not only makes decision making in a faster fashion possible but also curbs energy consumption and network congestion, which is critical when considering sustainability.

Predictive analytics, automated diagnostics, smart alert systems and other such capabilities have disrupted the healthcare IoT's ecosystem enough by artificial intelligence and machine learning. Today, even detection of anomalies in patient data, predicting disease outbreaks, and improving hospital workflow can be done using real time via AI based IoT platform. These capabilities have greatly improved the clinical decision making, facilitated better patient treatment and reduced the demand on the workers in the health care (Al Khatib, Shamayleh, & Ndiaye, 2024).

The convergence of IoT technology with blockchain has been another main disturbance, with special emphasis made on data security and interoperability. For health information confidentiality and data integrity, decentralized and tamper proof storage is what blockchain provides. Blockchain enables trust within patients so that they can share data from IoT devices and institutions without compromising on security or regulatory compliance.

Innovations also are starting to emerge that are targeted towards sustainability. Focused on these topics, the devices include energy harvest devices that can power sensors with body heat or ambient light, biodegradable materials in manufacturing, and modular IoT platforms that will last and be recyclable. These developments serve as a fundamental shift towards sustainable design thinking and a solid combination of high level of technological advancement with regards to environmental stewardship.

3.3 Transition from Traditional to Smart Healthcare

While technological transition from traditional to smart healthcare is a case of a step up, it is more than just a change of gears; it is a fundamental reorientation in the perception, practice and experience of the healthcare. Traditional healthcare models are reactive hospital focused and fragmented, uncoordinated organization among providers, patients and systems. Therefore, smart health care is purely proactive, patient focused and networked, using IoT and other digital tools in order to make personal, timely and efficient attentions.

The trend that remote and continuous technologies are gaining in relevancy is one of the most visible manifestations of this phenomenon. Most healthcare in the traditional sense is based on episodic visits and either paper record keeping or technology that serves to delay diagnosis and often suboptimal treatment outcome. The goal of smart healthcare systems is to monitor patients in a continuous manner by wearable or ambient sensors whose data is constituted in real time and either sent to cloud based platforms or edge devices (Kulkarni & Inamdar, 2025). This data can be accessed at any given time by the providers to make early interventions and better clinical decisions.

One dimension of the transition is about the operation efficiency. Such hospitals usually face problems in resource allocation, patient flow and inventory management. IoT enabled systems are used in smart hospitals to automating these functions to enhance the efficiency as well as reducing operational costs. For instance, real time location systems (RTLS) can track the movement of equipment, staff and patients to reduce search times and to enhance the process of running the business.

Patient engagement has also evolved. Typically in traditional settings patients play an inactive role in their own healthcare journey. Patient empowerment through smart healthcare systems enables patients to report their own health data, gain access to smart educational resources as well as smart communication tools through mobile apps and connected platforms. What it does is encourage accountability, adherence to treatment plans, satisfaction with care.

More importantly, the smart healthcare transition has made way for environmental sustainability. Smart healthcare models minimize resource use, waste through intelligent system, and reduce the need for physical visit by reducing ecological goals. This includes use of telehealth platforms like, that reduce travel related emissions; or, digital documentation which reduces paper use and administrative overhead.

4. EMERGING TRENDS IN SUSTAINABLE HEALTHCARE IoT

Healthcare sustainability and Internet of Things fusion are sparking in a marvelous yet top to bottom connected time frame which is about innovation, productivity, and the consideration towards environmental aspects. When global health systems are increasingly burdened to provide high quality care with negligible footprint on the environment, however, there are emerging trends of sustainable healthcare IoT that are setting the pace of the future of medical technology and service delivery. And they are a product of more values: markets getting smarter, cleaner and more inclusive systems of healthcare. The miniaturization of devices, improvement in wearable and implantable technologies, exponential growth of telemedicine and remote monitoring systems; and the convergence of IoT, AI and edge computing are some of the most influential trends. Taken together, they are shaping up as an entirely new landscape of digital health (Barwal & Rohilla, 2024).

4.1 Miniaturization and Energy-Efficient Devices

Improvements in energy efficiency and miniaturization of the devices are one of most important trends of sustainable IoT evolution in healthcare. With healthcare increasingly going towards more personalized and non invasive intervention, the quest for smaller and smaller, less power hungry and less visible devices has increased exponentially. Nowadays, miniaturised medical sensors and medical devices, sometimes small as a coin, can be used to collect a variety of physiological data; e.g., heart rate, temperature and blood glucose levels at high precision with minimal invasion.

It is driven by the need to make energy efficient design in order to extend the use of batteries and minimize an environmental impact. Slotted IoT systems are one way to reduce power consumption, especially when there are a large number of devices to be powered and they need to be sustained for long time periods without frequent maintenance or recharging. Therefore, in response, researchers and manufacturers have adopted low power electronics, new circuit design, and energy harvesting technologies (Alshammari et al., 2024). As a result of this, devices that can recharge from body heat, movement or ambient light are now in development, removing the need for change of batteries and therefore eliminating electronic waste.

Secondly, these energy conscious devices are frequently creating using sustainable materials and manufacturing procedures. Industry wide efforts to reduce ecological harm are reflected with the use of biodegradable polymers and recyclable components, along with lead free soldering techniques. Due to the maturity of these technologies, these will continue to be critical enablers of sustainable healthcare, especially in environments where there is a lack of or unreliability in traditional power sources.

4.2 Wearable and Implantable IoT Devices

A second defining trend in the sustainable healthcare IoT space is the proliferation of wearable and implantable IoT devices. These wearables—smartwatches, fitness bands, and biosensor patches—have now proven themselves as wide common to health track and chronic disease management. Continually, real time monitoring of vital signsts and lifestyle metrics: an important advantage of their bring both patients and clinicians actionable insights. Wearable devices enable detection of irregular heart rhythms as well as monitoring for sleep quality without invasive procedures and with a convenient user friendly form.

A more advanced category is implantable devices used mainly in critical care conditions, for example cardiac pacing, glucose monitoring and neurostimulation. They are used and embedded in the body to provide high fidelity data collection and intervention with minimal patient effort. Implantables have become miniature, wireless systems capable of real time communication with healthcare systems and has helped address emergency situations and reduce frequency of clinical visits (Ali & Ali, 2021).

Sustainability is being considered in both wearable and implantable devices. The effects of advances in materials science, with the development of biocompatible and biodegradable implants, has reduced the environmental impact over long periods of time as well as the pitfalls when rinsing end of life disposal. At the same time, modular design approaches make wearables easier to repair, with ease of replacing parts and recycled components. These innovations not only help us fulfill ecological goals but also enable accessibility of the device by lowering costs and improving usability.

Most importantly, these technologies help spread medical services to previously inaccessible populations simply because of the distance they live from medical providers, from the lack of economic resources, or due to infrastructural impediments. Wearable and implantable IoT devices are tantamount enablers of decentralized, patient centric care which comes as close as sustainable development and global health goals, respectively.

4.3 Telemedicine and Remote Monitoring Systems

Consequently, outcomes of telemedicine and remote monitoring systems are bang on the trends that require change in the healthcare industry with less cost and reduced cost alternative to traditional approaches. To address the hospital overcrowding, infection risk, and enhance patient satisfaction, the adoption of telehealth solutions has grown rapidly accelerated in recent months by the COVID pandemic. Telemedicine makes possible virtual consultations, electronic prescriptions, as well as digital follow ups, meaning that the carbon footprint related to healthcare services is reduced as the need of physical infrastructure and travel is lowered.

Remote Monitor has a profound role in the chronic disease and chronic illness management, processing health, and post-operative care. Patient data (collected in homes or in community centers) is transmitted from these systems to central health care platforms where clinician can retrieve, analyse and intervene when required. It monitors on a continuous basis so that there is early detection of the complication, there is less emergency room visits, and it is possible to have more timely and tailored interventions.

Remote care brings with it some excellent sustainability benefits. The reduction of patient travel results in reduced transportation emissions and fewer hospital visits mean less energy and resource use in medical facilities (Ahmad et al., 2022). Remote monitoring also allows for shorter hospital stays which saves us critical care resources to be used in other acute cases.

Remote care, however, is becoming more and more sustainable and effective with the technological advancements. With integration of mobile apps, AI analytics and cloud based platform, it becomes possible to deliver high quality care on scale in low resource settings. For telemedicine and remote monitoring to become viable characteristics of worldwide sustainable healthcare infrastructure, broadband access and digital literacy will continue to improve.

4.4 Integration with Artificial Intelligence and Edge Computing

The combination of IoT and artificial intelligence and edge computing is a powerful convergence that is putting new notions of healthcare on sustainable paths. Vast amounts of data generated by the IoT is not really information in the traditional sense, but rather signals—except that they are extraordinarily noisy. AI technologies such as machine learning, natural language, and predictive analytics are used to make sense of this data. In the real time these intelligent systems can detect patterns, detect anomalies, predict health risks and suggest personalized interventions. This results into a more proactive, efficient and precise health care experience. This increases dynamic capability by having data processing closer to its source and partially resolves the imprecision problem. Real time processing of data takes place on edge devices like smart sensors and gateways instead of sending all data to distant cloud servers for analysis. It lowers energy consumption, and this is a big advantage for sustainability because it reduces latency and conserves bandwidth. Additionally, edge computing helps in data privacy and security as the personal information does not have to travel over several networks.

Most impacts in using AI and edge computing together in healthcare IoT systems are seen in the time sensitive applications like emergency response, intensive care unit, and remote diagnostics. For example, wearable ECG monitor with AI algorithms can identify arrhythmias and notify health care providers in few seconds, which would save lives. In smart hospital environment, AI based IoT systems can be utilized in the form of energy use optimization, air quality management and waste cycle management to support greener operations. As well, it allows adaptive learning and system optimization (Hameed, Naha, & Hameed, 2024). Having AI in IoT devices allows them to self calibrate, update their firmware remotely and power consumption based on usage pattern and environment conditions. In addition to that, they improve the performance, decrease the maintenance needs, and minimize the impact on the environment. As AI and edge computing are increasingly part of the healthcare systems, they are going to be a key player in the triple aim of sustainable healthcare — better outcomes for patient, lower cost for operations, less environmental damage.

Table 1. Emerging trends in sustainable healthcare IoT

Trend	Key Features	Sustainability Benefits
Miniaturization and Energy-Efficient Devices	Compact, low-power sensors; energy harvesting from body heat/ movement/light	Reduced battery waste; longer device lifespans; biodegradable materials
Wearable and Implantable IoT Devices	Real-time monitoring wearables; wireless implantables for critical care	Less invasive diagnostics; improved accessibility; modular and recyclable design
Telemedicine and Remote Monitoring Systems	Virtual consultations; remote data collection and transmission	Lower carbon footprint; reduced hospital visits and patient travel
Integration with AI and Edge Computing	Smart algorithms for analysis; edge devices for local processing	Faster response; energy-efficient data handling; improved data privacy and security

Four key trends for the evolving agenda of sustainable healthcare IoT are displayed in table 1: miniaturization and the energy efficiency of the device design, wider usage of wearables and implanters of it, scaling telemedicine and remote monitoring systems, and the integration of AI with the edge computing. These are but to say the least, operational efficiency and patient care improvement, and indirectly, these are directly connected to the environmental sustainability goals which modern healthcare systems aim for by decreasing electronic waste potential, cutting carbon emissions, saving or using energy better and promoting decentralization and accessibility of medical services.

5. ENABLING TECHNOLOGIES FOR SUSTAINABLE MEDICAL IoT

In this context, we present the theory of sustainable medical IoT (Medical Internet of Things), which is based on the fusion of the technological innovation and environmental stewardship. Further clouding our understanding of the topic of enabling technology is that the healthcare sector is embracing digital transformation and needs to ensure that the technologies that we enable ecological and operational sustainability. Such technologies cannot only help in seamless data acquisition and analysis and data transfer, but also in reducing power consumption, carbon emissions and ensuring data security (Gupta & Singh, 2023). Several of these technologies are now considered as pillars of sustainable medical IoT. Some of these include the LPWANs, the energy harvester and self-powered devices, the green cloud and edge computing infrastructures as well as the blockchain enhanced cybersecurity frameworks.

5.1 Low-Power Wide-Area Networks (LPWANs)

For the IoT sector, and in particular the healthcare sector where it is of paramount importance to maintain continuous and real time monitoring over wide geographical areas, LPWANs are at the core. This small range in comparison to traditional wireless communication means that LPWAN technologies like LoRaWAN, NB-IoT and Sigfox are based on communication between IoT devices and network with minimal energy. Such low power requirements make them particularly suitable for use in battery powered medical devices with multi year battery life even in constant use scenarios.

LPWANs are in the context of sustainable healthcare supporting many applications ranging from rural telemedicine to remote patient monitoring to emergency alert systems. They enable connectivity in places where regular cellular infrastructure may not reach or is just too costly, and in the process create linkages between settlement patterns that previously were thought to be less densely settled. LPWANs deliver on the promise of low powered aspects of IoT, helping to reduce the costs and enable scalability of healthcare IoT ecosystems through energy efficiency and operational cost functions.

Additionally, the environment impact of LPWANs is smaller since they introduce simple and affordable communication channels for its deployment which is cost-effective. Due to their ability to function reliably in harsh environments or remote locations, they do not discriminate on their information and are inclusive of healthcare delivery for the greater goals of equity and accessibility. However, with enhanced encryption, bandwidth optimization as well as spectrum efficiency of LPWAN protocols, their role in promoting sustainable medical IoT platform will be even more significant.

5.2 Energy Harvesting and Self-Powered Devices

Energy harvest is an exciting approach to enabling the IoT through sustainable power. Self powered devices capture ambient energy from environment (e.g. solar, thermal, RF, etc.) and thus do not require conventional batteries or frequent recharges. In particular, this is of great importance in medical settings, where interruptions to device functionality can be a matter of life and death, and the replacement or service of the device might not be possible or easy logistically and environmentally.

Wearable sensors, implantable devices and environmental monitors are being also integrated with self powered medical IoT devices. For example, piezoelectric sensors can produce energy through bodily movements and thermoelectric generators can covert the body heat into electrical energy. These innovations make it possible to build ultra low power systems that can be left alone and run for long periods without any human input.

In terms of sustainability, energy harvesting breaks a dependency on disposable batteries, which are a large source of electronic waste containing hazardous materials. Besides, it supports deployment of devices in the remote or off grid environment where other power sources don't exist. In addition, energy efficient micro controllers coupled with smart power management systems allows the harvested energy to be exploited as effectively as possible improving performance as well as the length of life.

Integration of energy harvesting technologies into medical IoT devices represents a turning point that heralds the open new horizon of healthcare economy in the circuit, in line with the principles of wellness economy. It aids in resource conservation, declines environmental pollution and shapes a path for resilient, decentralized healthcare infrastructures which are both ecologically and economically sustainable.

5.3 Green Cloud and Edge Computing for Healthcare

Rapidly growing IoT generated data has created massive requirements on the traditional computing and storage infrastructures. Green cloud and edge computing have stepped up to become the major enablers of sustainable healthcare IoT systems in response. The goal of these computing paradigms is to lower the carbon footprint of data processing and storage through reduction of energy consumption, increased use of available resources, and decreased overheads in communication.

In green cloud computing, the data centers are run on energy efficient format by which energy waste is minimised with the help of renewable energy source along with virtualization and dynamic workload management. For storing electronic health records (EHRs), managing patient data, running diagnostics, delivery telemedicine services, cloud platforms are used in a healthcare setting. As for large scale data operations, such as the cloud, sustainable cloud practices như делит общное сервер компаковка, интеллигентные системы охлаждения,green data center design contribute to mitigating the environmental impact.

One of the features of the cloud is the ability to put data closer to the job by running the processing on edge devices. The decentralization on the other hand reduces the need of long distance data transmission thus conserving bandwidth and thus energy usage. Edge computing allows for near real-time analytics and make decisions for example by doing vital signs analytics locally on a wearable device and sending only critical alerts to the cloud for other applications in critical health care. This not only helps to improve response times, but also increases data security and privacy of patient's data.

Combination of green cloud infrastructures with edge intelligence provides healthcare system with the architecture that balances sustainability goals and high performance and reliability at the same time. Additionally, the ongoing development of the energy aware algorithms, adaptive scheduling, and containerized services places these technologies in a stronger position as key pillars for sustainable medical IoT.

5.4 Blockchain and Cybersecurity in Sustainable IoT

In healthcare where patient data is always being transmitted across inter connected devices and network, security and privacy is the most sensitive requirements. In the context of sustainable IoT, cybersecurity frameworks that are robust must be implemented such that the system's performance is not compromised nor the energy efficiency. One of the answers to this, blockchain technology, provides to manage the data in a decentralized and tamper-resistant way without requiring energy intensive centralized servers.

Blockchain helps to establish secure and transparent interchange of data between medical IoT devices, healthcare providers and patients. The cryptographically linked and stored transaction or data entry is across a distributed ledger, making it traceable, immutable and accountable. It reduces the risk of data breach, and increases trust to digital healthcare systems. In addition, smart contracts can be used to automate and enforce access controls, making it such that only well defined entities can interact with sensitive information.

Newer blockchain architectures (e.g., Proof of Stake (PoS), Directed Acyclic Graphs (DAGs), or lightweight consensus algorithms) impose much lower energy consumption than conventional Proof of Work (PoW). By integrating these innovations, it is possible to integrate blockchain into such resource-constrained IoT environments without adding to energy use.

Blockchain is only part of a sustainable cyber security strategy which also includes lightweight encryption, intrusion detection systems and anomaly detection based on Artificial Intelligence, all designed for low power power operation. Such measures ensure that the system vulnerabilities do not undo sustainability goals.

Combining blockchain and green cybersecurity practice lays a solid ground for sustainable and ethical operation of medical IoT systems. A more resilient and trustworthy digital health ecosystem can be enabled through the compliance with data protection regulations, protection of healthcare assets and healthcare rights they uphold.

Table 2. Key enabling technologies for sustainable medical IoT

Technology	Functionality	Sustainability Contributions
Low-Power Wide-Area Networks (LPWANs)	Facilitates low-power, long-range connectivity for IoT medical devices	Reduces energy consumption, enhances remote healthcare access, supports scalability
Energy Harvesting & Self-Powered Devices	Converts ambient energy (solar, thermal, kinetic) into electrical power	Eliminates reliance on disposable batteries, enables long-term deployment in remote areas
Green Cloud & Edge Computing	Decentralizes data processing and optimizes storage through eco-friendly methods	Minimizes carbon footprint, improves energy efficiency, boosts data security and real-time responsiveness
Blockchain & Cybersecurity	Ensures secure and tamper-proof medical data exchange	Provides decentralized data management with low-power protocols, ensures compliance, trust, and energy efficiency

The Medical Internet of Things (IoT), sustainable development relies on the dual objectives of the operational efficacy and environmental stewardship and, as such, the enabling technologies supporting the piggyback of the system. According to Table 2, the ecological sustainability in connected healthcare is driven by other core components whose combinations include Low Power Wide Area Networks (LPWANs), energy harvesting devices, green cloud and edge computing infrastructures as well as the blockchain and cybersecurity. LPWANs sacrifice energy in order to have long range communication, which is particularly suitable to rural and resource constrained deployments. Medical devices are developed that can operate without batteries using ambient energy, which reduces the dependence on the use of disposable batteries as well as enables long term deployment. Medical IoT offers massive data, and Green and edge computing paradigms meet the challenge of realizing such data demands by decentralizing processing and consuming energy as PEPs. Additionally, blockchain and state of the art security protocols are utilized to guarantee secure, transparent and low energy management of above keen medical data. Together, these technologies constitute the foundation of the smart healthcare ecosystem as a robust, energy aware, and equitable one.

6. APPLICATIONS AND USE CASES

Sustainable IoT in healthcare is now dealing with the practical deployment of this technology in healthcare whose impact is transforming delivery of medical care into intelligent data driven and patient-centric care models. For example, as smart hospitals and as remote rural healthcare outreach, such applications show promise of IoT technologies in dealing with critical health challenges and further sustainability. In combating rising needs, resource constraints, and environmental responsibility challenges across global healthcare systems, the IoT based solutions have became vital on various volumetric use cases.

6.1 Smart Hospitals and Connected Infrastructure

The integration of IoT technologies into almost all aspects of infrastructure and service provision in a smart hospital is a paradigm shift in hospital functioning. The combination of these institutions generates a response and intelligent environment with a web of interconnected devices (sensors, actuators, wearables, and analytics platform) with the intent of improving operational efficiency, patient experience, and sustainability.

In the case of hospitals, IoT can provide automated climate control, intelligent lighting and predictive maintenance of hospital equipment that reduces the energy consumption and carbon emissions. For example, smart beds can monitor patient vitals and automatically adjust patients to keep them from getting bed sores, thus improving patient comfort and decreasing the degree of reliance on inhuman intervention. Real time tracking of RFID tagged medical assets help optimize equipment utilization and prevent waste by not purchasing or losing the medical asset.

Additionally, the IoT enabled facility management system uses the real time monitoring and control to ensure efficient use of water, power and other kinds of resources. HVAC systems powered by smart sensors can reduce consumption of energy by adjusting ventilation in accordance with occupancy and air quality levels. Together, these systems can simultaneously monitor the cycle of medical devices at the tip of the customer use to guarantee proper recycling and disposal.

Smart hospitals also create data rich ecosystems which enable advanced analytics and artificial intelligence (AI) based diagnostics and help improve faster and more accurate decision making. Innovation is not only about supporting sustainability (through optimising resource use) but further leads to improved clinical outcomes, shortening waiting times and better quality care.

6.2 Chronic Disease Management and Personalized Care

Worldwide, several chronic diseases such as diabetes, cardiovascular conditions and respiratory disorders are significant burdens to healthcare systems. However, IoT technologies have taken center stage as instrumental devices of transformation in how these conditions could be purveyed and managed in a personalized manner; proactively and in a sustainable manner.

Both wearable devices and biosensors constantly monitor patient metrics of health such as glucose levels, heart rate, blood pressure, and oxygen saturation. The secure cloud platforms that this real time data is transmitted to healthcare providers on time frame for them to make the timely interventions and eliminate the frequency of hospital visits. Further integration of AI into this process acts as the means of spotting anomalies as well as deriving future disease progression predictions from longitudinal data patters.

Treatment plans become personalized when the patient history and lifestyle factors are combined with IoT generated data that would make sense of treatment plans. Connected mobile applications give patients individualized medication reminders, dietary advice and exercise recommendations. The personalisation to treatment improves adherence and health outcomes while reducing the need for unnecessary diagnostics and treatments and, consequently, the payment for the environmental and financial costs.

Importantly, IoT puts patients into a primary role in their own care, a change toward preventive health. Sustainability is achieved through reduced energy intensive operations of acute care facilities through reduction in hospital admissions and emergency intervention. Remote monitoring also decongests urban hospitals, which allow such hospitals to allocate resources more effectively.

Despite addressing disparities in access, IoT enabled sustainable chronic disease management both empowers patients and enables chronic disease management to occur outside the direct oversight of a clinician. A healthcare model that could be better and more resilient involves the fusion of connectivity, data intelligence, and user friendly interfaces.

6.3 Emergency Response and Critical Care Systems

The IoT technologies in emergency response systems are critical for generating speed, accuracy, and effectiveness of care for medical emergency. IoT will provide the solution to whether we are dealing with trauma incidents, cardiac arrests or pandemics — real time communication and coordination will be the difference between life and death.

Ambulances are equipped with telemedicine capabilities and connected diagnostic tools and hence are IoT-enabled ambulances that act as mobile ICUs. Patient data, including ECG readings, vital signs, travels through these units and transmit to emergency departments en route so medical teams can properly prepare for incoming cases. It significantly reduces the treatment delay upon arrival.

In critical care units, smart stretchers, emergency room monitors, and automated drug dispensers speed up workflows and enhance patient outcomes without increasing physical or cognitive stress on health care staff. The addition of IoT sensors to drones in disaster scenarios can evaluate damaged areas, provide first aid, find the victims, and improve situational awareness, as well as providing more efficient response.

IoT provides for efficiencies and redundancies less common in emergency care from a sustainability perspective. However, these efficiencies are only achieved by optimizing ambulance routes, resource allocation, interdepartmental coordination, globally, thanks to the use of IoT systems, which in turn results in faster responses with fewer emissions and less operational waste. On top of that, predictive analytic can predict the trends and surges of emergency visits to hospitals that may be prepared for such an event, thus planning the use of hospital resources and avoiding overstock or underutilization.

The intelligent systems that cope with patient changes dynamically can make the real time critical care greener due to power consumption optimum change there. In aggregate, these solutions build an environment wherein the care is concerned yet resource conservative, as is consistent with the foundations of viable health care.

6.4 Rural and Remote Healthcare Services

One of the most transformative IoEN in sustainable healthcare is to provide services to such rural and remote areas where medical infrastructure is scarce. In such a situation IoT enabled healthcare fills the gap by providing continuous care delivery, diagnosis and consultation without ensuring physical travel or huge infrastructure investments. IoT devices are used on the back end of telemedicine platforms to enable rural patients to photograph aspects, and video conference with specialists in urban centers. Through use of home based medical devices, real time vital parameters are fed into physicians who can be able to diagnose and prescribe real time treatments without being in the same place. The traffic time a patient spends in the car, the costs incurred in patient medical costs and emissions from transportation are reduced by this system.

Solar powered and low bandwidth IoT devices are used where no electricity or Internet access is available for uninterrupted service. By way of example, an

IoT sensor can be embedded in maternal health monitoring kits that can sense the well-being of the fetus and the vitals of the mother, then signal healthcare workers to potential risks during pregnancy and childbirth. The mobile health units, which have IoT tools incorporated onboard, visit remote communities once in a while to collect data that is then analyzed at the central place, but also, provides immediate care to the patients. They do help in tracking epidemiological, vaccination drives and chronic disease screening as all these are necessary for public health management.

Finally, IoT in rural healthcare also empowers the community. By training local health workers to use connected devices to respond quicker, these will improve the capability to respond as well as engaging local health workers to employ sustainable practices in hygiene, nutrition, and disease prevention. Such systems have the potential to evolve over time to resilient community based networks that could sustain long term health equity and environmental and sustainability.

7. CONCLUSION

Sustainable Internet of Things (IoT) technologies integration in healthcare is a revolutionary step towards no smarter, more effective, and environmentally friendly medical systems. In this chapter we presented the evolution, trends, technologies, applications and challenges of sustainable IoT in the world of modern healthcare. IoT has become an important part of patient centered care and real time monitoring as well as operational efficiency from the early stages of sensor based system to advanced AI enabled network. Despite the brutal uphill struggle towards a viable and just IoT driven healthcare ecosystem, there's still hope. Collaborative innovation and policy reforms are required to address power consumption issues, security, cost and regulation. Both the demand of the energy efficiency in device design as well as the demand of ethical data practices and long term ecological impact mitigation are part of sustainability focus. The role of sustain due IoT is becoming increasingly important with the healthcare sector being fully fledge into digital transformation. It opens an opportunity to promote health outcomes linked with planetary well being giving rise to systems that can respond to both individual and global challenges, and are resilient. Sustainable IoT can reshape healthcare delivery by enabling smarter, greener and more inclusive systems, by harnessing demands from the stakeholders across concerned groups to create and redefine this future.

8. FUTURE DIRECTIONS

And looking ahead, several new keys will enhance sustainable IoT healthcare in the future. The most drastic will be the preparedness of fully autonomous, energy harvest devices that can operate indefinitely without external power. With these innovations, companies could dramatically cut their dependency on batteries and waste in the environment, and extend IoT into restricted places lacking resources. For instance, integration with advanced artificial intelligence (AI), machine learning (ML), and edge computing will further strengthen IoT capabilities by providing predictive analytics, personalised care, and speed of decision making without the use of cloud infrastructures. In addition to improving efficiency, energy consumption and latency are both reduced. As the communication across the diverse healthcare systems and devices become more and more important, standardization and interoperability are going to become the main elements to help streamline the communication. At the same time, regulatory frameworks should remain in flux as well, guaranteeing their security, ethicality, and equity. The next wave of innovations, however, will be user centered, particularly for vulnerable populations, and in the process user centered, inclusion, accessibility and responsiveness to a range of diverse healthcare needs. Sustainability and digital healthcare will converge for moral and ecological reasons, and not only for technological ones, in order to build the health systems of the future.

REFERENCES

Ahmad, K. A. B., Khujamatov, H., Akhmedov, N., Bajuri, M. Y., Ahmad, M. N., & Ahmadian, A. (2022). Emerging trends and evolutions for smart city healthcare systems. *Sustainable Cities and Society*, *80*, 103695. DOI: 10.1016/j.scs.2022.103695

Al Khatib, I., Shamayleh, A., & Ndiaye, M. (2024, July). Healthcare and the internet of medical things: Applications, trends, key challenges, and proposed resolutions. *Informatics (MDPI)*, *11*(3), 47. DOI: 10.3390/informatics11030047

Alahi, M. E. E., Sukkuea, A., Tina, F. W., Nag, A., Kurdthongmee, W., Suwannarat, K., & Mukhopadhyay, S. C. (2023). Integration of IoT-enabled technologies and artificial intelligence (AI) for smart city scenario: Recent advancements and future trends. *Sensors (Basel)*, *23*(11), 5206. DOI: 10.3390/s23115206 PMID: 37299934

Ali, Z. H., & Ali, H. A. (2021). Towards sustainable smart IoT applications architectural elements and design: Opportunities, challenges, and open directions. *The Journal of Supercomputing*, *77*(6), 5668–5725. DOI: 10.1007/s11227-020-03477-7

Alshammari, H. K. F., Alqoud, H. M. H., Alrimal, M. M. T., Alanazi, H. F. M., Alshammari, M. S., Alharbi, F. S. J., & Almutairi, H. S. M. (2024). Emerging trends in healthcare and medical technology. *Gland Surgery*, *9*(2), 171–179.

Alsharif, M. H., Jahid, A., Kelechi, A. H., & Kannadasan, R. (2023). Green IoT: A review and future research directions. *Symmetry*, *15*(3), 757. DOI: 10.3390/sym15030757

Barwal, R. K., & Rohilla, N. (2024). Internet of Things (IoT): A comprehensive review, emerging trends, and future directions. *International Journal of Engineering Science and Humanities, 14*(Special Issue 1), 164–169.

Bhushan, B., Kumar, A., Agarwal, A. K., Kumar, A., Bhattacharya, P., & Kumar, A. (2023). Towards a secure and sustainable internet of medical things (IoMT): Requirements, design challenges, security techniques, and future trends. *Sustainability (Basel)*, *15*(7), 6177. DOI: 10.3390/su15076177

Bhutani, M., Lall, B., & Agrawal, M. (2022). Optical wireless communications: Research challenges for MAC layer. *IEEE Access : Practical Innovations, Open Solutions*, *10*, 126969–126989. DOI: 10.1109/ACCESS.2022.3225913

Bhutani, M., Lall, B., & Agrawal, M. (2023). An efficient and adaptive superframe structure for IEEE 802.15.7-based real-time sensor networks.

Bhutani, M., Lall, B., & Agrawal, M. (2023). A novel energy-efficient adaptive superframe structure for OWC-based real-time bio-sensor networks. *Peer-to-Peer Networking and Applications*, *16*(6), 3021–3031. DOI: 10.1007/s12083-023-01544-3

Chataut, R., Phoummalayvane, A., & Akl, R. (2023). Unleashing the power of IoT: A comprehensive review of IoT applications and future prospects in healthcare, agriculture, smart homes, smart cities, and industry 4.0. *Sensors (Basel)*, *23*(16), 7194. DOI: 10.3390/s23167194 PMID: 37631731

Gupta, A., & Singh, A. (2023). Healthcare 4.0: Recent advancements and futuristic research directions. *Wireless Personal Communications*, *129*(2), 933–952. DOI: 10.1007/s11277-022-10164-8 PMID: 36594000

Hameed, K., Naha, R., & Hameed, F. (2024). Digital transformation for sustainable health and well-being: A review and future research directions. *Discover Sustainability*, *5*(1), 104. DOI: 10.1007/s43621-024-00273-8

Krishnamoorthy, S., Dua, A., & Gupta, S. (2023). Role of emerging technologies in future IoT-driven Healthcare 4.0 technologies: A survey, current challenges and future directions. *Journal of Ambient Intelligence and Humanized Computing*, *14*(1), 361–407. DOI: 10.1007/s12652-021-03302-w

Kulkarni, M. H., & Inamdar, P. R. (2025). Healthcare unbound: Navigating emerging trends and future applications in IoT-based innovations for daily well-being. In *Evolution of Machine Learning and Internet of Things Applications in Biomedical Engineering* (pp. 238–250). Springer.

Kumar, M., Kumar, A., Verma, S., Bhattacharya, P., Ghimire, D., Kim, S. H., & Hosen, A. S. (2023). Healthcare Internet of Things (H-IoT): Current trends, future prospects, applications, challenges, and security issues. *Electronics (Basel)*, *12*(9), 2050. DOI: 10.3390/electronics12092050

Mishra, P., & Singh, G. (2023). Internet of medical things healthcare for sustainable smart cities: Current status and future prospects. *Applied Sciences (Basel, Switzerland)*, *13*(15), 8869. DOI: 10.3390/app13158869

Nti, I. K., Adekoya, A. F., Weyori, B. A., & Keyeremeh, F. (2023). A bibliometric analysis of technology in sustainable healthcare: Emerging trends and future directions. *Decision Analytics Journal*, *8*, 100292. DOI: 10.1016/j.dajour.2023.100292

Osama, M., Ateya, A. A., Sayed, M. S., Hammad, M., Pławiak, P., Abd El-Latif, A. A., & Elsayed, R. A. (2023). Internet of medical things and healthcare 4.0: Trends, requirements, challenges, and research directions. *Sensors (Basel)*, *23*(17), 7435. DOI: 10.3390/s23177435 PMID: 37687891

Settia, N., Bhutani, M., & Saini, V. (2025). Exploring the impact of artificial intelligence on healthcare: A comprehensive review. In *Digitalization and the Transformation of the Healthcare Sector* (pp. 115–138). Springer.

Chapter 2
Creating a Roadmap to 2030 for Achieving Sustainable Healthcare IoT

Jayanth Sai Pokkalla
https://orcid.org/0009-0009-6198-4561
Independent Researcher, USA

ABSTRACT

This chapter proposes a strategic definition of Smart Healthcare IoT solutions development and deployment strategies for the year 2030. It provides a roadmap which lays out policy recommendations, technological innovations, investment priority, collaborative frameworks. Global trends like climate change, aging populations, healthcare digitalization, drive the need for green and intelligent systems that are the subject matter of the chapter. It combines analysis of current gaps, as well as projections of future demands to provide an overall roadmap to stakeholders looking to create resilient, green, fair digital health ecosystems in line with South Africa's sustainable development goals.

1. INTRODUCTION

The healthcare sector is booming with many major technological advancements, provisions for patient requirements, environmental concerns, all accelerating the desired transformation. Healthcare like other industries around the world is embracing Internet of Things (IoT) as a critical enabler of more personalized, efficient, and data driven care in parallel with the broader digital transformation. Yet, as the number of

DOI: 10.4018/979-8-3373-4134-7.ch002

healthcare IoT systems grows, they must be made to fit within sustainability principles so they won't make existing environmental and social problems worse. This transformation is an urgency driven by multiple global trends including escalating effects of climate change, aging population with complex health needs, rising burden of chronic diseases, and others. For these challenges and to make a sustainable future for healthcare development and deployment of sustainable healthcare IoT requires a strategic, long term approach (Espinosa, López, 2021).

To achieve sustainable healthcare IoT by 2030, this chapter provides a comprehensive roadmap and what you should focus on: technological innovation, policy formulation, investment priorities, and cross sector collaboration. 2030 will be a healthcare ecosystem that is green, inclusive, digitally empowered, aligning the most sophisticated technology with reducing the environmental impact, improving patient outcomes and making systems of healthcare globally more accessible and affordable. As one of the critical new domains of IoT, healthcare, sustainability is required to be integrated into IoT design, deployment and operational aspects.

The concept of sustainability in healthcare IoT encompasses several dimensions: environmental, social, and economic. Beyond itself, bringing down energy use and carbon emissions, healthcare device lifecycle is committed to social equity, responsible resource use, and a circular economy model, which is also more broadly a commitment to sustainability. Wearable and remote patient monitoring devices as well as any other IoT system should not only focus on delivering optimal healthcare services, but also add or enhance value to the communities and environments in which IoT deployments are made. Holistic approach is required to achieve sustainability in healthcare IoT as it involves both design and disposal of devices at the design phase, procurement and their use.

Healthcare IoT in 2030 is defined by a number of key global drivers that, when combined, become the vision. For example, climate change has made the industries of healthcare, amongst others, put more pressure to reduce its carbon footprint. IoT solutions, encompassing remote patient monitoring systems and smart hospital management platforms, can have huge implications on aiding sustainability efforts by conservation of resources, and elimination of the same. Nevertheless, in order for the global environmental goals to be achieved, such as the Paris accord and the United Nations Sustainable Development Goals (SDGs), the healthcare sector will need to adopt a much greater scale of sustainable IoT practices. The growing digitalization of healthcare, spurned on by the COVID 19 pandemic, now has the desire for powerful, energy efficient and ecologically friendly healthcare technology ever more urgent (Kasinathan & Pugazhendhi, 2022).

Another important factor in deciding the future of healthcare is demographic changes. Rapidly, the global population is aging with the number of people 60 and over to double by 2050. This demographic shift is occurring in conjunction with an increasing population with chronic diseases that require long term management and personalized care. Healthcare IoT can tackle these challenges by providing continuous health monitoring, early detection of health conditions and personalized treatment regimens. The goal is for IoT where healthcare solutions are wide spread and promote support for populations aging, increasing quality of life and decreasing strain on health care systems by 2030. Consequently, these systems must be designed in such a manner that environmental impact is minimized, in particular, energy consumption and e-waste (Bhutani, Lall, & Agrawal, 2022).

Also, healthcare IoT landscape is driven by technological innovation. With advances in artificial intelligence (AI), advancement of machine learning, and low power high efficient devices for the IoT taking place today, much more cannot be done to in the field of sustainable healthcare solutions. Sensors on wearable devices, remote monitoring systems, and smart medical equipment will have to become more energy efficient and environmentally friendly, have much longer life cycles, and are capable of self powering with renewable energy sources. It also involves progressing AI driven optimization in order to minimize energy use and optimize overall efficiency of health care operations.

On the other hand, the demand for more sustainable practices is increasing in the healthcare sector, which is subject to mounting pressures from both regulatory bodies, governments and the worldwide organizations. The transition to a sustainable healthcare IoT ecosystem implies that there should be policies and regulations facilitating green procurement, promoting e-waste recycling, and the use of renewable energy sources. Additionally, international cooperation and alignment with global frameworks of sustainability, including the United Nations' SDGs (Mani & Goniewicz, 2024), will be required to ensure that the benefits of sustainable healthcare IoT are extended to all areas and populations, especially at the lowest resource levels, i.e., in low- and middle-income countries.

While technical and regulatory aspects are difficult, the challenge of sustainability in healthcare IoT lies essentially on a shift in the approach to healthcare delivery and assessing its impact on sustainability. To realize the transition to a sustainable healthcare IoT ecosystem, we will need to work closely by multiple stakeholders including governments, healthcare providers, technology firms, academia, and civil society organizations. Through this collaborative approach, it will also be possible to foster a shared commitment to sustainable development of healthcare IoT solutions and their deployment to everyone irrespective of where they are located or what their socioeconomic status is.

The aims of this chapter are to present the roadmap to sustain healthcare IoT to reach that target in 2030. In this roadmap, it would outline the key areas for the focus of stakeholders and that there would be the need to align technological, policy and investment priorities with the object of creating resilient, eco friendly, and equitable digital health ecosystems. In this chapter, the state of the art and current market of healthcare IoT will be reviewed, the gaps in this area will be identified, and a vision of the future of healthcare IoT that strongly relies on sustainability being incorporated into the design and the use of every healthcare technology will be presented. This chapter attempts to make a contribution to the continuing discussion on the future of health and its potential role in aiding the realisation of global sustainability goals, by offering actionable recommendations based on a clear timeline of implementation.

2. GLOBAL DRIVERS FOR SUSTAINABLE HEALTHCARE IoT

Of enormous significance for the healthcare systems of the world is the profound healthcare landscape transformation being driven owing to a range of global forces and driving forces mutually supporting each other. Thus, these forces are brought together at the point of need for healthcare technologies, in these cases, the Internet of Things (IoT), which has the power to usher in a new era of healthcare delivery. Yet the widespread adoption and deployment of IoT in healthcare must occur in the knowledge of such global issues as climate change, demographic shifts, technological disruption, and policy/regulatory pressures. These drivers are no also deciding the necessity of demand for healthcare IoT as well as how these technologies need to be designed, developed and deployed in order to contribute to the healthcare systems long run sustainability.

Given the crises happening in the world, the focus of sustainable healthcare IoT takes the center fold. The issue of climate change is not anymore in the far corner but a pressing issue around what we do in our day to day lives, including healthcare. As aging populations and the burgeoning pressure of chronic diseases, health care delivery needs to shift to more effective, scalable and personalized solutions. Alongside these, the accelerating digitization in healthcare and technological innovation offer an opportunity to increase the effectiveness of healthcare delivery and reduce healthcare's inefficiency. Yet, for proper exploitation of opportunities in healthcare IoT in a manner that leads to sustainability, a full understanding of the drivers of the future healthcare IoT is paramount (Albreem & Sheikh, 2023).

2.1 Climate Change and Environmental Imperatives

Sustainability in healthcare is one of the central global drivers for such an issue in the growing urgency of addressing climate change. In addition to being an environmental problem, climate change is an issue for public health, affecting health all around the globe through rising temperatures, change in what weather is seen as extreme, and spread of infectious diseases. Among the largest consumers of energy, the healthcare systems are also powerful carbon emitters. With the world accelerating on reaching net zero emissions by 2050, healthcare systems are making efforts towards reducing their environmental footprint. As digital healthcare technologies, such as IoT devices, have the ability to increase or lessen environmental stress, it is important to consider this extra element.

Healthcare IoT has significant roles to mitigate the environmental impact of healthcare systems. For example, remote patient monitoring systems can decrease necessary in person visits, and consequently decrease transportation related carbon emissions. Furthermore, IoT devices can aid in efficient energy usage in healthcare facilities by taking responsibility for smart management of resource such as heating and cooling as well as lighting. Using IoT devices which consume less energy and consume lesser power, makes healthcare systems consume a lot less energy which ultimately results in a more sustainable operational model. In addition, the design of IoT devices must be life cycle driven, including the energy use, the resource extraction and the disposal in order to lower their ecological footprint. The sustainability in healthcare IoT approach in this holistic way for sustainability is in concert with other world environmental licensing and the need to support responsible innovation in the context of climate change (Abaee, 2023).

2.2 Demographic Shifts: Aging Populations and Chronic Disease Burden

The demographic shift that is going on all over the world is another very major driver of sustainable healthcare IoT. Among the most important of the trends shaping healthcare today are demographic, primarily the aging population, especially in high income countries. It is estimated that the number of those over the age of 60 will double to 2.1 billion by 2050. Along with this, there's an increase in age related health condition: cardiovascular disease, diabetes, dementia etc. These conditions are long term, continuous care, which places extreme strain on the healthcare systems. The challenge of disease management will find its solutions in these IoT technologies ranging from wearable health trackers to smart home devices to remote monitoring tools, which make the patient able to manage his health by himself without frequent hospital visits.

It is also a growing burden of chronic diseases on healthcare systems all around the world. WHO states that chronic diseases kill more than 70 per cent of the world's deaths, with a large number of these conditions preventable and controllable through good monitoring and early intervention. IoT can help manage chronic diseases by real time monitoring of vital signs, medications, lifestyles factors, and it can detect early health issue before it becomes a big issue. Nevertheless, for such systems to be sustainable, they need to be energy efficient, cost effective and fast enough to deliver real-time actionable information. As the popularity of IoT based solutions to manage chronic diseases grow, it presents more need for technologies used to address the needs to complement sustainability efforts to avoid worsening the already stretched healthcare resources as well as climate issues.

2.3 Technological Disruption and Healthcare Digitization

Healthcare is entering into a digital transformation at an unprecedented rate, especially with the development of artificial intelligence (AI), machine learning, cloud computing, and IoT. These technologies are transforming how healthcare is consumed, with better patient outcomes and waste reduction. That said, the healthcare industry itself is also becoming very fast digitized, and these are obviously double edged swords when it comes to data privacy, security, and the environmental cost of setting up the digital infrastructure. In the digital health solution shift, there is high potential but there are also necessary steps part of which includes planning to ensure these new technologies are rolled out in a manner which is aligned with sustainability goals (Almuaythir & Singh, 2024).

IoT is one of the key stones of healthcare digitization that allows transition from patients care based on traditional estimates to more personalized, data driven, and remotely monitoring solutions. With the devices such as smart wearables, implantable sensors and connected medical equipment steadily becoming ubiquitous in healthcare settings, care of patients can be more continuously monitored outside of the standard clinical set up. These innovations bring tremendous benefits with regards to ways of delivering better healthcare and lowering the costs of healthcare at the same time, but they come with some sustainability challenges. In combination with higher energy densities and short lifecycles of many consumer grade devices contributions to an increase in the amount of electronic waste is the sheer volume of IoT devices in use. In addition, the ‘carbon footprint’ of that infrastructure that has to support these technologies, data centres and the like, needs to be considered too.

With the digital transformation in healthcare systems, the environment must ensure that it is eco-friendly. Animal and Human Health include adopting energy efficient technology, reducing waste and that digital health solution is accessible to all in particular for particularly underserved populations. The only way sustainable healthcare IoT solutions can exist with the least environmental impact is by using emerging technologies that will improve patient care while also making healthcare more efficient and less resource intensive in the first place.

2.4 Policy and Regulatory Pressures (e.g., SDGs, COP Commitments)

Another driving factor in the future of sustainable healthcare IoT is the policy and regulatory environment. Governments, international organizations, and regulatory bodies anywhere in the world are paying attention to bring in sustainability into healthcare systems. Healthcare systems are being held increasingly accountable for their contribution to the healthcare related and wider Sustainable Development Goals (SDGs) of the United Nations, which, as a global framework for dealing with such health, environmental, social challenges, are underpinned by the UN agreement to Complete the Decade for Action on Sustainable Development. Specifically, SDG 3 relates to the need to ensure healthy lives and well-being for all at all ages, whereas SDG 12 relates to the need for responsible consumption and production patterns which includes the reduction of waste and creeping towards a circular economy (Goyal, Elawadhi, & Sharma, 2024).

On top of SDGs, such international climate agreements as such as Paris Agreement (COP commitments) further compel healthcare systems to cut down their carbon footprints and to work more according to sustainability. Meeting these commitments requires the introduction of green procurement policies in the healthcare sector, the use of energy efficient technologies and adoption of circular economy principles. Also, healthcare IoT technologies are starting to be developed and deployed in accordance with stricter regulatory bodies standards in terms of e-waste management, device lifecycle sustainability and energy efficiency.

In the healthcare systems of tomorrow, regulators are poised to lead the charge under a very complex—notoriety—regulatory landscape, while simultaneously requiring that IoT technologies be developed and deployed within the context of local and global sustainability frameworks. Being able to drive the widespread adoption of sustainable healthcare IoT will require policy interventions to encourage innovation in sustainable healthcare technologies, ranging from good procurements with a green tint to fostering collaboration of governments, the private sector and international organizations.

Finally, it is concluded that these global drivers of the sustainable healthcare IoT are diverse and interrelated under environmental, demographic, technological and regulatory factors. These drivers to the healthcare sector present both challenges and opportunities and a holistic approach must be used to address them involving sustainability in every area that is addressed as it relates to healthcare technology development and deployment. As healthcare IoT aligns his technological innovation with global sustainability goals, healthcare IoT can positively impact healthcare system, outcome of patient, and making the global healthcare future sustainable and equitable(Abaee, Saeedi, & Taherdoost, 2024).

Table 1. Global drivers for sustainable healthcare IoT and their impact on healthcare systems

Driver	Description	Impact on Healthcare IoT
Climate Change	Rising temperatures, extreme weather, and infectious diseases impact public health	IoT can mitigate environmental impact by optimizing resource use and reducing carbon emissions in healthcare settings
Demographic Shifts	Aging populations and rising chronic disease burden drive demand for efficient healthcare solutions	IoT technologies enable remote patient monitoring, personalized care, and chronic disease management
Technological Disruption	Accelerating digital transformation driven by AI, ML, IoT, and cloud computing	IoT enables personalized, data-driven healthcare but raises concerns about data privacy, e-waste, and infrastructure sustainability
Policy & Regulatory Pressures	SDGs and international agreements (e.g., COP commitments) push for sustainable healthcare practices	Policies incentivize green procurement, energy-efficient devices, and responsible waste management in healthcare

Table 1 presents the main driving factors that affect the adoption of sustainable healthcare IoT at global level. Climate change presents both challenges and opportunities, and the healthcare systems are large users of energy and significant contributors of carbon emissions. The impacts of these tasks can be reduced through application of IoT in optimizing resource use in facilities for better facility management, and for enabling remote monitoring to reduce the need for in-person visits and resulting transportation related emissions. Ageing population and rising chronic disease burden create even more need for the IoT solutions like wearable devices and remote patient monitoring tools to deliver continuous, personalized care. But AI, machine learning and cloud computing are accelerating the digital transformation of healthcare and these are bringing data privacy concerns, e waste as well as the environmental cost of pouring tech support around digital infrastructure. Finally, healthcare systems are increasingly being influenced by frameworks such as SDGs and COP commitments since the policy and regulatory landscape has become shaped. To meet government and international organization's sustainability criteria in healthcare, the IoT solution needs to be energy efficient, waste reducing and support a circular economy. As a collective, these drivers necessitate the development and use of holistic healthcare IoT design and implementation alongside contributions to health and environmental sustainability (Fahal, Smith, Nyaoke, & 2024).

3. CURRENT LANDSCAPE AND GAPS

The use of Internet of Things (IoT) technology in healthcare has been rising rapidly in digital health, along with high prevalence of chronic disease and desire for more efficient, patient-centric care. Wearable devices to smart medical equipment, IoT systems have proved to be excellent tools that have transformed the way healthcare delivery takes place by allowing for remote monitoring, facilitating better patient outcomes and reducing the time that the healthcare operation takes. While the potential of IoT in healthcare is enormous, the same cannot be said for its sustainability, accessibility, standardisation and policy support within the healthcare setting. To maximize potential of IoT in healthcare, which is consistent with sustainability targets, there is a need to understand and tackle these gaps.

3.1 Overview of the Current Adoption of IoT in Healthcare

Already, IoT in healthcare has seen vast improvement in patient care and operational efficiencies on adoption. Wearable devices that monitor your health or fitness like heart rate monitors, glucose sensors, and fitness trackers can be considered as included in the healthcare IoT ecosystem, as well as more sophisticated application like remote patients monitoring systems, smart hospital equipment and the IoT enabled diagnostic tools. Real time data can be enhanced using these technologies and the real life outcome cites examples of how these technologies can be used in improving patient monitoring, reducing hospital readmission and improving overall healthcare delivery.

Along with chronic disease management, IoT adoption has experienced one of the key areas of growth in IoT. Remote monitoring of vital signs enabling better proactive care alleviates the patients' requirement for frequent visits to hospital. Also, IoT solutions are being embedded into hospital infrastructures like smart bed, auto medication dispensers, and energy efficient climate control systems though which hospitals are becoming efficient and more cost effective. Along the same lines, the advent of telemedicine enabled by IoT technologies has brought care to more people (recall the recent case of rural and underserved areas), especially in remote areas.

While IoT adoption is prevalent in some geographical areas, there is significant disparity in rate of adoption between country and healthcare placement. The leading countries in the IoT healthcare implementation are high income countries in Europe and North America, while many low and middle income countries are yet to bear witness to their introduction. However, its potential remains unfulfilled by many of the healthcare institutions that are faced with problems with its infrastructure, its technology integration and its financial investment to deploy the system on a broad basis.

3.2 Sustainability Challenges: E-Waste, Energy Use, Ethical Concerns

The viability of IoT devices for healthcare has been rapidly growing, rendering the sustainability challenges with them an increasingly pressing issue. Electronic waste (e-waste) is one of the major emerging issues in the healthcare world department. Lifetime of IoT devices, especially wearables and sensors is very short and they are obsolete or non functional within a few years. With the increasing discarding of these devices, e-waste is now growing, which is almost all of which is not disposed of properly, causing environmental pollution and adverse effects to human health. Healthcare devices contribute e-waste which also has hazardous materials such as heavy metals, which if not appropriately handled can contaminate the soil and water.

In addition, IoT devices can optimize the delivery of healthcare, but also increase the energy need for healthcare systems. The extra demand on energy resources are incurred by the devices which need constant connection, data transmission and processing like in data centers and hospitals where lots of data is processed. A major challenge with the healthcare IoT is that healthcare systems globally are looking to achieve sustainability targets like reducing carbon emissions and landing the net zero energy use. IoT devices will have a potential negative impact on the entire sustainability of healthcare systems if they do not pass through an energy efficiency and environment friendly consideration (Arora, Rao, & Bhutani, 2020).

There are also ethical concerns surrounding the spreading use of IoT in health care, and particularly privacy and security of data. Sensitive personal health data is collected by IoT devices that are commonly transmitted and stored in cloud-based systems. Thus, they create potential vulnerabilities, which call for the strong data protection measures from the side of healthcare providers and device manufacturers. Equity issues also become ethical on the subject of providing benefits of IoT equally is not distributed across the different socioeconomic groups. Access by a disadvantaged population to the most recent technology or skill to use the technology may lead to disparity in healthcare quality and outcomes.

Figure 1. Key gaps in healthcare IoT adoption

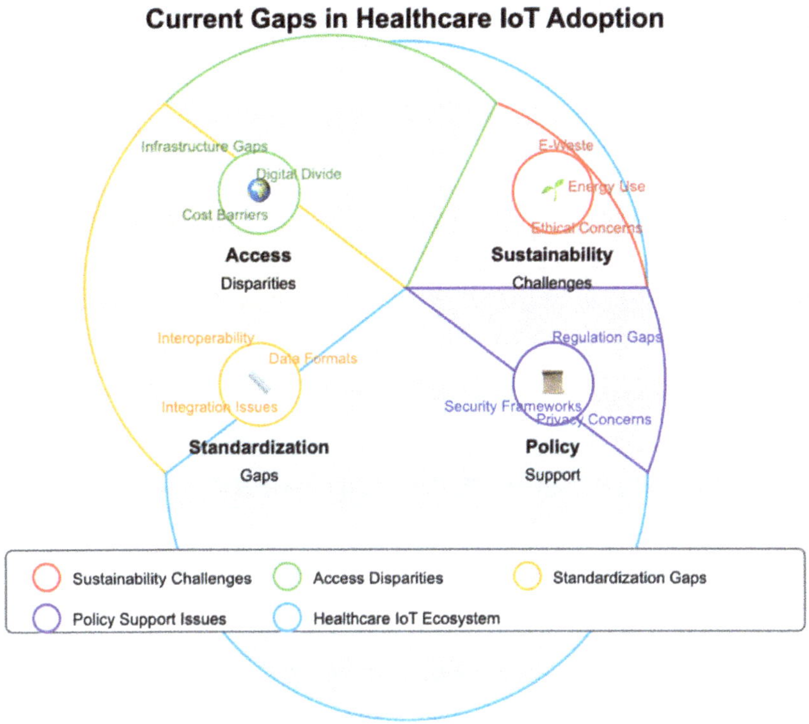

Figure 1 shows the main four vector gaps constraining the far-reaching adequate utilization of IoT in healthcare settings. I.e., the ecosystem of the healthcare IoT is shaped as a central circle connected with other challenges such as sustainability (e-waste, energy consumption, and ethical concerns) and access differences (developed and developing nations), standardization gaps for interoperability (lack of standardization) and system integration (lack of integrations), while lacking policy frameworks on privacy (lack of cryptographic security on health information), security, and regulatory compliance (lack of robustness in the system). For maximum value in healthcare through IoT technology, it is necessary to address these interdependent challenges while achieving equitable and sustainable scaling across a variety of healthcare settings (Settia, Bhutani, & Saini, 2025).

There are substantial gaps that currently prevent the widespread and sustainable adoption of IoT technologies in healthcare to improve patient care and thus the number of opportunities one gains in order to do so. The range of these gaps, as described in Fig. 1, encompassed four crucial domains: the sustainability concerns created by producing e-waste and utilizing energy, a lack of standardized wherein interoperability is affected, the access inequality between the developed & developing nations, and relative to the absence of policy frameworks. In the quadrant representation, these challenges are represented as they are interrelated, where one gap amplifies the others in a complex ecosystem. For example, the lack of sustainability standards contribute towards e-waste and the lack of infrastructure in the developing regions help widen the digital divide in healthcare access. To address these gaps, a coordinated approach with various health care provider, technology developer, policymakers, and environmental experts is needed to develop more sustainable solutions that will run across various health care settings and provide patient privacy, data security, and equitable access. As the healthcare IoT grows at accelerating speeds on a global scale, design from the get go with sustainable practices and having a holistic system of regulation for both minimization of environmental impact and healthcare disparities will be essential to maximize the benefits of healthcare IoT.

3.3 Disparities in Access and Infrastructure Between Developed and Developing Nations

Currently, the gap between developed and developing nations for adoption of healthcare IoT is a significant one. While IoT adoption is growing exponentially in high income countries with solid health care infrastructures, this is not the same story in low and middle income countries for which the adoption of IoT technology is hampered. It's a lack of infrastructure, such as internet connection, electricity, and cheap devices. Due to lack of resources, healthcare facilities in many rural or remote areas can not support the IoT technologies in these areas (Xames, & Topcu, 2025).

Most of developing nations suffer from shortage of skilled healthcare workers and technical expertise required to install and maintain the IoT systems. To deploy and manage IoT in health care settings, a workforce having the expertise and the knowledge of the technology and the healthcare application it supports is needed. Although such advancements are possible, many healthcare systems in these regions can not afford the training or resources required to accompany them. Additionally, the cost of IoT devices and required infrastructure are very high, thereby preventing many health care facilities in developing nations from owning them.

Similar to the access to healthcare IOT technologies, the digital divide also leads to health inequities. The IoT and healthcare devices that are IoT enabled are less likely to be present in rural populations and low income communities both in developed and developing countries. The lack of access to this state–of–the–art care compounds existing conditions of health disparities and, among these groups, the ones with the highest healthcare needs are the least likely to have access to it. The fullest potential of IoT for healthcare transformation will continue to be unrealized for a huge part of the global population without giving attention to filled up on these infrastructural, access and training gaps.

3.4 Lack of Standardization and Policy Support

A key gap in today's landscape of healthcare IoT adoption is the lack of standards and policy groundings there are for. As the healthcare IoT market continues to grow, multiple manufacturers developed numerous devices and platforms but do not have a unified method of guaranteeing interoperability in these systems. The lack of standardization prevents integration of different kinds of IoT devices within a functioning healthcare system. In order for healthcare IoT to fulfill its true potential, devices and systems need to communicate with one another easily throughout the sharing of secure, efficient data. However, standardization is not present in the integration of IoT technologies into healthcare settings which makes it expensive, time intensive and complex (Cassee, Bleeker, & Durand, 2024).

In addition, the governance environment where there is healthcare IoT is still developing. Most importantly, healthcare IoT is not legally regulated in many countries, and there are no comprehensive regulations or policies to regulate privacy, security, ethics of such systems. Without adequate regulatory frameworks protecting patient data, and safeguarding the safe and ethical use of IoT technologies, the deployment of these devices is a very high risk. In addition, there is a deficit of incentives and support for sustainable procurement activities like the advancement of energy efficient IoT devices or forming of policies for e-waste management.

Given the nature of IoT technology, the evolution of the healthcare industry will need to come up with more rigid and sustainable policy that will maintain these systems secure, private and health. Such policies should foster development of interoperable, energy efficient, and secure IoT solutions, which are not restricted to any of the predefined geographic areas and do not unnecessarily exclude people from great benefits owing to their social status, location, etc.

4. 2030 VISION: KEY GOALS AND OUTCOMES

By 2030, my wishes for a sustainable healthcare IoT vision have been built on the digital healthcare ecosystem to be innovative but also environmentally and socially responsible. Some core goals of this vision are to integrate digital health technology into healthcare in line with principles of sustainable development to contribute to the condition of the planet, people and health systems everywhere. To achieve this vision, competitive solutions in multiple domains, including emissions reduction, adoption of a circular economy, equitable access, and energy efficiency are needed in strategic action.

4.1 Net-Zero Emissions in Digital Healthcare Infrastructure

By 2030, net zero emissions in the health digital infrastructure is a highly critical goal for the healthcare IoT systems. Today, healthcare IoT is powered by the digital infrastructure that feeds that data center, servers, and networks, and each consume vast amounts of energy and contribute heavily to global carbon emissions. The environmental impact of data storage, transmission, and processing of the resulting IoT data will increase as more devices are deployed across healthcare settings unless steps are stronger well taken to mitigate its impact.

The objective in the 2030 vision is to create a healthcare IoT infrastructure with minimal carbon emissions. It will require the use of renewable forms of energy like solar, wind and hydroelectric power to supply healthcare data centers, hospitals and IoT devices. In addition, energy efficient computing and networking such as edge computing and low power sensors will reduce energy consumption by processing data closer to the source and they won't transmit the data to the centralize servers.

Additionally, carbon offsetting will be integral to making healthcare IoT net zero. To this, organizations will have to install carbon accounting mechanisms so that they can monitor and balance their greenhouse gas emissions through projects that cut or trap them. If healthcare systems establish and use emissions reduction targets and use of new technologies, they can drastically reduce their environmental footprint while maintaining the level of care and operations.

4.2 Circular Economy Adoption in IoT Device Design and Lifecycle

Circular economy principles are another cornerstone of the 2030 vision, with the intention to apply those principles to both the design and production of healthcare IoT devices as well as the disposal of them. While traditional linear models of production and consumption in the electronics industry have an enormous contribution on the environmental degradation, especially by excessive waste and resource extraction. On the other side, a circular economy aims to close the product life cycle loop by stressing on reuse, repair, refurbishment and recycling (waste reduction and reduction of raw materials need), not waste and discard.

It is expected that healthcare IoT devices used by 2030 will be designed with end-of-life considerations in mind, casting aside materials that will not give excess energy to pushing recyclable or reusable alloy from point A to point B. Manufacturers will concentrate on sustainable materials such as biodegradable plastics and/or less harmful to the environment metals. Also, such devices will be designed for longevity and upgradability so that healthcare facilities could extend the lifespan of IoT technologies by repairing rather than throwing away outdated equipment.

Establishment of take back programs and e waste recycling initiatives will be a key part of the healthcare IoT model of the circular economy. They facilitate the proper disposal of old, broken devices with hazardous materials disposed safely, and with valuable components recovered for reusing. As the healthcare sector continues to drive towards a circular economy, so it steps away from the environmental impact of IoT devices and towards broader sustainability goals that limit the ecological footprint of the healthcare industry.

4.3 Resilient, Equitable Digital Health Systems Accessible to All

By 2030, the healthcare IoT ecosystem will be based on equity and resilience. A resilient system takes challenges whether from social and economic adversities, pandemics or natural disasters and still provide quality care. An equitable system allows for all people, from any particular location, income, or social status to have equal access to healthcare service and technologies.

By 2030, the vision is to provide IoT based healthcare solutions to bridge the digital divide for the underserved and marginalized populations towards enjoying the digital health innovations. In order to realize this, policies and initiatives aimed at making IoT systems inclusive in their design and deployment will have to reflect inclusivity in their design and deployment of healthcare devices and digital platforms, making them affordable, accessible and usable by all people. They further asserted that the healthcare sector in low resource areas will also have to make investments in capacity building, training healthcare workers and strengthening the infrastructure, which would help these communities to use these technologies effectively (Alasiri & Mohammed, 2022).

Additionally, healthcare IoT solutions will be highly resilient to work within different contexts, that is rural, urban and emergency environments. In other words, this has to do with making sure IoT devices work on low bandwidth networks and without power, as well as with creating back up systems and contingency plans in case of crisis. The use of digital healthcare systems is increasing, and the advances in digital healthcare systems should ensure their reliable and uninterrupted operation, even under such challenges as climate-related disruptions or health care emergencies.

4.4 Intelligent, Patient-Centric IoT Systems Powered by Green Energy

In 2030, the healthcare IoT ecosystem will be based on intelligent systems focused on the care of the patient powered by renewable energy sources. These systems will be energy efficient and will use the most advanced technology in artificial intelligence (AI) and machine learning to provide the best personalized care to the patients. Patient centric IoT systems include those focus on patient care and allow real time data to monitor patient health, predict medical condition, and even suggest personalization interventions.

AI and data analytics will play into intelligent IoT systems for offering more exact diagnostics, superior medical treatment alternatives, and live updates on patients' health condition. This type of wearable devices will sense patients' vital signs at all times and analyse data to make early alerts on early signs of health issues such as heart attack and diabetic crises. Future health condition can also be predicted with these systems using historical data, which can lead to prevention and decrease of emergency care (Mondejar, 2021).

In these patient-centric systems, the integration of green energy will ensure that the healthcare delivery will be sustainable by using energy efficient devices and renewable powered networks. This will also lead in reducing healthcare costs because of energy efficiency and predictive analytics to reduce the amount of waste and make the resource utilization optimal. Moreover, AI driven optimization techniques will further decongest the hospital operations, such as optimizing the energy use in healthcare facilities to enhancing the efficiency of IoT devices.

Understandable, by this time next decade, healthcare IoT systems will have integrated with green energy such as solar powered wearable devices and energy efficient data centers to form a future proof as well as sustainable healthcare ecosystem. Through this, the times will change, and we will come up with a healthcare environment which not only improve the outcomes of a patient but also lessen a burden on the environment, thereby becoming an example of how technology is advancing a health whiles promoting global sustainability.

5. STRATEGIC PILLARS OF THE ROADMAP

To achieve the vision of sustainable healthcare IoT by 2030, a comprehensive and multi-faceted approach is required. The roadmap for this transformation will be grounded in several strategic pillars that address technological, regulatory, financial, and social dimensions. These pillars will guide stakeholders through the complex process of creating an eco-friendly, resilient, and equitable healthcare IoT ecosystem. The following sections outline the key pillars essential to this transformation.

5.1 Technological Innovation

The foundation of any sustainable healthcare IoT system is technological innovation. As the healthcare industry continues to digitize, it is essential to focus on developing technologies that minimize energy consumption, reduce e-waste, and optimize operational efficiency. The evolution of green hardware design, low-power IoT devices, and energy harvesting technologies will be crucial in ensuring that the environmental impact of digital health technologies is minimized (Gagneja, Lall, & Bhutani, 2024).

Green Hardware Design

Healthcare IoT devices must be designed in such a manner so that its lifecycle takes account of the minimal environmental footprint. It involves using eco-friendly materials, maximizing manufacturing processes as well as making the devices for easy recycling. Using materials such as the biodegradable plastics or reducing the use of rare earth metals, manufacturers can considerably lessen the ecological impact of healthcare IoT devices. In addition, products should be designed to have high longevity, to be modular and upgradable, which will help reduce the frequency of replaceable devices, thereby reducing waste and extending the possible operational life of the devices.

Low-Power IoT Devices

Because of continuous operation, healthcare IoT systems consume a lot of energy. The main purpose behind the low-power devices is to extract maximum output while consuming minimum energy to minimize the use of the environmental impact of IoT systems. IoT devices using low power protocols (such as LPWAN: Low Power Wide Area Network) can operate for an extended period without the need to be charged or powered with batteries or other power supplies often. It will be very important in reducing the energy consumption of wearable devices, patient monitoring systems and environmental sensors in healthcare facilities.

Energy Harvesting Technologies

Energy harvesting is a viable solution for powering the health care IoT devices independent of the conventional energy sources. Healthcare IoT devices that can operate autonomously by capturing and converting ambient energy from the environment— e.g. perform biological analysis, collect environmental data such as patient vital signs, harvest environmental energy. However, due to their valuable characteristics, these technologies are particularly well suited to remote environments or where there is no assurance that electricity is secure, as there can be self functioning IoT systems created to monitor patient wellness, monitor medical gear or take care of hospital foundations.

AI-Driven Optimization

Healthcare IoT systems can have a huge artificial intelligence (AI) impact on their operation. AI algorithms employed on IoT data from real time can enable the healthcare provider to predict patient outcomes, optimize resource usage, and streamline their operations. AI can also suggest improvements on the energy efficiency of healthcare systems by analyzing energy usage patterns to spot energy waste and its associated causes. An example is how IoT devices or infrastructure could be benefited with AI-driven predictive maintenance to predict when they are going to need repair to minimize downtime and extend the life span of critical systems (Kumar & Singh, 2025).

5.2 Policy and Regulation

Therefore, it is pivotal to have more comprehensive policies and regulatory framework in order to make the adoption of such sustainable healthcare IoT techs more widespread. It will be precisely governments and international organizations that will have to play a crucial role in incentivising the development of green technologies, ensuring proper electronic waste disposal and instigating the global cooperation on sustainability issues.

Incentives for Sustainable Procurement

Policy incentives, for example, that promote procuring eco-friendly devices and systems, are one of the best methods available to promote sustainability in healthcare IoT. Tax incentives, grants and subsidies can be introduced by governments to healthcare providers who undertake the purchase of energy efficient and environmentally friendly IoT devices. These policies would encourage companies to include sustainability into the product design, and health care organizations would have a fiscal reason to adopt greener technologies.

E-Waste Legislation

However, the healthcare industry holds the electronic waste (e-waste) in high regard as old IoT devices and medical equipment often end up in landfilling that causes environmental pollution as well as reduction of precious resources. Healthcare IoT devices will require governments to tighten up their e-waste legislation demanding proper recycling and disposal of these devices. This can include setting up e-waste recycling facilities, take back programs for old devices, and requiring manufacturers design products that are easier to recycle. It will also include provisions for the proper handling and disposal of hazardous material such as lead or mercury which is used in medical devices.

International Cooperation

As this is a global issue of the healthcare IoT, international cooperation will be needed for harmonizing regulation, exchange of best practices and alignment of sustainability goals. To develop shared sustainable healthcare IoT standards, multinational health organizations like WHO, UN will have to cooperate with national governments, industry stakeholders and environmental organizations. Fostering a cohesive approach to green ecosystem for health will be driven with combined international sustainability target, carbon emission, and circular economy principles.

5.3 Investment and Infrastructure

Investment in sustainable healthcare IoT solutions will be pivotal in achieving the 2030 vision. A combination of public and private investments is necessary to drive the development, deployment, and scaling of green IoT technologies. Strategic funding, infrastructure development, and sustainable supply chains will help overcome the barriers to widespread adoption (Qwaider & Al-Ramadan, 2023).

Public-Private Partnerships

There will be a need for collaboration between governments and private enterprises to accelerate the transition of the healthcare IoT towards a sustainable one. PPPs can help to bridge the gap needed for providing support for the development and deployment of sustainable technologies through the sharing of resources, expertise and funding. Through PPPs, governments can use incentives and financial help so that the private companies involved in green innovations in healthcare do so. These partnerships also contribute to scaling successful pilot projects, or in putting an end to healthcare IoT solutions towards a mass audience.

Smart Funding for Green R&D

In order to create sustainable healthcare IoT systems, investment in research and development (R&D) of necessary technological innovations will be crucial. The R&D on green in energy efficient devices, green in manufacturing techniques and green in energy harvesting should be directed by governments, philanthropic organizations and private investors. For smart funding strategies, time is of the essence;

usually, long term sustainability goals must be the main priorities and resources should not be distributed to the projects that have less potential to impact or scale.

Sustainable Supply Chains

However, the supply chain of healthcare organizations must be sustainable at every stage, starting from source of raw materials up to device manufacturing and disposal of end of life. It includes working with suppliers following sustainability standards, reuse of materials and reduction of carbon footprint in transportation and logistics. Building sustainable supply chains mitigates the environmental impact on healthcare providers as they use technologies that match healthcare providers' sustainability goals.

Table 2. Strategic pillars for achieving sustainable healthcare IoT by 2030

Strategic Pillar	Description	Key Focus Areas
Technological Innovation	Focus on developing IoT devices that minimize energy consumption, reduce e-waste, and optimize efficiency	Green hardware design, low-power devices, energy harvesting, AI-driven optimization
Policy and Regulation	Implementation of policies and regulations to promote sustainability in healthcare IoT	Incentives for sustainable procurement, e-waste legislation, international cooperation
Investment and Infrastructure	Investment in sustainable healthcare IoT through public-private partnerships and smart funding	Public-private partnerships, green R&D funding, sustainable supply chains

Table 2 outlines the strategic pillars necessary to achieve a sustainable healthcare IoT ecosystem by 2030. Technological innovation is fundamental, focusing on reducing energy consumption and e-waste through green hardware design, low-power IoT devices, energy harvesting technologies, and AI-driven optimization. Policy and regulatory frameworks must incentivize sustainable procurement, enforce e-waste management laws, and foster international cooperation on sustainability goals. Finally, investment and infrastructure development, particularly through public-private partnerships and targeted funding for green R&D, are critical for scaling sustainable healthcare IoT solutions. These pillars, collectively, will guide the healthcare sector towards a more eco-friendly, resilient, and equitable future in digital health technology.

6. MILESTONES AND TIMELINES (2025–2030)

The journey toward achieving a sustainable healthcare IoT ecosystem by 2030 will require strategic milestones, carefully planned timelines, and consistent progress monitoring. These milestones will serve as key indicators for the advancement of sustainable digital health systems and the broader integration of IoT technologies within healthcare infrastructures globally. The period from 2025 to 2030 will encompass three distinct phases: short-term, mid-term, and long-term. Each phase will focus on specific actions and achievements that contribute to the overarching goal of a greener, more resilient, and equitable healthcare IoT landscape.

Figure 2. Sustainable healthcare IoT implementation roadmap (2025-2030)

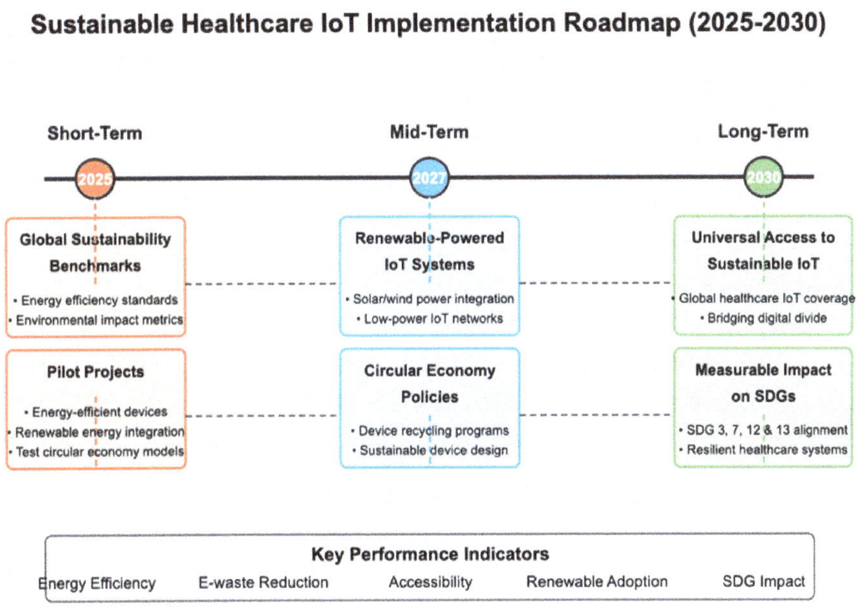

Figure 2 demonstrates the strategic implementation roadmap of sustainable healthcare IoT from 2025 to 2030, and contains three phases: short term (2025), mid term (2027) and long term (2030). The milestones of each phase focus on the establishment of global sustainability benchmarks, pilot projects at the short term, renewable powered IoT system scaling over the mid term and universal access to sustainable IOT systems by 2030 which will directly impact SDGs. This figure shows how milestones develop into the progressive and interconnected way of a fully integrated, intelligent and environmentally sustainable healthcare IoT ecosystem.

To accomplish the goal of implementing sustainable healthcare IoT, an approach involving milestones that identify and delineate progress toward environmentally accountable and equitable digital health systems is necessary. Figure 2 displays a three phase strategy (building on one another) to achieve a holistic change in the health care IoT from 2025 through 2030. This short term phase is to establish the global sustainability benchmarks of energy efficiency and its environmental impact while also launching pilot projects within real world healthcare setting to test the integration of renewable energy and circular economy principles. The actions performed in the first part of the lifecycle create the requisite knowledge base and standards for which the final part formalizes a circular economy policy and an IoT system in renewable power in mid term, by 2027. By 2030, these additive efforts should, in total, result in universal access to environmentally sustainable healthcare IoT systems with verifiable contributions to several Sustainable Development Goals, specifically SDGs 3, 7, 12 and 13. This timeline reflects that healthcare IoT implementation can only proceed with technological transformation with the concurrent emergence of policy, infrastructure, and stakeholder involvement, the outcomes of which can contribute to both environmental and social aspects of the sustainability agenda as accomplished by addressing both environmental and social dimensions of sustainability at the same time (Panchal, S. 2022).

6.1 Short-Term Milestones (2025)

The short term phase or from the present time to 2025, will prepare the grounds for a sustainable ecosystem for healthcare IoT. In this phase, major goals will be to set up global sustainability benchmarks as well as launch pilot projects that demonstrate innovative and green application in real healthcare settings.

Establish Global Sustainability Benchmarks

The step in the short term of creating global reference benchmarks for sustainable healthcare IoT will be the first crucial milestone. These will become the barometer for energy efficiency, environmental impact and sustainability of IoT devices, healthcare infrastructure and operations. The benchmarks for these concepts will be defined by international organizations like WHO, UNDP and relevant standardization bodies who will make them applicable internationally to various healthcare systems all over the world. These will be used as benchmarks by governments, healthcare providers and manufacturers in times when they may consider adopting greener, more sustainable roadmaps. This will create the opportunity for an involvement in global comparison of the efforts the sustainability and tracking the progress towards the specific goals.

Pilot Projects and Proof-of-Concept Initiatives

During the short term phase pilot projects will be initiated for testing and refinement of sustainable healthcare IoT solutions. As part of these pilot programs, energy efficient devices, renewable energy sources and circular economy principles will be integrated into healthcare IoT systems. The pilot projects will be launched half in developed region and half in the developing regions therefore ensuring the possible adaption to various healthcare infrastructures. Programs piloting successful sustainable IoT technologies will help with obtaining necessary data on the feasibility, cost effectiveness, and scalability of such technologies so that, later on these can be provided on a larger scale. What lessons were learned from these projects will also inform the development of future policy, regulation, and best practice for sustainable healthcare IoT (Hira, F. A., Khalid, H., & Moshiul, A. M. 2024).

6.2 Mid-Term Milestones (2027)

From 2025 to 2027, the mid term phase will concentrate on scaling throughout successful innovations and framework sustainable healthcare IoT systems. Efforts will be made during this period to deploy renewable powered IoT systems on a large scale and embed circular economy policies in its practice.

Scaling Renewable-Powered IoT Systems

By 2027, the use of renewable energy sources in healthcare IoT systems will be the focal point of any integration. This will encourage hospitals, clinics, as well as healthcare providers to transition from usually used energy sources to renewable, like solar or wind power, in order to provide the operation of the IoT devices and infrastructure. Specifically, in areas that would have limited access to a reliable energy grid, renewable powered IoT systems will be deployed on a larger scale. Switching to renewable energy for healthcare IoT systems will not just cut the carbon footprint but also improve the sustainability and resiliency of a digital health infrastructure. It will aim to get a large proportion of IoT devices operating on renewable energy sources by 2027.

Implement Circular Economy Policies

Then we need to get to the next critical milestone of implementing circular economy policies which support energy efficient, circular device design, and reuse as well as recycling within the healthcare IoT sector. Healthcare organizations and manufacturers will adopt circular design principles by 2027, which will force the creation of devices built to last, be recyclable and have a minimum negative impact on the environment. This will also force the take-back and recycling of end of life devices under national and international regulations resulting in a closed loop system, creating an environment of minimal e waste and maximum resource efficiency. Second, refurbished or recycled devices will be encouraged to be used by the healthcare providers, thus minimizing the demand for new raw materials and waste generation.

Increased Focus on Data Governance and Ethics

Between now and 2027 there will be more defined how this data generated by these healthcare IoT devices be handled not only technically but ethically. Healthcare organizations and policymakers will develop robust data governance frameworks to protect patient privacy, secure data storage and appropriate sharing of data. The ethical implications of consent, ownership and use of healthcare data will be emphasized with minimization of the environmental charges of data centers and data transfer. There will be more concrete guidelines of how to intermationally handle data from people and how to do so in a ethical way and with it will become a standard of doing so in healthcare across the globe.

6.3 Long-Term Milestones (2030)

The long-term milestones, set for 2030, represent the culmination of the efforts undertaken in the previous phases. By 2030, the goal is to achieve universal access to sustainable healthcare IoT systems, with measurable impacts on the Sustainable Development Goals (SDGs) and global health equity.

Universal Access to Sustainable Healthcare IoT

By 2030, sustainable healthcare IoT should be available everywhere, at any geographic location, or anywhere at any economic status. The first benchmark on this list will be to close the digital divide and make sure that even the most underserved and remote populations have access to the advantages of digital health technology. In healthcare facilities around the world, as well as in low resource settings, the standard will be healthcare IoT systems running on renewable energy. The digital health solutions will be equitable—one that benefits all the segments of the population as well as the marginalized ones—with this global access to sustainable technologies.

Measurable Impact on SDGs

The long term goal is to substantially improve the healthcare IoT systems and widely use sustainable healthcare IoT systems to measure progress on the SDGs. Healthcare systems will play a direct contribution to multiple SDGs by 2030, such as to good health and well being (SDG 3), affordable and clean energy (SDG 7), responsible consumption and production (SDG 12), and action on climate (SDG 13). The health care IoT systems will play an important role in making progress in these areas by promoting sustainable health practices, reducing resource consumption and minimizing environmental impact.

Fully Integrated, Intelligent, and Resilient Healthcare IoT Ecosystems

The healthcare IoT ecosystem is integrated into the broader healthcare infrastructure by 2030 where intelligent patient centric system will drive the healthcare delivery. Using AI and data analytics to optimize patient care, resource allocation, operational efficiency and minimise environmental impact, these systems will exist. Green energy will power healthcare IoT devices, and all systems will be compliant with the circular economy principles considering, that sustainability will be a touchstone during and after the complete lifecycle. Such healthcare ecosystem is one that is integrated with sustainable technologies to the extent that it can be resistant to environmental, social and economic challenges.

Continuous Monitoring and Improvement

Continuous monitoring and improvement of the healthcare IoT systems will serve as the last 2030 milestone with the creation of the mechanisms. The sustainability targets and SDG impact will be monitored by governments, international organisations and industry stakeholders who will reflect on the progress made on a regular basis. However, healthcare IoT systems will continue to evolve by using ongoing data collection, research and feedback from stakeholders so that they will still adapt to the emerging challenges and opportunities.

In conclusion, the milestones set for 2025, 2027, and 2030 provide a clear roadmap for achieving a sustainable healthcare IoT ecosystem. The successful realization of these milestones will require a concerted effort from governments, industry leaders, healthcare providers, and other stakeholders. As progress is made toward these goals, healthcare IoT systems will become more energy-efficient, accessible, and aligned with global sustainability and health equity objectives.

7. COLLABORATIVE FRAMEWORKS AND GLOBAL ALLIANCES

No entity can make the transition to sustainable healthcare IoT ecosystem by itself. They must involve various stakeholders from global organizations, governments, private bodies, academic institutions, and civil society, to name a few. In order to achieve sustainable healthcare transformation, it's important to bring in the complexities in multilateral collaborations and alliances to make sure the required resources, expertise and innovative solutions are in all regions, particularly in underserved populations. This section looks at the roles of key global institutions, the need for multilateral cooperation and the possibilities in open source platform for fostering democracy.

Major global health organizations like World Health Organization (WHO), United Nations Development Programme (UNDP), World Economic Forum (WEF) guide the global conversation on healthcare sustainability and the digital health transformation. They help put in place frameworks for collaboration, shape policy and mobilize resources for sustainable healthcare solution. They can use their authority and global distribution to promote international standards, guidelines and action plans(Jamoussi, B., Abu-Rizaiza, A., & AL-Haij, A. 2022).

The WHO is a key actor to define the global health priorities and to integrate the sustainable healthcare technologies into them. By defining health related sustainability goals for the organisation, promoting the use of green technologies or advocating for collaboration between the governments and the private sector players can guide the development of the sustainable healthcare IoT systems. The WHO is positioned to lead global health initiatives and convene stakeholder from disparate sectors to such an extent that the healthcare IoT systems maintain alignment with international health and environmental policies where possible. The WHO can also aid in efforts to develop a global infrastructure for giving IoT solutions to healthcare to low income and remote areas(Ariffin, N. H. M., Yusoff, F. H., 2024).

The second key player in the global drive for sustainability, both in general and in the integration of such into healthcare, is the UNDP. In the broader effort towards the achievement of the Sustainable Development Goals (SDGs), the UNDP can contribute to generate financing, technical assistance and policy expertise to support the wide dispersion of healthcare IoT systems. As an ideal partner to create inclusive and green digital health ecosystem, its focus on addressing inequalities, delivering healthcare access and environmental sustainability remains its strong point. Sustainability in the healthcare IoT systems can be accelerated by the UNDP through its partnership with governments, businesses and civil societies.

8. CONCLUSION

With an eye toward the future of sustainable healthcare IoT, we realize that if we are to be on track to reach 2030 on this vision, that there are a number of technological innovations that need to be made, as well as strong policy of what and how things will be done to come to pass and the collaboration of innovators globally. Healthcare is ominated for structutral transformation, which is driven by climate change, demographic changes, and digitisation of health care services. If healthcare IoT is appropriately used responsibly and sustainably, it has the potential to alleviate many of these issues by enhancing patients' outcomes, improving care, and having a lesser environmental impact. In this chapter, we discuss these drivers, challenges and opportunities of the sustainable healthcare IoT system development. The 2030 road map toward an integrated and inclusive approach to address healthcare's most pressing issues comes from global pressures of environmental sustainability and demographic shifts, the policy and regulatory frameworks shaping digital health innovation, and what and how healthcare really can, can and may need to become. Based on the technological innovations on green hardware design, energy efficiency, AI driven optimization and policies, investment, and capacity building efforts described below, this transformation will be built upon the pillars of technological innovations, policy incentives, investment on infrastructure, and capacity building.

9. FUTURE DIRECTIONS

While the healthcare IoT starting 2030 provides for modification, evolution will continue in the future, driven by development of technology driven and global collaboration. The next stage in this transformation will probably witness the continuation of the acceleration of green IoT innovations and the extent to which AI, machine learning and edge computing will be engaged in the healthcare applications. Rising economies, aging pandemics, the strengthening expandability, and the increasing cost of employing staff will lead to novel approaches to improve patient care, reduce costs, and improving the environmental outcomes of facilities. The future paths will see the development of fully autonomous IoT healthcare systems of themselves, which do not only need to save energy, but are also intelligent enough to predict the needs of its patients and provide personalized and proactive care. Such systems, based on artificial intelligence and powered with renewable energy will be able to control patient cases in real time and cut the number of hospital visits and medical emergencies. In addition, wearable and remote monitoring devices connected to real time data analytics will assist in monitoring the environmental conditions that help valuable information regarding patients and healthcare providers. One other thing is, in the future, there will be greater focus on ethical data governance and patient privacy as well. With that comes a massive amount of sensitive data being generated about everyone's bodies by the healthcare IoT system, which requires safe, secure, and transparent data handling practices to be adhered to. Decentralizing data models of patients will solve the concern regarding privacy and gain trust in digital healthcare. In addition, further developments in data transactions in the health care area law are possible thanks to the advancement of blockchain technology.

REFERENCES

Abaee, M. (2023). A systematic review of how cloud computing (CC) and the Internet of Things (IoT) can contribute towards achieving the sustainable development goals (SDGs). *Internet of Things : Engineering Cyber Physical Human Systems*, *22*, 100805. DOI: 10.1016/j.iot.2023.100805

Abaee, M., Saeedi, M., & Taherdoost, H. (2024). IoT and cloud computing for sustainable development goals in Industry 4.0: A bibliometric analysis. In Tyagi, A. K., & Rajest, S. S. (Eds.), *Reshaping environmental science through machine learning and IoT* (pp. 98–118). IGI Global., DOI: 10.4018/979-8-3693-2351-9.ch006

Alasiri, A. A., & Mohammed, V. (2022). Healthcare transformation in Saudi Arabia: An overview since the launch of Vision 2030. *Health Services Insights*, *15*, 1–7. DOI: 10.1177/11786329221121214 PMID: 36081830

Albreem, M. A., Sheikh, A. M., Bashir, M. J. K., & El-Saleh, A. A. (2023). Towards green Internet of Things (IoT) for a sustainable future in Gulf Cooperation Council countries: Current practices, challenges, and future prospects. *Wireless Networks*, *29*(2), 539–567. DOI: 10.1007/s11276-022-03133-3

Almuaythir, S., Singh, A. K., Alhusban, M., & Daoud, A. O. (2024). Robotics technology: Catalyst for sustainable development—Impact on innovation, healthcare, inequality, and economic growth. *Discover Sustainability*, *5*(1), 486. DOI: 10.1007/s43621-024-00744-y

Ariffin, N. H. M., Yusoff, F. H., Abu Mansor, N. N., Javed, S., & Nazir, S. N. (2024). Assessment of sustainable healthcare management policies and practices in Oman: A comprehensive analysis using an AI-based literature review. *Multidisciplinary Reviews*, *7*(11), 2024257. DOI: 10.31893/multirev.2024257

Arora, A., Rao, A., & Bhutani, M. (2020). A MATLAB simulation model for MAC layer of visible light communication. In *2020 7th International Conference on Signal Processing and Integrated Networks (SPIN)* (pp. 941–945). IEEE. DOI: 10.1109/SPIN48934.2020.9071254

Bhutani, M., Lall, B., & Agrawal, M. (2022). Optical wireless communications: Research challenges for MAC layer. *IEEE Access : Practical Innovations, Open Solutions*, *10*, 126969–126989. DOI: 10.1109/ACCESS.2022.3225913

Cassee, F. R., Bleeker, E. A. J., Durand, C., Exner, T., Falk, A., Friedrichs, S., Heunisch, E., & Wijnhoven, S. W. P. (2024). Roadmap towards safe and sustainable advanced and innovative materials. *Computational and Structural Biotechnology Journal*, *25*, 105–126. DOI: 10.1016/j.csbj.2024.05.018 PMID: 38974014

Fahal, A., Smith, D. J., Nyaoke, B., Asiedu, K., Falves, F., Warusavithana, S., Argaw, D., & Hay, R. (2024). Towards enhanced control of mycetoma: A roadmap to achieve the UN's sustainable development goals by 2030. *Transactions of the Royal Society of Tropical Medicine and Hygiene*, *118*(10), 646–651. DOI: 10.1093/trstmh/trae016 PMID: 38530874

Gagneja, A., Lall, B., & Bhutani, M. (2024). Unveiling CM-Det: Leveraging ConvMixer architecture for advanced object detection. *International Journal of Information Technology : an Official Journal of Bharati Vidyapeeth's Institute of Computer Applications and Management*, *16*(7), 4273–4278. DOI: 10.1007/s41870-024-02069-5

Goyal, R., Elawadhi, O., Sharma, A., Bhutani, M., & Jain, A. (2024). Cloud-connected central unit for traffic control: Interfacing sensing units and centralized control for efficient traffic management. *International Journal of Information Technology : an Official Journal of Bharati Vidyapeeth's Institute of Computer Applications and Management*, *16*(2), 841–851. DOI: 10.1007/s41870-023-01527-w

Hira, F. A., Khalid, H., & Moshiul, A. M. (2024). Blockchain-based electronic health records: Revolutionizing healthcare in Malaysia. *Asia Pacific Journal of Health Management*, *19*(1), 226–233. DOI: 10.24083/apjhm.v19i1.2621

Holt, S. G., Koornneef, E., Al Obeidli, A. A. K., Hubbert, L., & Nicholson, L. (2024). Decarbonisation of kidney care in the United Arab Emirates: A roadmap to an environmentally sustainable care. *International Journal of Nephrology and Renovascular Disease*, *17*, 241–253. DOI: 10.2147/IJNRD.S481121 PMID: 39430398

Jamoussi, B., Abu-Rizaiza, A., & Al-Haij, A. (2022). Sustainable building standards, codes, and certification systems: The status quo and future directions in Saudi Arabia. *Sustainability (Basel)*, *14*(16), 10314. DOI: 10.3390/su141610314

Kasinathan, P., Pugazhendhi, R., Elavarasan, R. M., Ramachandaramurthy, V. K., Ramanathan, V., Subramanian, S., Kumar, S., Nandhagopal, K., Raghavan, R. R. V., Rangasamy, S., Devendiran, R., & Alsharif, M. H. (2022). Realization of sustainable development goals with disruptive technologies by integrating Industry 5.0, Society 5.0, smart cities and villages. *Sustainability (Basel)*, *14*(22), 15258. DOI: 10.3390/su142215258

Kumar, R., Singh, A., Ahmed Kassar, A. S., Humaida, M. I., Joshi, S., & Sharma, M. (2025). Leveraging artificial intelligence to achieve sustainable public healthcare services in Saudi Arabia: A systematic literature review of critical success factors. *Computer Modeling in Engineering & Sciences*, *142*(2), 1191–1228. DOI: 10.32604/cmes.2025.059152

Mani, Z. A., & Goniewicz, K. (2024). Transforming healthcare in Saudi Arabia: A comprehensive evaluation of Vision 2030's impact. *Sustainability (Basel)*, *16*(8), 3277. DOI: 10.3390/su16083277

Mondejar, M. E., Avtar, R., Baños Diaz, H. L., Dubey, R. K., Esteban, J., Gómez-Morales, A., & Hallam, B. (2021). Digitalization to achieve sustainable development goals: Steps towards a smart green planet. *The Science of the Total Environment*, *794*, 148539. DOI: 10.1016/j.scitotenv.2021.148539 PMID: 34323742

Panchal, S. (2022). A road map to achieve sustainable goals in the healthcare system: A perspective of a pharmacist. *Nirma University Journal of Pharmaceutical Sciences*, *9*(1–2), 61–65.

Qwaider, S., Al-Ramadan, B., Shafiullah, M., Islam, A., & Worku, M. Y. (2023). GIS-based progress monitoring of SDGs towards achieving Saudi Vision 2030. *Remote Sensing (Basel)*, *15*(24), 5770. DOI: 10.3390/rs15245770

Settia, N., Bhutani, M., & Saini, V. (2025). Exploring the impact of artificial intelligence on healthcare: A comprehensive review. In Tyagi, A. K., & Rajest, S. S. (Eds.), *Digitalization and the transformation of the healthcare sector* (pp. 115–138). IGI Global., DOI: 10.4018/979-8-3693-5577-0.ch006

Verdejo Espinosa, Á., López, J. L., Mata Mata, F., & Espinilla Estevez, M. (2021). Application of IoT in healthcare: Keys to implementation of the sustainable development goals. *Sensors (Basel)*, *21*(7), 2330. DOI: 10.3390/s21072330 PMID: 33810606

Xames, M. D., & Topcu, T. G. (2025). How can digital twins support the economic, environmental, and social sustainability of healthcare systems: A systematic review focused on the triple-bottom-line. *IEEE Access : Practical Innovations, Open Solutions*, *13*, 1594–1613. DOI: 10.1109/ACCESS.2025.3559502

Chapter 3
Sustainable Healthcare Through IoT and Circular Economy

Himanshu Tyagi
https://orcid.org/0009-0001-8435-9832
Quantum University, India

Anju Bhatt
https://orcid.org/0009-0002-3617-7644
JB Institute of Technology, India

Pawan Goel
https://orcid.org/0000-0003-3601-102X
RKGIT, India

Rohan Chaudhary
https://orcid.org/0009-0003-1386-3470
Quantum University, India

Himani Tyagi
https://orcid.org/0000-0002-5005-9954
Quantum University, India

ABSTRACT

Environmental impact, depleting resources, and the production of wastes pose significant challenges to the healthcare industry. This chapter explores how a healthier, more sustainable healthcare system can be constructed through the integration of the Internet of Things (IoT) and the principles of the circular economy. Enhanced patient outcomes, minimization of wastes, and the optimization of resources are all

DOI: 10.4018/979-8-3373-4134-7.ch003

achievable for healthcare professionals who employ IoT technology for data collection, monitoring, and analysis. A closed-loop system minimizing environmental impact is developed by the circular economy model, which promotes the development of resilient, reusable, and recyclable medical devices and gadgets. How IoT and circular economy strategies have the potential to transform healthcare, minimize its ecological footprint, and enhance its sustainability in the long term is analyzed in this chapter.

1. INTRODUCTION TO THE NEED FOR SUSTAINABLE HEALTHCARE

1.1 Environmental Impact of Healthcare

The healthcare sector accounts for about 4.4% of global greenhouse gas emissions, the equivalent of the emissions from 514 coal-fired power plants each year, qualifying it as an important global contributor to environmental degradation. The impact arises from energy-consuming operations, transportation, and production and disposal of medical products. Hospitals, being the major healthcare providers, are dependent on fossil fuel-burning energy to support lighting, heating, ventilation, and medical equipment demands (Eckelman & Sherman, 2016).

The sector also produces a lot of waste, such as biohazardous waste, single-use plastics, and outdated equipment, resulting in landfill and pollution. The global supply chain—resource extraction, production, and cross-continent shipping of medical equipment—also increases healthcare's carbon footprint (Lenzen et al., 2020).

Healthcare's environmental footprint goes beyond emissions to ecological disturbance. For example, chemical reactions in the production of pharmaceuticals tend to discharge contaminants into water bodies, negatively impacting aquatic ecosystems. Inappropriate disposal of unused drugs can pollute water and soil, threatening human and environmental health (World Health Organization, 2021). Combating these issues means a paradigm change towards sustainable procedures that give much importance to environment conservation but sustain high-quality care for patients (Eckelman & Sherman, 2016; Lenzen et al., 2020).

1.2 Challenges of Resource Depletion and Waste

The health sector's dependence on limited resources, including the use of rare earth materials for imaging machinery, polymers for syringes and packaging materials, and water for cleaning and sterilization, is primarily responsible for depleting resources (Eckelman & Sherman, 2016). According to the World Health Organization (2021), more than 13 pounds of waste are produced per patient per day by hospitals, with much of this being disposable items such as needles, gloves, and gowns. Toxic materials, like infectious waste and sharps, have special disposal needs, adding expense and environmental hazards. The prevailing linear "take-make-dispose" healthcare model, where little is recycled or reused, exacerbates scarcity by creating a need for virgin materials (Lenzen et al., 2020).

The large-scale nature of global healthcare operations complicates waste management. In resource-poor environments, poor infrastructure tends to result in improper disposal or open burning, which puts communities at risk of health effects and produces toxic emissions. In industrialized countries, the heterogeneity of medical waste streams—isolation of biological materials, electronics, and plastics—presents logistical and regulatory obstacles. The build-up of waste and the depletion of resources highlight the need for immediate adoption of sustainable practices that lengthen material lifecycles and reduce environmental degradation (Eckelman & Sherman, 2016; Lenzen et al., 2020).

1.3 The Role of Sustainability in Modern Healthcare

In health, sustainability is the incorporation of social, economic, and environmental factors into the provision of healthcare. Environmentally, it emphasizes minimizing waste, pollution, and resource use. Socially, it provides access to healthcare while safeguarding communities from resource degradation and pollution. Economically, it advocates for cost savings through long-term resilience and optimal resource utilization (Eckelman & Sherman, 2016). Sustainable healthcare is consistent with international goals, including the United Nations' Sustainable Development Goals (SDGs), specifically SDG 3 (Good Health and Well-Being), SDG 12 (Responsible Consumption and Production), and SDG 13 (Climate Action).

Both ethical necessity and practical necessity compel the implementation of sustainable practices. Growing pressure from patients, stakeholders, and regulators forces healthcare providers to reduce their environmental impact. In addition, sustainability can provide financial returns, e.g., in the form of lower operating costs through waste reduction and energy conservation (Eckelman & Sherman, 2016). Adopting principles such as the circular economy and capitalizing on technologies like the Internet of Things, healthcare can evolve from a system that is high in resource usage and waste production to one that is regenerative and resilient (Lenzen et al., 2020).

Figure 1. Key applications of IoT in healthcare, highlighting their contributions to efficiency and waste reduction.

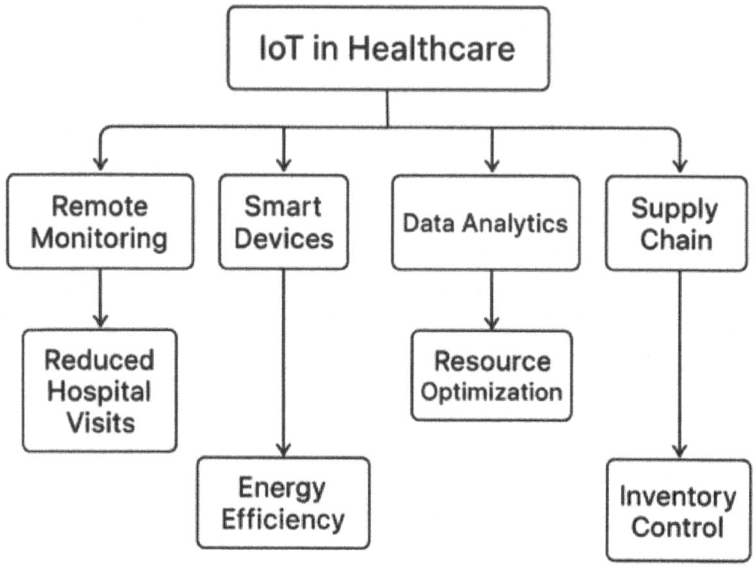

2. IoT IN HEALTHCARE: ENHANCING EFFICIENCY AND REDUCING WASTE

The Internet of Things (IoT) is revolutionizing healthcare by enhancing efficiency and minimizing waste. IoT devices like smart sensors and wearable monitors make the best use of resources through monitoring equipment, inventory management, and real-time monitoring of patients' health (Kelly et al., 2020). IoT systems, for example, can automate supply chains, eliminating excess stock and lowering waste from spoiled medical supplies (Deloitte, 2018). Intelligent energy management systems in hospitals reduce energy use, minimizing environmental footprint. IoT also improves patient treatment by enabling remote monitoring, minimizing unnecessary hospital trips and resulting emissions (Lenzen et al., 2020).

IoT deployment is not without challenges, such as data security and initial expenses (Kelly et al., 2020). But in the long run, there are economic and environmental dividends in terms of minimal operational expense and minimized resource utilization (Deloitte, 2018). With the inclusion of IoT, healthcare can move towards a circular economy paradigm, lengthening material lifecycles and reducing waste

(Lenzen et al., 2020). As in line with Sustainable Development Goals, IoT encourages sustainable healthcare approaches balancing efficiency, patient care, and environmental management.

2.1 Remote Patient Monitoring and Telemedicine

Remote patient monitoring (RPM) and telemedicine are transforming healthcare by increasing access, enhancing patient outcomes, and minimizing environmental footprint. RPM leverages IoT-enabled devices, including wearable sensors and mobile health applications, to monitor vital signs and health metrics in real time, enabling providers to remotely monitor patients (Kelly et al., 2020). Telemedicine supports RPM by enabling virtual consultations, minimizing the necessity for in-person visits (Smith et al., 2020). These technologies enhance care access, especially for underserved and rural populations, supporting the United Nations' Sustainable Development Goal 3 (Good Health and Well-Being).

Environmental considerations, telemedicine and RPM minimize the carbon footprint of healthcare. By lessening patient travel, these technologies lower transportation-related emissions (Holmner et al., 2014). Virtual consultations also minimize the energy and resources used by hospital visits, like electricity for heating and lighting (Lenzen et al., 2020). For instance, one study estimated that telehealth would cut healthcare-related travel emissions by as much as 20% in some areas (Holmner et al., 2014). Moreover, RPM maximizes resource utilization by facilitating early interventions, minimizing hospital readmissions, and lessening dependency on resource-heavy emergency care (Kelly et al., 2020).

Challenges notwithstanding, these have advantages. Cybersecurity and data privacy concerns are major risks since RPM devices and telemedicine platforms deal with sensitive patient data (Smith et al., 2020). High upfront expenditures on infrastructure and training can also impede take-up, especially in low-resource environments. Still, long-term economic gains through decreased healthcare spending and enhanced operating efficiency outweigh them (Kelly et al., 2020). In addition, the incorporation of RPM and telemedicine into healthcare systems facilitates a move toward sustainable practice by lowering waste and encouraging effective resource utilization (Lenzen et al., 2020).

Healthcare systems need to overcome regulatory and technological hurdles while providing equal access to digital health solutions in order to maximize impact. With the use of RPM and telemedicine, healthcare can become more resilient, patient-focused, and environmentally friendly, helping to achieve global sustainability objectives.

2.2 Smart Medical Devices and Equipment

Intelligent medical equipment and devices, driven by the Internet of Things (IoT), are revolutionizing healthcare through increased efficiency, better patient outcomes, and sustainability. Such devices, such as smart inhalers, wearable biosensors, and networked imaging systems, allow for real-time monitoring and data-driven decision-making (Kelly et al., 2020). By optimizing the use of equipment and avoiding unnecessary procedures, intelligent devices reduce resource utilization and waste in line with Sustainable Development Goal 12 (Responsible Consumption and Production). For example, Internet of Things (IoT)-connected inventory systems avoid overstocking of medical materials, limiting waste from expired products (Deloitte, 2018).

Smart technology also helps reduce energy consumption. Intelligent diagnostic systems can go into low-power modes when not in use, lowering hospital energy loads (Lenzen et al., 2020). Predictive maintenance through IoT also minimizes equipment downtime and maximizes device lifecycles, reducing the demand for energy-intensive replacements (Kelly et al., 2020). Such innovations enable a circular economy system, reducing the environmental footprint.

Challenges are high implementation costs, interoperability problems, and cybersecurity threats (Deloitte, 2018). In spite of these challenges, long-term advantages—cost savings and minimized environmental impact—render smart devices a necessity for sustainable healthcare (Lenzen et al., 2020). Through the integration of these technologies, healthcare systems can attain resilience and environmental stewardship without sacrificing care quality.

2.3 Data Analytics for Resource Optimization

Data analytics is transforming healthcare through optimization of resource usage, minimization of waste, and sustainability maximization. Leveraging big data, machine learning, and predictive modeling, the healthcare system is able to integrate operations, cut down on environment impact, as well as reduce patient complications (Bates et al., 2014). Data analytics promotes accurate inventory control, ensuring optimum levels of medicine supplies, preventing wastage that comes with expiring or not using items (Deloitte, 2018). For example, predictive algorithms will be able to predict demand for resources such as syringes or medicines, avoiding overstocking and supporting Sustainable Development Goal 12 (Responsible Consumption and Production).

Energy usage, one of healthcare's most substantial environmental contributions, is another sector where analytics optimizes efficiency. Live data from intelligent sensors in hospitals can maximize lighting, heating, and cooling systems, decreasing energy consumption and carbon emissions (Lenzen et al., 2020). An analysis projected that data-driven energy management would reduce hospital energy expenses by as much as 15% (Deloitte, 2018). In addition, analytics enables predictive maintenance of medical equipment, lengthening device lifespans and minimizing the demand for resource-consuming replacements (Bates et al., 2014). This strategy promotes a circular economy model, reducing waste and resource consumption.

Data analytics further improves patient care effectiveness, decreasing resource-consuming interventions. Through analyzing patient data, healthcare professionals are able to pinpoint high-risk patients and institute preventive interventions, minimizing hospital admissions and related resource consumption (Bates et al., 2014). For instance, predictive analytics can minimize wasteful diagnostic procedures, decreasing energy and material use (Kelly et al., 2020). Analytics also maximizes staff planning, which saves on human resource utilization, ultimately decreasing operational expense and environmental degradation (Deloitte, 2018).

Although it has the potential, the adoption of data analytics is challenging. High initial infrastructure costs, difficulty in integrating data, and the requirement of trained staff can dissuade adoption (Kelly et al., 2020). Data security and privacy are important as well because health data is sensitive in nature and comes under stricter regulation (Bates et al., 2014). Yet the long-term advantages—cost-effectiveness, better patient outcomes, and environmental impact reduction—overrule these impediments (Lenzen et al., 2020). To maximize impact, healthcare systems need to invest in interoperable platforms and workforce training while maintaining strong cybersecurity arrangements.

Through data analytics, healthcare can shift from resource-consumptive practices to sustainable, high-efficiency systems. This is congruent with sustainability agendas worldwide, such as SDG 3 and SDG 13 (Climate Action), and allows healthcare to harmonize quality care with environmental responsibility.

2.4 Supply Chain Management and Inventory Control

The Internet of Things (IoT) is revolutionizing healthcare supply chains by providing real-time information on inventory levels, demand forecasts, and product expiry dates, enhancing efficiency and sustainability. IoT technologies, including RFID tags and sensors, monitor vital medical supplies such as vaccines and pharmaceuticals, allowing for timely replenishment and avoiding overstocking, thereby minimizing waste and financial losses (Jayaraman,2020). This is especially critical for perishable products, whose spoilage comes with hefty environmental and financial burdens. IoT devices can inform managers when the expiration dates of supplies approach, allowing them to act ahead by donating excesses to clinics or redistributing within other facilities. Apart from managing stock, IoT simplifies logistics through the reduction of delivery miles and transportation emissions, making healthcare supply chains into circular economy compliant sustainable systems.

In medicine, effective supply chain management is critical to guarantee the availability of essential supplies while reducing waste. IoT-based RFID tags and sensors enable real-time monitoring of inventory, monitoring the location, quantity, and condition of products such as medicines and vaccines. Real-time visibility enables hospitals to have optimal levels of stock, neither running short and disrupting care nor overstocking and causing spoilage. For instance, IoT systems can track the temperature of vaccines to keep them viable during storage and transportation, which is important for perishable goods with stringent cold-chain requirements. By avoiding spoilage, IoT minimizes the environmental cost of wasted medical supplies and the economic losses of replacing expired products. Jayaraman et al. (2022) emphasize that such IoT uses increase supply chain transparency, facilitating data-driven decisions by healthcare providers that favor sustainability.

IoT's capability to track expiry dates is a revolution for perishable medical items. IoT sensors can provide automated alerts to stock managers when products, like blood products or medicines, are near their expiry dates. This enables hospitals to utilize these products in priority, donate them to community clinics, or shift them to other hospitals, avoiding wastage. For example, a hospital that uses IoT to monitor insulin supplies can divert approaching expiry stocks to an adjacent clinic to prevent any useful medication from being waste. This not only saves resources but also facilitates fair access to healthcare, especially in deprived communities. Such methodologies align with the objectives of the circular economy by ensuring maximum use of resources and least contribution to landfills.

Table 1. IoT in healthcare: Enhancing efficiency and reducing waste

Application Area	Technology Used	Description	Primary Benefit	Secondary Benefit	Examples
Remote Monitoring	Wearables, IoT Sensors	Remote monitoring of patient vitals	Fewer hospital visits	Early issue detection	Fitbit, HealthPatch
Telemedicine	Video, IoT Integration	Allows for remote consultations	Facilitates accessibility in rural regions	Lower costs	Teladoc, Amwell
Smart Devices	IoT-Connected Equipment	Automates and monitors device usage	Enhances accuracy	Eliminates human error	Smart IV pumps
Smart Beds	Sensor-enabled beds	Tracks movement, vitals in bed	Improves patient comfort	Prevents injuries	Hill-Rom Smart Beds
Data Analytics	AI, Big Data Platforms	Analyzes patient and operational data	Optimized staffing	Predictive maintenance	IBM Watson Health

continued on following page

Table 1. Continued

Application Area	Technology Used	Description	Primary Benefit	Secondary Benefit	Examples
Workflow Management	Analytics Dashboards	Streamlines clinical operations	Resource efficiency	Time saving	Tableau, Power BI
Inventory Management	RFID, IoT Tags	Monitors medical supplies in real time	Eliminates stockouts	Trims waste	Terso Solutions
Supply Chain Tracking	Cloud + IoT Logistics	Real-time tracking of pharma logistics	Better planning	Avoids expiries	SAP, Oracle SCM

3. CIRCULAR ECONOMY PRINCIPLES IN HEALTHCARE

The concept of a circular economy is becoming more popular as a sustainable substitute for the conventional linear economy in a number of industries, including healthcare.

Healthcare companies may save waste, maximize resource utilization, and promote innovation in medical procedures and technology by adopting the concepts of the circular economy.

Figure 2. Circular economy cycle in healthcare

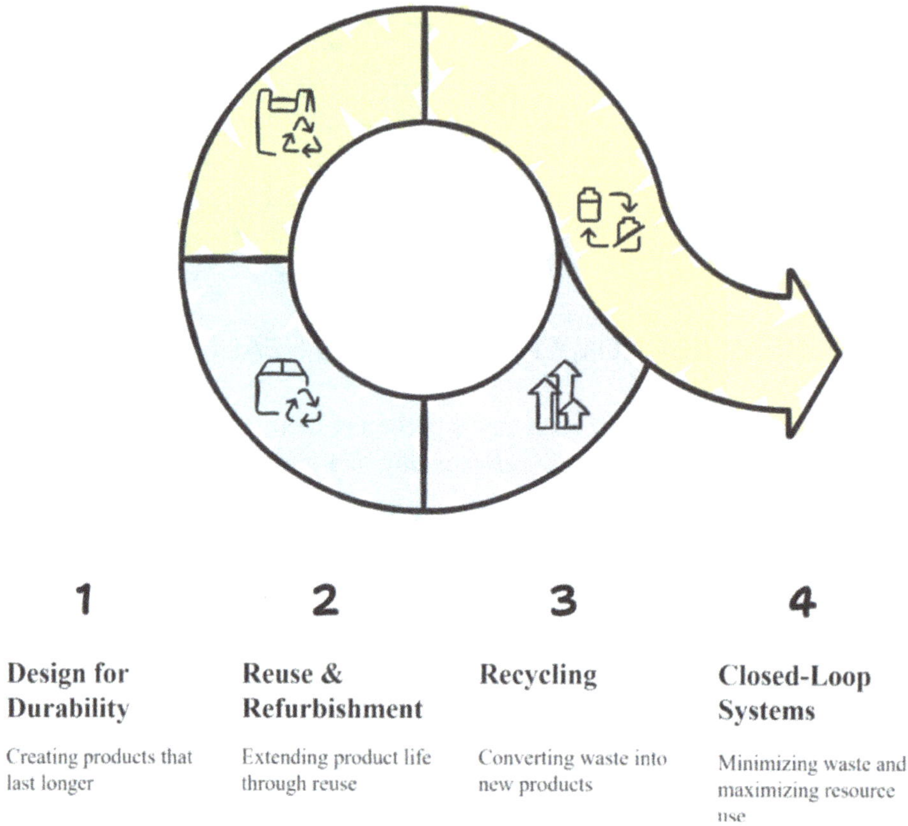

3.1 Design for Durability and Longevity

The circular economy promotes designing products to last to reduce resource use and waste, a concept that has revolutionary potential for the healthcare sector. By focusing on modular, repairable, and upgradable medical devices, manufacturers can prolong device lifetimes, minimize material usage, and decrease environmental footprint. For instance, modular ultrasound devices enable single components to be replaced or upgraded, avoiding the necessity to scrap entire systems and hence minimizing material waste (Ellen MacArthur Foundation, 2020). GE Healthcare has adopted modular designs for imaging machines, lengthening device lifetimes by ten years and improving sustainability. Also, using sustainable materials, including recyclable metals and biodegradable polymers, further reduces the environmental impact of manufacturing and disposal. With durability in mind, manufacturers both aid environmental preservation and provide cost savings for healthcare providers, bringing economic and ecological value into a circular economy system.

Under conventional healthcare manufacturing, the linear "take-make-dispose" pattern creates routine equipment replacement, utilizing finite resources and producing massive waste. The circular economy reverses this by promoting designs that are long-lasting and flexible. Modular medical devices, like diagnostic equipment or surgical instruments, are made up of replaceable parts that can be repaired or refurbished separately. This minimizes the necessity for complete system replacement, saving materials such as metals and plastics that are costly to manufacture. For example, a modular ultrasound system could enable a hospital to replace its software or swap out a worn transducer without having to replace the underlying system, greatly extending its useful lifespan. GE Healthcare's modular imaging systems, such as MRI and CT scanners, are a prime example of this approach, with designs that facilitate upgrades and repairs, with lifespans up to a decade longer than standard models. This not only minimizes waste but also decreases the rate of resource extraction, including mining for rare earth metals contained in imaging technology.

Material selection that is sustainable is another key element of durable design. By using recyclable metals, including stainless steel or aluminum, producers make sure equipment parts can be reclaimed and reused when their life is over. Biodegradable polymers, produced from renewable materials such as corn or sugarcane, provide a sustainable option for disposable or short-term medical products, e.g., sutures or packaging. They naturally break down, minimizing landfill space and the negative environmental impact of incineration, which discharges toxic pollutants such as dioxins. Biodegradable polymers in medical devices, for instance, reduce the carbon footprint of production and disposal in line with circular economy principles. Moreover, using sustainable materials reduces dependency on fossil fuel-derived plastics, which contribute to greenhouse gas emissions during manufacturing. By integrating these materials, manufacturers can create equipment that is both functional and environmentally responsible.

3.2 Reuse and Refurbishment of Medical Equipment

Reuse and refurbishment of medical devices are bedrock strategies in the circular economy, extending the life of equipment, minimizing waste, and providing substantial cost savings to healthcare providers. Refurbishment recycles used devices, including CT scanners, ventilators, and surgical tables, to like-new status through exhaustive testing, repair, and sterilization, making them safe and functional. Hospitals that implement refurbished equipment can save 20–40% of the cost of new devices, which is a cost-saving option compared to buying new equipment (Kane et al., 2018). Reuse programs for single-use devices (SUDs), such as catheters, entail reprocessing and resterilization, reducing waste by up to 50% in participating hospitals (Practice Greenhealth, 2020). Through the adoption of reuse and refurbishment, healthcare facilities can limit their dependence on virgin materials, lower waste inputs, and conform to sustainability targets without compromising high standards of patient care.

Refurbishment is a detailed procedure that rejuvenates high-value medical equipment, allowing hospitals to prolong the utility of expensive assets. An example is a refurbished used CT scanner that receives extensive diagnostics, component upgrade, and calibration to original manufacturer standards. This provides performance on par with new equipment at a fraction of the expense. Financial advantages are considerable: a remanufactured ventilator, which may cost $50,000 versus $80,000 for a new machine, frees up hospital budget for other essential uses, like employee training or patient care. Aside from expense, refurbishment saves resources in minimizing the need for raw materials, including metals and plastics, to produce new devices. It also reduces environmental pollution from disposal, keeping functioning equipment out of landfills and incinerators, where potentially toxic components like batteries might unleash toxins. SUD refurbishers like firms partnered with GE Healthcare have brought thousands of imaging systems back online around the world, proving scale and reliability.

Reuse of SUDs like catheters, endoscopes, and laparoscopic tools further move towards sustainability. Previously regarded as disposable because of infection hazards, most SUDs can be reprocessed for safe reuse by employing state-of-the-art cleaning, sterilization, and testing processes. Regulatory agencies, such as the FDA, monitor reprocessing to ensure safety standards are met, so devices can be reused several times without diminishing patient outcomes. According to Practice Greenhealth (2020), hospitals that have employed SUD reprocessing programs have eliminated 50% of hospital waste, sending tons of plastics and metals from being sent to landfills. For instance, reprocessing one catheter can save the equivalent of 0.5 kg of waste, and when extrapolated to a hospital system, the effect is significant. These programs also minimize the environmental impact of production, which is based on energy-intensive operations and finite materials such as petroleum for plastics. By maximizing the lifecycle of SUDs, healthcare providers minimize costs and environmental damage.

3.3 Recycling and Waste Management Strategies

Healthcare recycling is instrumental in reducing environmental effect by treating plastics, metals, and electronics derived from healthcare products and devices. Specialized recycling programs collect recyclable material from products like syringes, IV packs, and printed circuit boards for reusing in making new goods (Stahel, 2019).

They are part of circular economy policies, recycling waste into value-added materials and decoupling virgin material dependence. A good example is a UK hospital recycling scheme that managed to cut landfill waste by an impressive 500 tons per year by recycling 70% of its plastic waste, which was used for non-medical purposes (NHS, 2021). Efficient waste management, facilitated by employee training and cutting-edge technologies such as IoT-based systems, is crucial to achieve high recycling rates. By incorporating recycling into hospital processes and utilizing IoT for automatic waste sorting, healthcare facilities can minimize their ecological impact and support sustainable material cycles.

The medical industry produces high volumes of waste, such as disposable plastics, metal equipment, and electronic parts from diagnostic equipment. If not managed, these materials contribute to landfill waste and ecological contamination. Recycling initiatives mitigate this problem by converting medical waste into recyclable raw materials. For example, IV bag and syringe plastics are chemically recycled to yield premium-quality polymers for non-medical or industrial applications, like packaging and construction materials. Likewise, metal from surgical devices and electronics from infusion pumps or monitors are salvaged through special processing facilities, eliminating the necessity for resource-scarce mining. These programs not only save limited resources but also reduce the emission of toxic substances, like dioxins from incinerated medical waste, that can harm ecosystems and human health.

Proper waste management starts at the source, where healthcare personnel must carefully sort recyclables, biohazardous waste, and general waste. This is important to avoid contamination, which can make recyclable materials unusable. Training programs ensure that doctors, nurses, and support staff know about proper segregation procedures, like keeping sharps, infectious waste, and non-hazardous plastics separate. Proper labeling and special bins make it easy, but human mistakes are always a problem. To counter this, hospitals are increasingly implementing IoT-based technologies to make waste management more efficient. Sensor-enabled smart garbage cans employ sophisticated detection technology to detect and sort waste streams in real-time, sorting and reducing dependence on manual segregation. Such systems are able to identify plastic, metal, and biohazardous materials and send them to respective recycling or disposal streams. With enhanced accuracy, IoT technologies increase recycling percentages and reduce contamination, thereby ensuring that materials of value are efficiently recovered.

3.4 Closed-Loop Systems and Product-as-a-Service Models

Healthcare sustainable supply chain management utilizes circular economy concepts, like closed-loop systems and product-as-a-service (PaaS) models, to minimize environmental degradation and maximize resource efficiency. Linear supply chains in the healthcare sector extract raw materials, produce goods, and dump waste, thus accelerating resource depletion and pollution. Conversely, closed-loop systems encourage material reuse and recycling, for instance, reprocessing one-time use medical devices or recycling plastics from packaging, thus minimizing landfill waste and virgin resource demand (Lenzen et al., 2020). For instance, hospitals can have programs to sterilize and reuse some devices, maintaining safety while reducing waste.

Product-as-a-service models revolutionize supply chains by enabling healthcare facilities to rent equipment, e.g., ventilators or diagnostic equipment, instead of buying them. Manufacturers upgrade, maintain, and recycle the devices, making their lifecycles longer and their resource usage lower (Stahel, 2016). This reduces providers' costs and is in line with Sustainable Development Goal 12 (Responsible Consumption and Production). Yet, challenges to overcome are regulatory compliance, logistical complexity, and the requirement for strong recycling infrastructure. By embracing these circular economy approaches, healthcare supply chains can be made more sustainable, aligning environmental stewardship with operational efficiency.

4. SYNERGIES BETWEEN IOT AND CIRCULAR ECONOMY

The fusion of the Internet of Things (IoT) and circular economy concepts in healthcare promotes sustainability by maximizing resource utilization and reducing wastage. Smart devices and internet-connected medical equipment made possible through IoT promote closed-loop systems by monitoring and tracking material flows and enabling recycling or reuse of resources (Kelly et al., 2020). For example, IoT systems may track the life cycle of one-use medical devices, determining potential for safe reprocessing, lowering landfill waste and virgin material demand. This supports Sustainable Development Goal 12 (Responsible Consumption and Production).

IoT also caters to product-as-a-service (PaaS) models wherein healthcare facilities rent equipment such as imaging machines, with manufacturers holding responsibility for servicing and end-of-life recycling (Stahel, 2016). IoT offers real-time information on machine performance, adding lifecycles and lowering consumption of resources (Deloitte, 2018). In addition, IoT-based analytics refine supply chains through demand forecasting, avoiding overstocking, and minimizing waste caused by expired inventory (Lenzen et al., 2020). These synergy effects minimize the environmental impact of healthcare while making operations more efficient.

Challenges are massive implementation expenses, security risks for data, and regulatory barriers for reprocessed devices (Kelly et al., 2020). Notwithstanding these, the synergistic IoT-circular economy solution enables a regenerative health system, wherein patient care is matched by environmental responsibility.

Figure 3. IoT and circular economy principles work together to enhance sustainability in healthcare.

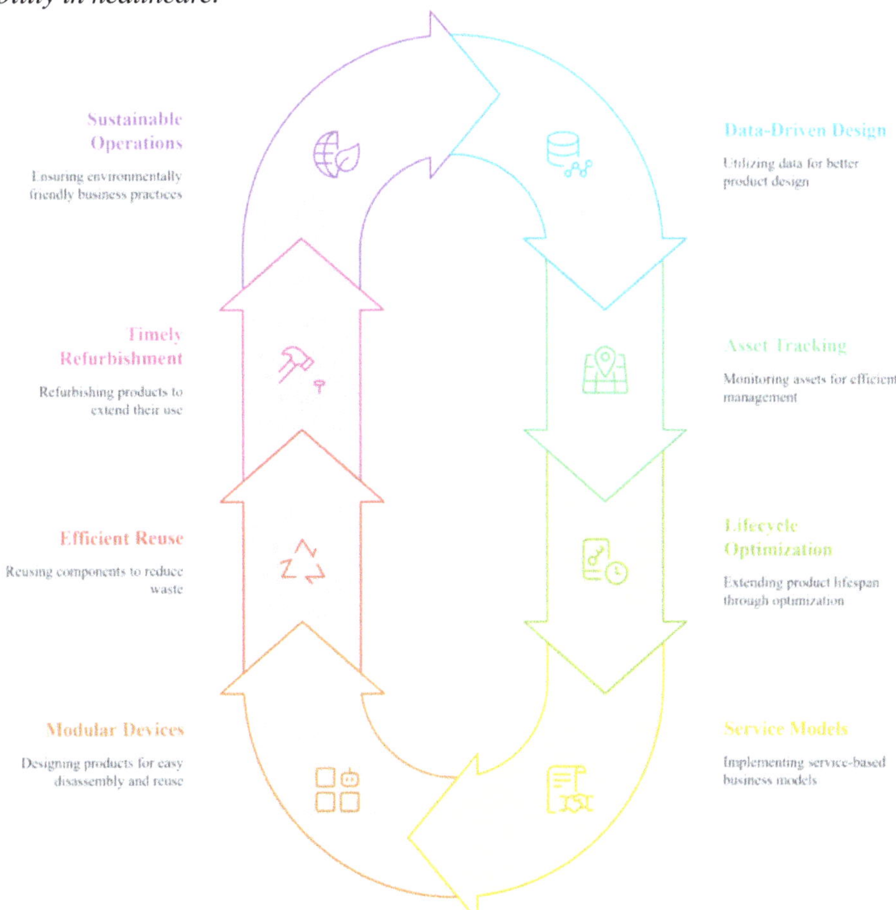

4.1 Data-Driven Insights for Circular Design

The incorporation of Internet of Things (IoT) data in product design is transforming the health sector by allowing the development of circular products according to the dictates of the circular economy. Based on data of device performance, usage patterns, and failure points, designers are able to make medical equipment durable, modular, and easy to repair, towards Sustainable Development Goal 12 (Responsible Consumption and Production). For example, IoT sensor data from infusion pumps can identify repeated wear points so engineers can strengthen critical parts for longer life (Moreno et al., 2016). Further, IoT analysis informs the use of sustainable materials, including recyclable composite for surgical instruments, with minimal environmental footprint but continued functionality (Kelly et al., 2020). These data-based innovations reduce waste, save resources, and encourage a regenerative economy that allows the health sector to focus on sustainability without sacrificing performance.

In the circular economy, the emphasis is placed on creating products that use the most efficient use of resources and least amount of waste, an objective that IoT data greatly improves. Medical equipment, for example, infusion pumps, ventilators, or diagnostic equipment, produce real-time information regarding operating performance, e.g., stress on mechanical components, usage patterns, and failure rates. When this data is aggregated and analyzed, design flaws become apparent to engineers. For instance, whenever IoT data show that a valve in a pump fails after every 10,000 cycles, designers can reenforce the valve with heavy materials or restyle it so that it becomes simple to replace, increasing the life of the device. With this strategy, as noted by Moreno et al. (2016), products are designed to last long so that there are fewer instances for replacements and accompanying use of raw materials such as metals and plastics. Modular design further promotes circularity in that individual components can be refurbished or upgraded, with no need to discard entire systems, and with alignment with sustainable manufacturing practices.

IoT information also has a significant role in material choice, with manufacturers able to select sustainable materials that meet performance requirements. Conventional medical devices tend to utilize non-recyclable plastics or metals that have high resource requirements, leading to destruction of the environment. IoT analytics also offer comprehensive performance data on materials in actual working conditions, leading to the uptake of sustainable alternatives. For example, Kelly et al. (2020) observe that IoT-based insights have accelerated the deployment of recyclable composites for surgical instruments, which have equal strength and accuracy to traditional materials but are recoverable at the lifecycle's end. These composite materials minimize environmental impact of production and waste disposal since they can be recycled to create fresh products, as opposed to disposable plastics that are sent to landfills or incinerated. Manufacturers become circular economy aligned by giving preference to materials facilitating closed-loop systems, thus reducing depletion of resources and pollution.

The advantages of IoT-enabled circular design are numerous. Long-lasting, modular equipment reduces healthcare providers' lifecycle costs, with repairs and updates being cheaper than buying new equipment. Less waste and resource usage make them environmentally sustainable, tackling the healthcare industry's high level of emissions and landfill waste. Additionally, these technologies increase system resilience by making equipment operational in times of supply chain disruptions. But to scale these practices, issues like the necessity of sophisticated data analytics capabilities and standardized design procedures need to be overcome. Cooperation between healthcare providers, manufacturers, and technology developers can spur adoption of IoT-enabled design methodologies.

4.2 IoT-Enabled Tracking and Traceability of Medical Assets

IoT-based real-time monitoring of device performance transforms healthcare by facilitating proactive maintenance and timely refurbishment, drastically improving sustainability. IoT sensors placed in key equipment, for example, ventilators or MRI machines, continuously track performance indicators such as operational efficiency and component wear. These sensors can identify subtle indications of reduced performance and automatically initiate maintenance alerts prior to failures, avoiding expensive downtime and guaranteeing patient safety (Geissdoerfer et al., 2017). This predictive maintenance practice increases device life, lowers the rate of replacements, and decreases waste directly contributing to Sustainable Development Goal 12 (Responsible Consumption and Production).

Healthcare facilities can utilize IoT data to deploy closed-loop systems, in which equipment is refurbished or reused instead of being thrown away. For example, IoT analytics can detect reusable parts in diagnostic equipment, making it possible to reintegrate them into new systems, saving resources and minimizing virgin material demand. An analysis put the potential reduction of equipment replacement costs by up to 25% and of associated waste by 20% due to predictive maintenance driven by IoT in hospitals. This supports circular economy, which promotes a regenerative healthcare system that emphasizes resource efficiency.

IoT also improves energy efficiency in healthcare devices. Intelligent sensors can minimize power consumption by putting devices into low-energy states when not in use, further minimizing the environmental impact of healthcare (Lenzen et al., 2020). Nevertheless, issues like high deployment costs, data security threats, and interoperable systems need to be resolved to increase IoT uptake (Kelly et al., 2020). In spite of these challenges, the long-term advantages—cost savings, longer equipment lifecycles, and less waste—make IoT a pillar of sustainable healthcare. By combining IoT with circular economy principles, healthcare systems can align operational excellence with environmental responsibility, supporting global sustainability objectives.

4.3 Optimizing the Lifecycle of Medical Devices

IoT real-time monitoring of device performance helps ensure proactive repair and timely refurbishment, furthering healthcare sustainability. For example, IoT sensors in ventilators can identify compromised performance and cause repairs prior to failure, protecting operational continuity and patient safety (Geissdoerfer et al., 2017). This strategy helps extend device longevity, minimize the necessity for replacement, and minimize waste, upholding Sustainable Development Goal 12 (Responsible Consumption and Production).

IoT also increases end-of-life management by sensing when devices can no longer be refurbished, assuring correct recycling and minimizing the footprint on the environment. A case in point, a hospital in Germany using IoT to enhance diagnostic equipment lifecycle optimization realized 10% savings from waste and a 20% boost in mean device lifespan (Siennes Healthineers, 2022). Implementing IoT with lifecycle management maximizes asset value, saves resources, and drives circular economy philosophies for healthcare providers. This data-driven approach promotes a regenerative healthcare system, where efficiency is balanced with environmental stewardship.

4.4 Implementing Circular Economy Models with IoT Support

The application of circular economy concepts with Internet of Things (IoT) technologies in healthcare promotes sustainability through waste reduction and maximization of resource utilization, in line with Sustainable Development Goal 12 (Responsible Consumption and Production). IoT-supported technologies, including RFID tags and intelligent sensors, improve closed-loop systems by tracking medical equipment throughout its life cycle, enabling effective reuse, refurbishment, and recycling (Bressanelli, Adrodegari, Perona, & Saccani, 2018). For example, IoT sensors on medical devices monitor usage and maintenance requirements, allowing for on-time refurbishment to maximize device lifecycles and minimize equipment loss. Blockchain integration also boosts transparency, providing assurance of adherence to ethical disposal standards. Furthermore, IoT enables product-as-a-service (PaaS) strategies, where hospitals rent equipment such as MRI scanners, with the manufacturers handling maintenance and end-of-life recycling. These practices make devices last longer, minimize resource utilization, and turn healthcare into a regenerative system that harmonizes patient care with environmental sustainability.

Circular economy values transform healthcare from the linear "take-make-dispose" to a regenerative system that is based on resource effectiveness. IoT is key to enabling real-time information to inform closed-loop systems, where devices are reused, refurbished, or recycled instead of being disposed of. RFID tags and intelligent sensors track medical device condition and utilization across their lifespan. For instance, IoT sensors deployed on laparoscopic equipment can identify wear or performance problems, and send maintenance notices to avoid premature waste. This anticipatory measure, according to Bressanelli et al. (2018), cuts equipment loss by as much as 30% and increases reuse levels by 15% in hospitals implementing

IoT asset management. Blockchain technology augments IoT by establishing secure, transparent histories of a device's life cycle, guaranteeing recycling and disposal procedures adhere to regulatory requirements and ethical standards, such as those avoiding hazardous waste pollution.

IoT also powers new PaaS models, restructuring equipment ownership toward sustainability. Through PaaS, hospitals rent value-intensive devices such as MRI or CT scanners and manufacturers are accountable for maintenance, upgrades, as well as recycling at the end of life. This model spurs manufacturers toward designing long-life, modular products that can simply be repaired or refurbished, improving lifecycles and minimizing utilization of resources. For example, a rented MRI scanner may get software patches or part replacement, keeping it in working order for a decade or more without requiring a new system. IoT facilitates PaaS by enabling manufacturers to get real-time information about equipment performance and usage, maximizing maintenance schedules and facilitating timely recycling. This strategy serves to support circular economy objectives in limiting material usage and waste and benefitting the healthcare sector in acquiring advanced technologies without the ownership-capital expense.

The effect of IoT integration along with the concept of circular economy is a turning point toward healthier sustainability within the healthcare industry. IoT-assisted asset management facilitates improved levels of reuse and refurbishment and curtails the eco-footprint in the manufacture of medical devices using resource-using material such as metal and plastics. Higher rates of reuse and lower equipment loss equate to savings for healthcare organizations, since refurbished equipment can be 20–40% less expensive than new equipment. Additionally, PaaS models decrease the need for virgin materials, helping to drive a regenerative economy. Nonetheless, issues such as high IoT deployment costs, data privacy issues, and standardized recycling procedures must be overcome to drive these solutions at scale. Joint action by manufacturers, healthcare professionals, and policymakers can stimulate innovation and regulatory harmonization.

Through the combination of IoT and circular economy thinking, healthcare systems can reach a sustainable equilibrium between patient treatment and environmental stewardship. IoT real-time monitoring and blockchain transparency strengthen closed-loop systems, and PaaS models ensure resource efficiency. These innovations minimize waste, save resources, and enable a regenerative healthcare model, guaranteeing long-term sustainability and conformance to international environmental objectives.

Table 2. Synergies between IoT and circular economy

Subsection	Description	Circular Economy Principle	Benefits	Challenges	Examples	Future Potential
4.1 Data-Driven Insights for Circular Design	Uses IoT data on device performance to design durable, modular medical equipment.	Design for durability and modularity.	Reduces waste, extends device lifespans.	Requires advanced analytics infrastructure.	IoT data informing modular infusion pump designs.	AI-enhanced design for predictive durability.
4.2 IoT-Enabled Tracking and Traceability	Tracks medical assets via RFID and GPS for efficient reuse and recycling.	Closed-loop systems and material recovery.	Minimizes asset loss, improves recycling rates.	High costs of tracking systems.	RFID tracking in hospital equipment management.	Blockchain for enhanced traceability.
4.3 Optimizing Device Lifecycle	Monitors device performance to enable timely maintenance and refurbishment.	Reuse and refurbishment.	Extends equipment life, reduces disposal.	Needs robust IoT integration.	Connected ventilators alerting for filter changes.	Predictive maintenance via edge computing.
4.4 Circular Economy Models with IoT	Supports Product-as-a-Service (PaaS) models with usage and condition data.	Product-as-a-Service and resource efficiency.	Lowers material use, promotes sustainability.	Regulatory and adoption barriers.	Philips' PaaS for imaging systems.	Scalable PaaS with 5G connectivity.

5. CASE STUDIES AND EXAMPLES

5.1 Successful Implementations of IoT in Healthcare

The Internet of Things (IoT) is transforming healthcare through increased efficiency, better patient outcomes, and a lower environmental footprint. IoT technologies, by allowing real-time integration of data and sophisticated analytics, help healthcare providers make the best use of available resources and limit waste. Philips' HealthSuite platform, Medtronic's pacemakers with IoT integration, and Apollo Hospitals' energy management system demonstrate how IoT influences sustainability in a variety of healthcare environments. HealthSuite is a platform that brings together data from wearables, medical devices, and electronic health records (EHRs) to enable remote monitoring, lowering hospital readmissions for patients with chronic disease by 38% while minimizing emissions and resource use (Philips, 2020). Medtronic pacemakers send data to doctors, reducing doctor visits by 40% and limiting related environmental footprint (Medtronic, 2022). In the meantime, Apollo Hospitals in India installed an IoT-based energy management system and recorded a 22% decrease in energy consumption across 50 buildings (Apollo Hospitals, 2021). Such instances illustrate the power of IoT to transform healthcare systems into sustainable and efficient healthcare systems.

The Philips' HealthSuite platform is a foundation of IoT-based healthcare innovation. By combining information from wearable devices, including glucose monitors and fitness trackers, with medical equipment and EHRs, the platform allows for real-time remote monitoring of patients suffering from chronic conditions like diabetes or heart disease. This minimizes the frequency of hospital visits, subsequently lowering patient travel emissions and facility resource utilization. The platform's 38% reduction in readmissions reflects its ability to deliver personalized care plans based on real-time data, preventing complications that necessitate resource-intensive hospitalizations. Additionally, by minimizing in-person care, HealthSuite reduces the energy consumption of hospital operations, such as lighting, heating, and equipment usage. Philips' incorporation of IoT not only improves patient outcomes but also supports sustainability objectives by reducing the healthcare industry's carbon footprint, which contributes 4.4% of global greenhouse gas emissions.

Medtronic's pacemakers with IoT capabilities further demonstrate the ecological advantages of remote monitoring. Cardiac information is sent to physicians directly by the devices, allowing timely interventions without necessitating patients' visits to clinics. By cutting in-person consultations by 40%, Medtronic's pacemakers reduce carbon emissions from travel as well as consumption of disposable healthcare supplies, e.g., gloves and sterile drapes, conventionally used in consultations. Reduction in physical sessions also reduces pressure on healthcare premises, decreasing energy and material consumption. Medtronic's strategy illustrates how IoT can be taken beyond patient care to meet environmental challenges, ensuring the sustainability of healthcare delivery without compromising high levels of medical supervision.

Apollo Hospitals' energy management system based on IoT illustrates the use of IoT to optimize healthcare infrastructure. Installed at 50 locations in India, the system utilizes sensors to track and manage energy usage in real-time, focusing on high-use facilities such as operating rooms and imaging suites. Analyzing lighting, HVAC, and medical equipment data, the system realized a 22% energy reduction, which meant substantial cost savings and reduced carbon emissions. This project has a particularly large impact in India, where hospitals are both high-usage and grid-variability challenged. Apollo's success highlights the scalability of IoT solutions in addressing operational inefficiencies, providing a model for hospitals globally to minimize environmental impact without sacrificing care quality.

5.2 Circular Economy Initiatives in Medical Device Manufacturing

The incorporation of circular economy in the manufacturing of medical devices is revolutionizing the healthcare sector by minimizing waste, saving resources, and reducing costs. Some of the top players in the industry, including Medtronic, Johnson & Johnson, and GE Healthcare, have launched revolutionary initiatives that focus on material recovery, biodegradable material use, and refurbishment. Medtronic's cardiac device recycling program returns 80% of materials, including titanium and batteries, to be used in new product development (Medtronic, 2021). Johnson & Johnson's Ethicon has also created biodegradable sutures to reduce the environmental burden of surgical waste (Johnson & Johnson, 2022). While GE Healthcare's imaging devices refurbishment program has refurbished more than 10,000 products across the world, saving $500 million in hospital costs and preventing 15,000 tons of waste (GE Healthcare, 2023). These efforts illustrate the potential of circular economy approaches to develop sustainable, cost-saving measures in medical device manufacturing, merging environmental responsibility with high-quality health care delivery.

Medtronic's cardiac device recycling initiative showcases the circular economy's ability to reclaim valuable material. Through collaborative arrangements with experienced recycling firms, Medtronic provides safe and effective processing of materials such as titanium, which is used in pacemakers, and lithium-ion batteries in defibrillators. These reclaimed materials are reused as new medical devices or other industrial goods, minimizing the need for virgin resources and landfill waste. The success of the program is its closed-loop system, where devices are collected after use, dismantled, and processed under rigorous regulatory guidelines to avoid contamination. This strategy not only saves limited materials but also reduces the environmental damage resulting from improper disposal, including soil and water pollution due to battery leakage. Medtronic's move underscores the significance of industry collaborations in upscaling circular economy practices, making recycling processes both efficient and meeting international safety standards.

Johnson & Johnson's Ethicon business solves surgical waste by creating biodegradable sutures that break down naturally within the body, never needing to be removed and minimizing harm to the environment. In contrast to conventional sutures, which can remain as waste or need to be incinerated, biodegradable sutures are created with materials such as polylactic acid, a substance sourced from renewable resources such as corn. This technology lessens the carbon footprint of surgery and supports circular economy principles by favoring materials that safely and harmlessly feed into natural cycles. Ethicon's work illustrates how green design for products can solve the environmental problems of single-use medical devices, which have a significant impact on healthcare's waste stream. Through investment in research and development, Johnson & Johnson is charting the path for increased use of environmentally friendly materials in medical use.

5.3 Integration of IoT and Circular Economy in Hospitals

The convergence of IoT and circular economy concepts in healthcare facilities promotes sustainability through efficient utilization of resources and minimizing waste. IoT technologies like RFID tags and intelligent sensors support real-time tracking of medical assets, which drive closed-loop systems that encourage reuse, refurbishment, and recycling (Bressanelli et al., 2018). For instance, IoT-enabled surgery equipment can be tracked for utilization and maintenance purposes, ensuring that refurbishment takes place on schedule and loss is avoided. Blockchain integration improves transparency and adherence to regulatory and ethical end-of-life disposal standards.

IoT also makes product-as-a-service (PaaS) business models possible, enabling hospitals to rent equipment such as diagnostic devices, with producers handling maintenance and end-of-life recycling. Such a model maximizes equipment lifespan and reduces resource use, contributing to Sustainable Development Goal 12 (Responsible Consumption and Production). Hospitals that adopt IoT-based asset management have recorded a 15% boost in reuse rates and a 30% decrease in equipment loss. By combining IoT with circular economy principles, hospitals can develop regenerative systems that reconcile high-quality patient care with environmental sustainability (Kirchherr et al., 2017).

6. CHALLENGES AND BARRIERS

6.1 Data Privacy and Security Concerns

The adoption of IoT technologies in healthcare, although revolutionary, presents profound concerns over data security and privacy. IoT devices, including wearable sensors and medical gear with connectivity, gather and transfer sensitive patient information and are at risk of cyberattacks and data breaches (Bressanelli et al., 2018). Health records or device tampering by unauthorized personnel may compromise patient safety and trust. Blockchain and encryption technology are sometimes used to protect data, but they are difficult and expensive to implement.

Regulatory systems, like the Health Insurance Portability and Accountability Act (HIPAA), require stringent data protection guidelines, but compliance is not easy with the amount of and variety in IoT-created data. Inadequate security measures in low-resource environments increase risks, with the potential to leave patient data open to exploitation (Kirchherr et al., 2017). Solving these concerns involves effective cybersecurity mechanisms, personnel training, and interoperable systems for maintaining data integrity. By placing privacy and security first, healthcare systems can use IoT safely to enable sustainable practices while ensuring patient confidentiality.

6.2 Regulatory and Policy Frameworks

The use of IoT technologies in the healthcare sector, as it enhances sustainability and efficiency, is hindered by huge regulatory and policy framework issues. IoT devices like smart medical equipment and wearable sensors have to be compliant with tough regulations such as the Health Insurance Portability and Accountability Act (HIPAA) and the General Data Protection Regulation (GDPR) of the European Union in order to secure patient data privacy and security (Aceto et al., 2018). Nevertheless, the exponential growth of IoT devices makes it challenging to comply due to heterogenous data streams and issues of interoperability. Lacking uniform global standards, deployment of IoT solutions becomes more challenging in low-resource environments where there could be weak regulatory oversight.

Policy structures also fail to handle the ethical implications of IoT-enabled circular economy frameworks like reuse of medical devices. Safety and effectiveness in remanufactured devices need to be ensured through guidelines, but most areas lack well-developed policies (Kirchherr et al., 2017). Furthermore, administrative barriers will slow down the use of product-as-a-service models since leasing contracts for medical devices tend to undergo legal examination. Designing harmonized, adaptive policies that facilitate innovation with patient protection is essential to incorporating IoT and circular economy principles into the healthcare sector.

6.3 Technical and Infrastructure Limitations

The incorporation of IoT technology into the healthcare sector, which is important for promoting circular economy frameworks, is plagued by immense technical and infrastructure challenges. IoT devices, including smart sensors and networked medical devices, necessitate strong network connectivity and compatibility to operate optimally (Aceto et al., 2018). Yet, most hospitals, particularly those in low-resource environments, do not have the high-speed internet connectivity and sophisticated IT infrastructure required to facilitate real-time processing of data and device communication. This restricts the scalability of IoT solutions such as asset tracking and predictive maintenance.

Interoperability issues stem from the heterogeneity of IoT devices and platforms, which tend to run on incompatible protocols, making data integration and system efficiency challenging (Kirchherr et al., 2017). Moreover, the intensive computational requirements of IoT analytics put pressure on current infrastructure, necessitating expensive upgrades to servers and cloud infrastructure. Technical constraints, including battery capacity limitations in wearable devices, also hinder continuous monitoring. Overcoming these challenges requires high investment in infrastructure, standardized protocols, and technical training to facilitate smooth IoT uptake while accommodating sustainable healthcare practice.

6.4 Economic and Financial Considerations

The implementation of IoT technologies in health care, while promising for embedding in circular economies, is underpinned by major economic and financial factors. Deploying IoT systems, such as smart sensors and networked medical devices, necessitates a large initial outlay in terms of hardware, software, and infrastructure upgrades (Aceto et al., 2018). Hospitals, especially in low-income settings, cannot possibly pay these expenses, so the scalability of IoT-based solutions such as asset tracking and predictive maintenance is undermined. Recurring costs, such as maintenance, cybersecurity, and employee training, further stretch budgets.

Although IoT has high up-front costs, long-term financial savings are available. Predictive maintenance and the optimization of resources can lower equipment replacement costs by as much as 25% and minimize waste-related costs. Product-as-a-service models, enabled by IoT, enable hospitals to rent equipment, which reduces capital spending while providing access to new technology (Kirchherr et al., 2017). Nonetheless, economic risks, including ambiguous returns on investment and regulatory compliance expenses, present challenges. Strategic funding frameworks and public-private partnerships are necessary to offset these economic factors, allowing sustainable IoT uptake in healthcare.

7. FUTURE DIRECTIONS AND RECOMMENDATIONS

7.1 Advancements in IoT Technologies for Healthcare

The use of Internet of Things (IoT) technologies in the healthcare sector has tremendous potential for promoting circular economy values like efficiency in the use of resources and minimizing waste but poses tremendous economic and financial implications. Installation of IoT systems like intelligent sensors, networked health devices, and data analysis platforms involves considerable initial investments in hardware, software, and infrastructure enhancement (Aceto et al., 2018). To hospitals, especially low-resource settings, such costs become exorbitant, reducing the scalability of IoT-enabled solutions such as asset tracking, predictive maintenance, and remote monitoring. Regular expenses such as system upkeep, cybersecurity features, and personnel training put pressure on budgets and become hindrances to broad-based uptake. Beyond such hindrances, IoT also yields long-term cost benefits via efficient operations and groundbreaking business models such as Product-as-a-

Service (PaaS). Mitigating the financial risks and regulatory compliance expenses calls for strategic funding and public-private partnerships to facilitate sustainable IoT deployment in healthcare.

The upfront cost of IoT implementation is a significant barrier. Implementing smart sensors on medical equipment, enhancing network infrastructure to handle real-time data transmission, and integrating analytics software require substantial capital. For instance, installing IoT devices in a mid-sized hospital costs millions, apart from the cost of cloud storage and cybersecurity measures to secure sensitive patient information. In resource-constrained settings, like rural clinics or hospitals in developing nations, these are usually unaffordable, limiting access to life-changing technologies such as IoT-based asset tracking, which can optimize equipment usage, or predictive maintenance that avoids expensive breakdowns. Furthermore, ongoing expenses contribute to the economic burden. Maintenance of IoT networks, software upgrades, and education for healthcare workers to use and analyze IoT information demand ongoing expenditures. Cybersecurity is especially essential, as IoT devices are subject to hacking, requiring strong encryption and surveillance systems to meet the requirements of laws such as HIPAA.

In spite of these high initial and recurrent expenditures, IoT provides enormous long-term savings that are aligned with the principles of circular economy. Predictive maintenance based on IoT data can save up to 25% of equipment replacement costs by detecting problems before they cause failures (Kirchherr et al., 2017). For example, IoT sensors in imaging devices can alert when components require maintenance, prolonging device lifetimes and curbing waste through early disposal. Smarter resource use, e.g., IoT-based inventory monitoring, reduces overstocking of perishable materials, reducing waste-related costs. In addition, IoT enables pioneering PaaS models, whereby hospitals rent equipment rather than buying it outright. This strategy, practiced by firms such as Philips, reduces capital outlay while guaranteeing access to advanced technology, with manufacturers taking responsibility for upkeep and upgrades. These models encourage resource effectiveness by rewarding long-lasting, recyclable designs, in accordance with circular economy tenets.

7.2 Policy Recommendations for Promoting Circular Economy

The implementation of IoT technologies within healthcare, although leading to increased sustainability and efficiency, raises major regulatory and policy framework issues. IoT technologies, including smart sensors and Internet-of-Things-enabled medical devices, have to be in accordance with strict regulations such as the Health Insurance Portability and Accountability Act (HIPAA) and the European Union's General Data Protection Regulation (GDPR) to safeguard patient information and ensure device safety (Bressanelli et al., 2018). Nevertheless, the swift proliferation of IoT devices surpasses current regulatory structures, leaving gaps in oversight and standardization. Varying worldwide regulations make compliance challenging for multinationals offering healthcare services, raising operational expenses and risks.

Policy structures also fail to support the ethical recycling and disposal of IoT-enabled medical devices, paramount for circular economy models. Ineffective guidelines on refurbishment and end-of-life management can result in environmental degradation and regulatory transgressions (Kirchherr et al., 2017). Furthermore, low-resource environments tend to lack the infrastructure to support compliance, further increasing risks of inappropriate device disposal. Resolution of these issues necessitates harmonized global standards, strong policy enforcement, and incentives for sustainable practices to ensure IoT integration promotes both patient care and environmental objectives.

7.3 The Role of Collaboration and Stakeholder Engagement

Collaboration and stakeholder involvement are pivotal in executing sustainable healthcare practices, especially in the adoption of IoT and circular economy paradigms. Involving stakeholders—hospitals, manufacturers, regulators, patients, and technology suppliers—guarantees convergence on sustainability objectives, for example, minimizing waste and optimizing resources (Ghisellini et al., 2016). For example, collaborations between hospitals and device manufacturers enable product-as-a-service (PaaS) models, whereby IoT-supporting equipment is rented, serviced, and recycled, increasing lifecycles and reducing environmental impact. Collaborative efforts also fuel the use of closed-loop systems, which facilitate the recycling and reuse of medical materials.

Stakeholder engagement that is effective stimulates innovation and compliance with the regulations. Multi-stakeholder platforms, such as healthcare professionals, policymakers, and environmental groups, create standards for green practices in order to ensure consistency with Sustainable Development Goal 12 (Responsible Consumption and Production). These partnerships have increased equipment reuse rates by 15% in certain hospitals. Through a process of shared responsibility, stakeholder engagement ensures that sustainability programs initiated by IoT find a balance between patient care and caring for the environment (Domínguez et al., 2021).

7.4 The Vision of a Sustainable Healthcare Ecosystem

The vision of a sustainable healthcare ecosystem unites environmental, social, and economic sustainability to provide quality care with minimal ecological footprint. Circular economy concepts and IoT technologies can be used in healthcare systems to minimize waste, optimize resource use, and increase resilience. IoT-based devices like smart sensors and RFID tags enable closed-loop systems through monitoring equipment lifecycles to allow for reuse, refurbishment, and recycling (Bressanelli et al., 2018). For instance, IoT-sensed diagnostic equipment can be rented via product-as-a-service offerings, with manufacturers taking responsibility for maintenance and end-of-life recycling, minimizing resource usage.

This ecosystem is in line with the United Nations' Sustainable Development Goals, specifically SDG 3 (Good Health and Well-Being), SDG 12 (Responsible Consumption and Production), and SDG 13 (Climate Action). Socially, it provides fair access to care, while economically, it generates cost savings through effective resource utilization. Environmentally, it reduces waste and contributes to pollution, creating a regenerative system. This poses challenges, including data security and significant implementation costs, with the need for strong cybersecurity and scalable infrastructure (Kirchherr et al., 2017). Healthcare can gain a sustainable, patient-focused ecosystem balancing the quality of care and environmental stewardship by merging IoT and circular economy approaches.

8. CONCLUSION: TOWARDS A SUSTAINABLE HEALTHCARE FUTURE

The health sector confronts considerable ecological challenges, among which are excess carbon emissions, natural resource usage, and intensive generation of wastes. By deploying the Internet of Things (IoT) through the application of circular economy ideas, healthcare has the potential to move toward the sustainability of tomorrow without compromising environmental accountability and effective patient care. This complementary strategy uses IoT's strengths in data aggregation, monitoring, and analysis to increase efficiency and minimize waste, while the circular economy fosters the conservation of resources using long-lasting, recyclable, and reusable systems. Though there are challenges like data security issues, regulatory issues, and upfront expenditure, the potential advantages—lower emissions, cost efficiency, and increased system robustness—make this strategy strong. Through partnership, investment in technology, and policy coordination, healthcare can create a sustainable system that advances environmental responsibility as well as patient health.

Internet of Things technologies are transforming healthcare by streamlining resource usage and reducing the footprint on the environment. IoT-supported devices, like wearable sensors and intelligent medical devices, allow for remote patient monitoring and telemedicine, cutting back on hospital visits and energy use. For example, ongoing monitoring of chronic diseases such as diabetes or hypertension reduces readmissions to hospitals, reducing patient travel and facility carbon footprints. Intelligent devices, such as networked infusion pumps and imaging equipment, utilize real-time data to maximize performance and maintenance cycles, minimizing waste and prolonging equipment life. IoT-based analytics also improve efficiency by anticipating equipment failure and automating inventory management, avoiding overstocking of perishable items. These developments not only enhance patient outcomes but also support sustainability objectives by conserving resources and minimizing operational inefficiencies.

Circular economy supports IoT by promoting closed-loop systems with a focus on resource preservation. In contrast to the conventional linear model of "take-make-dispose," the circular economy encourages designing medical devices for durability, modularity, and recyclability. Modular structures enable parts to be refurbished or updated, prolonging product life and minimizing replacement requirements. Refurbishment and reuse schemes for expensive equipment, e.g., MRI scanners, reduce expenses and save materials, while recycling processes harvest plastics and electronics from medical waste streams. Product-as-a-Service (PaaS) solutions, in which producers own and service devices, encourage sustainable design and optimal resource utilization. With IoT integrated, healthcare professionals are able to monitor assets in real-time, streamline maintenance, and enable recycling, offering a hassle-free closed-loop system with minimal environmental footprint.

The combination of IoT and circular economy is highly rewarding. IoT data drives the design of recyclable and long-lasting devices, and tracking of assets ensures timely reuse and recycling. This strategy can minimize healthcare emissions, with estimates of a possible 30% reduction by 2050. Cost savings from less waste, efficient use of energy, and reused equipment boost financial stability, especially for hospitals with limited resources. Sustainable practices also enhance system resilience, guaranteeing uninterrupted care in the event of resource shortages or climate disruptions.

Despite this, challenges remain. Security of data is a paramount issue since IoT devices gather sensitive patient data, which requires compliance with standards such as HIPAA. Regulatory environments differ across the world, making it challenging to implement circular practices like equipment refurbishment. The high upfront costs of IoT infrastructure and circular systems can discourage implementation, especially in low-resource environments. These challenges need to be addressed through concerted efforts. Governments can encourage sustainability through tax incentives and coordinated regulations, and public-private partnerships can finance infrastructure and pilot initiatives. It is vital that collaboration exists between policymakers, healthcare providers, and manufacturers in order to achieve scaling up of innovations and synchronizing the interest of the stakeholders.

REFERENCES

Aceto, G., Persico, V., & Pescapé, A. (2018). The role of Information and Communication Technologies in healthcare: Taxonomies, perspectives, and challenges. *Journal of Network and Computer Applications*, *107*, 125–154. DOI: 10.1016/j.jnca.2018.02.008

Bates, D. W., Saria, S., Ohno-Machado, L., Shah, A., & Escobar, G. (2014). Big Data In Health Care: Using Analytics To Identify And Manage High-Risk And High-Cost Patients. *Health Affairs*, *33*(7), 1123–1131. DOI: 10.1377/hlthaff.2014.0041 PMID: 25006137

Bauer, W., & Dorn, J. (2016). Requirements for Product-service Description at e-marketplaces in the Manufacturing Domain. *Procedia CIRP*, *47*, 406–411. DOI: 10.1016/j.procir.2016.03.040

Bressanelli, G., Adrodegari, F., Perona, M., & Saccani, N. (2018). Exploring How Usage-Focused Business Models Enable Circular Economy through Digital Technologies. *Sustainability (Basel)*, *10*(3), 639. DOI: 10.3390/su10030639

Eckelman, M. J., & Sherman, J. (2016). Environmental Impacts of the U.S. Health Care System and Effects on Public Health. *PLoS One*, *11*(6), e0157014. DOI: 10.1371/journal.pone.0157014 PMID: 27280706

Fu, L., Pan, R., Gu, T., Cheng, X., & Zhang, H. (2024). IoT-Enabled Supply Chain Management From A Customer Perspective: Challenges and Opportunities. *IEEE Internet of Things Journal, PP, 1*. DOI: 10.1109/JIOT.2024.3505412

Geissdoerfer, M., Savaget, P., Bocken, N. M. P., & Hultink, E. J. (2017). The Circular Economy – A new sustainability paradigm? *Journal of Cleaner Production*, *143*, 757–768. DOI: 10.1016/j.jclepro.2016.12.048

Ghisellini, P., Cialani, C., & Ulgiati, S. (2016). A review on circular economy: The expected transition to a balanced interplay of environmental and economic systems. *Journal of Cleaner Production*, *114*, 11–32. DOI: 10.1016/j.jclepro.2015.09.007

Ho, H.-J., Iizuka, A., Shibata, E., Tomita, H., Takano, K., & Endo, T. (2021). Utilization of CO2 in direct aqueous carbonation of concrete fines generated from aggregate recycling: Influences of the solid–liquid ratio and CO2 concentration. *Journal of Cleaner Production*, *312*, 127832. DOI: 10.1016/j.jclepro.2021.127832

Holmner, Å., Ebi, K. L., Lazuardi, L., & Nilsson, M. (2014). Carbon Footprint of Telemedicine Solutions - Unexplored Opportunity for Reducing Carbon Emissions in the Health Sector. *PLoS One*, *9*(9), e105040. DOI: 10.1371/journal.pone.0105040 PMID: 25188322

Jayaraman, P. P., Forkan, A. R. M., Morshed, A., Haghighi, P. D., & Kang, Y. (2020). Healthcare 4.0: A review of frontiers in digital health. *Wiley Interdisciplinary Reviews. Data Mining and Knowledge Discovery*, *10*(2), e1350. DOI: 10.1002/widm.1350

Kelly, J. T., Campbell, K. L., Gong, E., & Scuffham, P. (2020). The Internet of Things: Impact and Implications for Health Care Delivery. *Journal of Medical Internet Research*, *22*(11), e20135. DOI: 10.2196/20135 PMID: 33170132

Kirchherr, J., Reike, D., & Hekkert, M. (2017). Conceptualizing the circular economy: An analysis of 114 definitions. *Resources, Conservation and Recycling*, *127*, 221–232. DOI: 10.1016/j.resconrec.2017.09.005

Lenzen, M., Malik, A., Li, M., Fry, J., Weisz, H., Pichler, P.-P., Chaves, L. S. M., Capon, A., & Pencheon, D. (2020). The environmental footprint of health care: A global assessment. *The Lancet. Planetary Health*, *4*(7), e271–e279. DOI: 10.1016/S2542-5196(20)30121-2 PMID: 32681898

Smith, A. C., Thomas, E., Snoswell, C. L., Haydon, H., Mehrotra, A., Clemensen, J., & Caffery, L. J. (2020). Telehealth for global emergencies: Implications for coronavirus disease 2019 (COVID-19). *Journal of Telemedicine and Telecare*, *26*(5), 309–313. DOI: 10.1177/1357633X20916567 PMID: 32196391

Stahel, W. R. (2016). The circular economy. *Nature*, *531*(7595), 435–438. DOI: 10.1038/531435a PMID: 27008952

Yang, L., Nguyen, H., Bui, X.-N., Nguyen-Thoi, T., Zhou, J., & Huang, J. (2021). Prediction of gas yield generated by energy recovery from municipal solid waste using deep neural network and moth-flame optimization algorithm. *Journal of Cleaner Production*, *311*, 127672. DOI: 10.1016/j.jclepro.2021.127672

Chapter 4
Adopting a Circular Economy Approach to IoT Healthcare Devices

Jeevithesh Reddy Narravula Reddy
https://orcid.org/0009-0006-1041-1139
SimplyTurn Technologies, USA

Pratheesh Manikonda
Independent Researcher, USA

ABSTRACT

In this chapter, the idea of the circular economy is proposed as a transformative strategy on how IoT based healthcare devices can be designed for sustainable use and disposal. However, as connected medical technology exponentially grows, CycleTimber's environmental and economic implications of devices all the way through to their lifecycles are emerging as an important issue. Models in the chapter strive towards resource efficiency, reuse, refurbishment and recycling to decrease its electronic waste while maintaining performance and safety standards. It provides case studies and emerging policy supports the ways a circular economy can help to aid in both ecological responsibility and long term cost savings in the healthcare sector.

1. INTRODUCTION

Changes to the medical service are taking place in the healthcare industry due to technological innovations that are rewriting how patient services are rendered, tracked and handled. The Internet of Things (IoT) is at the forefront of this evolu-

DOI: 10.4018/979-8-3373-4134-7.ch004

tion, it is a network of interconnected devices that process and transmit real time health related data. Wearable health monitors, smart infusion pumps and remote patient management systems are becoming more and more essential parts of modern healthcare delivery, and these devices have evolved from these just a few years ago. The process they validate for patient care delivery improves patient outcomes, decreases hospital readmissions, and aids in the delivery of proactive and personalized care. But as these IoT enabled technologies are being deployed widely fast, there are growing concerns over them being an environmental foot print and their economic sustainability.

Having IoT healthcare devices proliferate also introduces new challenges when it comes to the lifecycle of these products. These devices were designed for high performance, typically with a short useful life as a result of rapid technological obsolescence, only limited upgradeability, strict regulatory constraints and power requirements. This results in a burgeoning pile of electronic waste (e-waste) that not only poses a major threat to the environment, but also in increased expenses for healthcare systems through e-waste procurement, maintenance and disposal. Due to this growing concern, there should be a paradigm shift in the way healthcare technologies are designed, utilized, and retired.

In response to the challenges posed by these, the notion of the circular economy is becoming a compelling concept for sustainable innovation in the medical device market. A central concept of circular economy is to decouple the economic activity from the consumption of finite resources and instead redefine growth. Through effective product design, they support systemic change by encouraging devices designs that enable reuse, refurbishment and recycling, and thereby extend device life and reduce waste. Unlike the linear model of 'take make dispose', the circular economy is intended to be regenerative by design, maintaining products, components and materials in use time as much as possible (Fernandes, 2023).

In this chapter, the author explores how the circular economy model can be best put into practice for IoT healthcare devices to produce economically feasible and ecologically responsible healthcare technology ecosystem. IoT and circular economy principles in the intersection are an ideal opportunity to rethink the medicine machine. Resource efficiency and eco design when brought into the early stage of product develop will help to reduce material input, simple maintenance, and support end of life management. It also means that we can use digital technologies like IoT to help circularity by improving tracking, predictive maintenance and lifecycle management of medical devices based on data.

In the context of healthcare IoT, adopting a circular approach is of great value. It environmentally contributes to reducing e-waste, greenhouse gas emissions, and raw material conservation, which play an important role in the war against climate change and environmental degradation. It also contributes towards making healthcare providers more economically efficient, potentially resource new business models that include leasing, device as a service or modular upgrades on top of the existing model. From a health systems perspective, the trade the reuse of and rehabilitating of devices so that these can safely be utilized aids in greater accessibility and equity in delivery of care, especially in settings with a lack of resources.

Additional support for the necessity of integrating circular economy practices in healthcare is provided by emerging regulating and policy frameworks adopting sustainability. International organizations, as well as governments and healthcare institutions are realizing to transition to greener practices. Systemic reforms are being set in place through initiatives such as the European Union's Circular Economy Action Plan and the United Nations Sustainable Development Goals (SDGS). One of the sectors that is also expected to be of critical importance in meeting these global sustainability targets is the one of healthcare (Rejeb & Suhaiza, 2022).

However, there are challenges to the circular economy approach in healthcare IoT. Often the complexity of many medical devices, the stringent safety and hygiene standards, the proprietary nature of some IoT systems, make difficult to 'sell' circular principles. In addition, healthcare delivery systems are fragmented and standard protocols for refurbishment and recycling are not developed. Changing these issues require collaboration between developers of technology, healthcare providers, policymakers and researchers.

This chapter can stand in two roles. It first gives a general idea of how we can maximize circular economy to achieve the sustainability of IoT enabled devices in healthcare. Second, it features real world, strategies, models, and case studies that describe tangible benefits and challenges in this transition. The chapter contributes through an interdisciplinary lens to bridging gaps between environmental science, medical technology, economics and policy studies.

The structure of the chapter is as follows:

- Second, Section 2 introduces the core principles of the circular economy and the IoT in healthcare by contextualizing them. The difference with circular and linear economic models is demonstrated and the former is pointed out as a driver of innovation and resilience in healthcare systems.
- In section 3 we look at IoT healthcare devices impacts on environment and economy in order to understand the necessity of sustainable practices.
- In section 4 we study circle economy models for the IoT healthcare sector by giving special attention to resource efficient design and lifecycle management.
- Specifically, Section 5 discusses usable technology pathways for reuse, refurbishment and recycling as well as enabling technology and supportive systems.
- Section 6 discusses the implication of circularity for health care provider in terms of Cost Effectiveness, Performance and Regulated Compliance.

- Section 7 discusses the practical limitations related to implementation in terms of technical and regulatory, cultural and organisational barriers.
- Lastly, Section 8 concludes the chapter by summarizing key findings and complementing it with the potential of circular economy adoption in transforming the healthcare sustainability.
- Section 9 concludes by suggesting future directions for this agenda, specifically research and policy objectives to continue this line of work.

To conclude, this chapter provides a rationale for why healthcare IoT is an environmental imperative for healthcare institutions and a strategic imperative for innovation balancing with sustainability. NOW is the time for the healthcare sector to promote itself as a leader in a global movement toward a new way of doing responsible technology use: Quality care, economic resilience, environmental stewardship all on the same terms.

2. UNDERSTANDING THE CIRCULAR ECONOMY IN THE CONTEXT OF HEALTHCARE IoT

The healthcare sector, adapting to smart technologies that enable better patient care and operational efficiency, continues to drive the demand for sustainable frameworks to oversee lifecycle of connected devices. A framework for the circular economy exists—the one based in energy and material flow, and environmental stewardship, economic viability, and systemic resilience (Turskis & Šniokienė, 2024). This section seeks to explore the theoretical and practical basis of the circular economy in healthcare IoT.

2.1 Defining the Circular Economy Model

The circular economy is a substitute model for linear production and consumption economy. On the whole, it is a regenerative system that seeks to reduce waste and make as good use of available resources as possible. Three main pillars upon which the model is built include designing out waste and pollution, keeping products and materials in use and regenerate natural systems.

Firstly, in the context of our healthcare IoT in a circular economy model, device manufacturer and healthcare providers are encouraged to step forward from a take-make-dispose paradigm. Instead, devices are built to be long, modular, and repairable. The materials selected also try to be easy to disassemble and recycle. Additionally, hardware is configured to enable return, refurbishing, or repurposing of the hardware at the end of its first use cycle.

2.2 Key Principles: Reduce, Reuse, Refurbish, Recycle

The implementation of a circular economy uses four fundamental principles which direct both product design and system-level conceptualization. The development and deployment of IoT healthcare devices requires minimizing their virgin material usage and energy consumption under the Reduce principle. Medical device manufacturers design multiple functionality into one compact system which decreases the essential product count both for patients and clinical facilities (Jauhar, Pratap, Lakshay, & Paul, 2023).

When designing devices and components the principle of Reuse involves creating products which can operate multiple times or serve different purposes after their initial use. Sensor modules and enclosure systems can be regenerated for new equipment uses after the original program requirements expire.

Devices that need refurbishment first receive restoration services to become operational and compliant and safe for use. The complete workflow involves part replacements and software updates and extends to severe testing for maintaining performance requirements. Healthcare institutions can use refurbished devices either for their internal needs or transfer them to facilities that provide insufficient medical technology to patients.

Recycling serves as the definitive option when devices turn outdated since both refurbished and recharacterized options have been exhausted. The principle includes extracting valuable materials including rare earth metals along with plastics and circuit components for re-entry into manufacturing circles.

2.3 Contrasting Circular and Linear Economy Models in Healthcare Devices

In healthcare device manufacturing and management the common linear economic system operates through a single path starting with raw materials extraction into device production before devices reach their end-of-life and get disposed. The traditional one-way product cycle causes fast depletion of resources while generating more hazardous waste materials that increase operational expenses for healthcare organizations (D'Alessandro, 2024).

A systems-level redesign of complete device lifecycles represents the core principle of circular economy operations in healthcare. Healthcare institutions achieve cost reductions while building resilience in their supply chains along with sustainability compatibility by using this system approach which maintains patient security and treatment standards.

Table 1. Linear and circular economic models, relevant to sustainable healthcare technology and medical IoT device management.

Feature	Linear Economy	Circular Economy
Design Focus	Performance & cost	Performance, longevity & recyclability
Product Lifecycle	Short-term, disposable	Extended through reuse and refurbishment
End-of-Life Strategy	Disposal	Recycling and reintegration into production
Environmental Impact	High (waste, emissions)	Lower (resource conservation, e-waste reduction)
Economic Model	One-time purchase	Service-based or leasing models

Table 1 demonstrates how a linear economy transition creates a powerful solution to sustainability problems in healthcare services. The linear model achieves short-term performance combined with cost efficiency but produces extensive environmental consequences through regular disposals whereas the circular economy sources durability alongside recyclability features and resource management protocols. The circular approach cuts down e-waste creation and material waste through renewable product use and refurbishment and leasing service economics. The medical IoT field requires this directional change because it enables more efficient healthcare solutions which combine sustainability with cost-effective practices and environmentally friendly procedures.

2.4 Benefits of Adopting Circular Principles in Healthcare IoT

Properties involved in healthcare IoT when applied to circular economic principles help achieve various advantages which exceed single environmental benefits:

- Medical device production that utilizes modular elements and recyclable materials helps organizations avoid finite supply chain dependencies on rare earth elements used in sensors along with chips.
- The healthcare industry can minimize its e-waste impact from toxic item remnants when it extends product utility through product renewal and system restorations.
- Leasing combined with device-as-a-service and refurbished device programs under circular strategies help reduce costs in healthcare technology thus making technology more accessible for developing countries.
- Circular-principled organizations demonstrate superior competitive ability in product design together with material science research and business model development. Such initiatives create business advantages that produce access to fresh market opportunities.

- Increasing global sustainability consciousness leads to new regulations alongside improved reputation among the public. Organizations taking environmental responsibility seriously will receive incentives along with escaping penalties and they will build superior brand reputations.
- The integration of IoT enables real-time monitoring of device condition as well as usage trends and environmental data through data-based life cycle management practices. The collected data helps to both predict device maintenance needs and maximize equipment use before decision-making about end-of-life supports the circular economy model.

3. ENVIRONMENTAL AND ECONOMIC IMPACT OF HEALTHCARE IoT DEVICES

The Internet of Things (IoT) has undergone tremendous growth in healthcare thus driving changes to global medical service delivery. IoT devices serve the healthcare industry by enhancing both patient diagnostics and treatment while improving their medical engagement through smart infusion pumps and wearable biosensors and remote patient monitoring technology. The digital transformation process presents urgent problems regarding its negative environmental and economic impact. Connected medical technologies now saturate the market thus creating substantial life cycle challenges which span from extraction of raw materials to waste disposal for healthcare sector operators (Rønn & Wieland, 2023).

Rapid technology advancement and increasing needs for digital healthcare practices have created a substantial rise in electronic medical device output as well as market demand. These advanced innovations drive healthcare efficiency but produce increasing electronic waste volumes while using more natural resources and raising total energy requirements. Modern healthcare facilities face economic challenges because they must pay expensive acquisition prices and maintain their systems and follow complex disposal guidelines which expenses hospitals a lot of money and disrupts operational activities.

The resolution of these problems calls for an entirely new method of operation. The healthcare sector needs to develop a comprehensive evaluation system for IoT devices throughout their entire lifecycle so they can see how environmental sustainability closely connects to economic success. The examination and measurement of present operations enables stakeholders to discover critical areas which allow them to develop circular economy strategies that reduce negative consequences and maximize economic value.

Figure 1. Environmental and economic impact dimensions of healthcare IoT devices

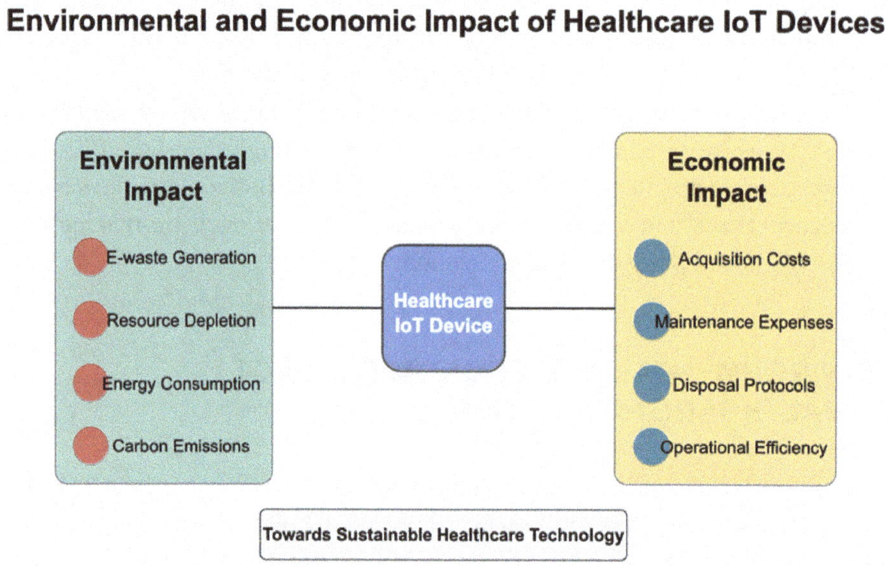

The illustration in Figure 1 presents healthcare IoT devices as agents which affect sustainability standards of the environment in addition to positively influencing economic stability. Human health benefits from these technological advancements in medical service delivery although these systems produce major difficulties for patient care systems according to section 1. Healthcare IoT technology generates environmental damage through the production of e-waste and natural resource depleting effects as well as higher energy use and carbon dioxide emissions during its complete operational life. Acquisition costs along with maintenance fees and indefinite disposal procedures and operating requirements result in financial consequences for healthcare establishments. A unified view underscores the necessity to handle healthcare IoT implementations through complete solutions which merge environmental durability assessment with financial stability analysis in order to direct strategy toward circular economic models that counter these dual issues.

3.1 Growing Use of Connected Medical Technologies and the Challenges They Pose

Healthcare institutions now experience an unexpectedly quick rapid adoption of internet of things technology. Hospitals as well as home care environments now utilize wireless ECG monitors with smart thermometers and digital insulin pens and AI-enabled imaging equipment. Recent market research shows the healthcare

IoT market will surpass $400 billion by 2030 because it is driven by the aging population and increases in chronic diseases as well as telemedicine service expansion.

Such renowned advantages of this technology include better diagnostics and real-time monitoring and personalized healthcare yet its exponential growth presents sophisticated difficulties. The combination of sensors along with batteries microprocessors and communication modules inside IoT healthcare devices results in high functionality but also creates major environmental impact. Regulations and safety standards as well as hygiene regulations often restrict device lifespans to only one-time use which produces widespread device replacements (Akbari, 2024).

The implementation of disposability features allows for faster accumulation of electronic waste when combined with regular hardware replacements and outdated software versions. Healthcare providers who lack sustainable procurement and end-of-life strategies expose themselves to both environmental ruin and economic waste as well as non-compliance issues.

3.2 Environmental Impacts: E-Waste, Resource Depletion, and Carbon Footprint

Multiple aspects make up the environmental impact of healthcare devices that use Internet of Things technology. Manufacturing electronic components as well as batteries requires essential raw materials including lithium together with cobalt and rare earth elements which creates harmful effects on ecosystems and causes pollution to water resources and demanding energy usage during the mining operations. Before reaching hospitals or patients the devices contribute to various environmental issues.

Created IoT devices draw power without interruption since they operate on batteries or connect to the electric grid. The individual power usage of healthcare devices remains minimal but when spread among numerous healthcare facilities across the country the combined impact becomes major. Devices rarely incorporate features for abled recycling because of which critical materials remain unrecoverable after their lifespan ends (Moutaoukil & Mahfoud, 2024).

Healthcare generates substantial electronic waste that presents a major difficulty because of its large volume. Healthcare serves as an important source of e-waste production because data from the Global E-Waste Monitor indicates that 53.6 million metric tons of e-waste emerged worldwide in 2019. Electronic devices disposed as waste contain dangerous substances which include lead, mercury, and cadmium that have the potential to enter soil and groundwater unless they receive appropriate disposal treatment.

Table 2. Environmental burden of common IoT healthcare device components

Component	Environmental Concern
Batteries (Lithium-ion)	Resource depletion, fire hazard, groundwater contamination
Circuit Boards	Toxic metals (lead, mercury), complex recycling
Plastic Housings	Non-biodegradable, high carbon footprint in production
Sensors and Transmitters	Use of rare earth metals, disposal complexity
Wi-Fi/Bluetooth Modules	Energy use, component miniaturization limits recyclability

The major environmental impacts from IoT healthcare devices core components can be found in Table 2. The environmental burdens affect the entire lifecycle of objects from the manufacturing stage to post-consumer waste generation thus it requires environmental design considerations and management strategies to combat this issue.

3.3 Economic Challenges: High Device Costs, Maintenance, and Disposal Expenses

Aside from environmental aspects, economic impact associated with healthcare IoT adoption is also quite high. Acquisition costs of advanced IoT devices can be quite prohibitive to a resource constrained healthcare system, therefore this is the first challenge. The costs in terms of upfront investment are often accompanied with ongoing costs, including software licensing and firmware updates, along with preventative maintenance.

In addition, the fragmented nature of the IoT device market, featuring different standards, limited interoperability as well as rapid product cycle creates hidden costs which include integration and staff training. In addition, the frequent replacement of devices that are not easily serviceable and upgradable adds to long term capital expenditure (Venkateswaran & Khosravi, 2024).

And the economic cost goes on at the end of life. On the other hand, disposal of e-waste is a loss of embedded value, e.g. raw materials, energy, and labor invested in a good, is a cost due to compliance with regulations and the handling of hazardous materials. However, with the lack of reuse or recycling when the devices are discarded, healthcare providers lose out on possible cost recoveries and sustainability benefits.

However, these financial stressors are especially pronounced in public healthcare systems and rural hospitals, where budget constraints often require a betting double orphan: technology versus fiscal prudence. Unless healthcare institutions have robust lifecycle planning, they are left in a potentially untenable view of unsustainable spending cycles with little environmental responsibility.

3.4 The Need for a Holistic Approach to Mitigate These Impacts

Apart from environmental impact, economic impact related with adopting healthcare IoT is also quite high. First, the acquisition costs of advanced IoT devices are quite prohibitive to the resource constrained healthcare system. Ongoing costs include preventative maintenance, as well as software licensing and firmware updates, and these can often be pushed along with the costs of upfront investment.

Moreover, the IoT as a device market is fragmented, leading to different standards on devices and limited Interoperability, and short product life cycle with hidden costs such as integration, staff training. Furthermore, the high frequency of replacing the devices that are not designed for easy service and upgrade continues to increase long term capital expenditure.

And there is an economic cost at the end of life. On the other hand, e waster disposal is loss of embedded value, (e.g. raw materials, energy and labours invested in a good) is a cost as it constitutes of spending on compliance with regulations and hazardous materials handling. But when devices are discarded, there is presently no reuse or recycling of the devices, resulting in missed opportunity to recuperate cost as well as sustainability gains for the healthcare providers.

Budget constraints are particularly intense in public healthcare and rural hospitals, where the betting double orphan is public healthcare and fiscal prudence. However, without robust lifecycle plan the healthcare institutions are stuck in an unsustainable view of spending cycles that are neither responsible to the environment.

4. CIRCULAR ECONOMY MODELS FOR IOT HEALTHCARE DEVICES

Knowing that the healthcare sector is adopting Internet of Things (IoT) technologies, it is imperative that the principles of circular economy be integrated into the design, production, usage and disposal of devices. The waste and resource efficiency are minimized and maximized respectively through closed systems in circular economy. This transition is also a strategic opportunity to improve the economic performance and reduce the environmental impact of IoT healthcare devices, many of which are loaded with expensive but quickly outdated resources (Cavalieri, Reis, & Amorim, 2021).

The idea of circular economy models in this context, aims to ensure that sustainability is also supported at each stage of the device life cycle. This required eco design principles at the development stage, the use of durable reusable materials, modular configuration to ease repair and improvement, as well as recovery at end of life.

This approach is based on lifecycle thinking and innovative business models, which offer a road map for ecological stewardship and cost efficiency at the same time.

4.1 Resource Efficiency and Eco-Design in the Development of IoT Healthcare Devices

The basis of a circular economy model is how devices are designed. Eco Design is a method of reducing products environmental footprint by choosing the best material, designing more efficient manufacturing processes and supporting product recovery at the end of its life. This translates to designing healthcare IoT devices to be made out of biodegradable or nonhazardous materials, be safe enough to not compromise performance or safety and also be energy and material efficient for a useful, long life.

Judicious use of raw materials is the starting point of the development of resource efficiency. They decide and select components that have fewer environmental costs (like biodegradable polymer or recyclable metals) and also design to optimize for less material use. In addition, it is also critical to reduce energy consumption, since IoT devices are usually powered continuously. Huge reductions in lifetime emissions can be achieved by incorporating low energy communication protocols (BLE or ZigBee), energy harvest capabilities and power saving modes.

Eco-design also emphasizes product longevity. They should build the devices in such a way that they last longer and degrade their performance little over time. With robust materials, modular assemblies (that make replacing components easy) and firmware that increases functionality without hardware changes, this can be made possible.

4.2 Examples of Sustainable Design Practices

Numerous sustainable design strategies have been successfully applied to IoT healthcare devices. Key among them are:

- Modular design: It allows for the modular ability to disassemble and replace defective parts, (e.g. sensors, batteries and communication modules).
- High quality plastics, aluminum alloys and biodegradable casings provide durable, long lasting, and recyclable devices.
- Energy-Efficient Components: Incorporation of low-power microcontrollers, efficient battery systems, and smart energy management circuits.
- Design for Disassembly: which attempts to simplify disassembly through elimination of metal adhesives and the use of snap fit schemes to facilitate the easy separation of material sub assemblies.

These design choices ensure devices are easier to maintain, adapt to evolving standards, and recover value at the end of their service life.

Table 2. Circular design practices and their functional benefits in healthcare IoT devices

Design Practice	Functional Benefit	Circular Economy Advantage
Modular Architecture	Easy repair, upgrade, and customization	Extends product lifespan
Biodegradable Materials	Reduces landfill accumulation	Minimizes environmental harm
Low-Energy Components	Decreases power consumption and carbon footprint	Enhances energy efficiency
Standardized Interfaces	Promotes interoperability and reuse	Facilitates part replacement
Firmware Upgradability	Updates features without hardware replacement	Reduces need for new devices

As shown in **Table 2**, applying circular design practices not only serves sustainability goals but also enhances device usability, flexibility, and cost-effectiveness throughout the lifecycle.

4.3 Lifecycle Management: From Production to Disposal

The production is the starting point of lifecycle management under the circular economy framework, during which sustainability goals are defined and design principles are imbedded. It includes all the stages from distribution, through usage, to maintenance, and finally disposal or reuse. Each phase in this model is designed to waste as little and maintain as much value as possible (Rejeb & Keogh, 2022).

Manufacturers can be encouraged at the production stage to use cleaner technologies, use less raw material and implement traceability so that, post use, the device is easier to recover. Leasing or product as a service is something that distribution models can embrace to keep ownership and have the device return to the manufacturer when it is done being used.

Predictive maintenance systems during the use phase may assist users in anticipating the potential of failures ahead of time, rather than replacing it. In clinic environments requiring the frequent change as technology evolves, devices should be readily re calibrated and easy to clean, as well as repurpose.

End of life management is done by take back programs, incentivized returns, or working with recycling firms to recover valuable materials. They advocate for circular flow of resources and cut down on dependency on virgin raw materials.

4.4 Case Studies of Companies Implementing Circular Economy Principles in Healthcare IoT Devices

Several companies are already pioneering circular economy principles in the healthcare technology sector:

- To address the challenges of reprocessing FDA regulated medical equipment, Philips Healthcare has developed its medical equipment as a circular-ready product through the implementation of modular designs and offering equipment as a service solution. To extend life of diagnostic and monitoring devices and to minimize their well known environmental footprint, they have take-back and refurbishment programs.
- Remanufactured imaging devices are available through GoldSeal program from GE Healthcare. Large truck bodies are disassembled, thoroughly tested and rebuilt with new or refurbished parts to the same quality and performance of a new unit, at a fraction of the cost and environmental burden.
- ResMed, a leader in respiratory care devices, employs recyclable packaging, modular ventilator designs, and energy-efficient firmware updates. They collaborate with recycling partners to guarantee responsible reuse of the materials in their designed end of life devices.

These examples demonstrate that circular models are not just theoretical constructs but practical, scalable solutions that can be embedded into business operations without sacrificing profitability or innovation.

5. STRATEGIES FOR REUSE, REFURBISHMENT, AND RECYCLING

In a circular economy for Iot healthcare devices, products are made to last further than their first lifecycle, for using and obtaining a resource value as long as possible. Traditional approach involves discarding devices used whereas circular approach involves recovering, renewing and reintegration components and materials. It is an essential mechanism of reducing environmental impact and the cost of production, as well as a mechanism to increase sustainability in medical technology, reuse, refurbishment, and recycling. Regardless of this, these strategies not only contribute to ecological aims, but also represent a business opportunity based essentially on cost recovery or added value (Settia, Bhutani, & Saini, 2025).

The Healthcare IoT devices typically contain valuable materials like rare earth elements, advanced semiconductors and special sensors. Reimagining end of life device handling as a process of strategic reuse and refurbishment enables key stakeholders to greatly decrease reliance on virgin materials, decrease electronic waste, and improve efficiency of the supply chain. In this section, we discuss the core strategies needed to attain these goals, as well as the technology that allows them.

Figure 2. Circular economy strategies for healthcare IoT devices

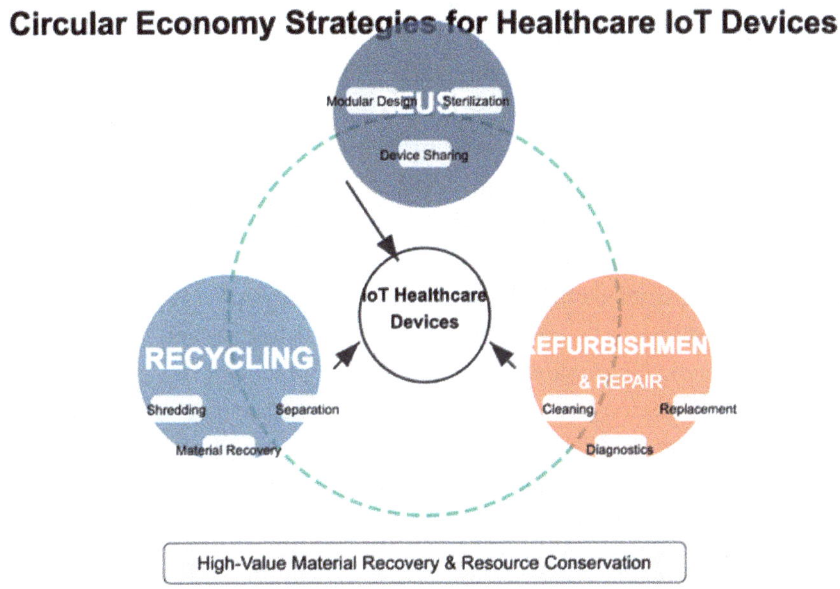

The figure 1 shows three main circular economy approaches for elongating the life of Healthcare IoT devices. Section 1 describes these strategies that integrate to recover, renew, and reintegrate valuable components and materials from medical technology. Reuse is all about extending the device lifespan through device sharing, proper sterilization techniques as well as modular design. Refurbishment involves diagnostics, replacing of components and professional cleaning, done to bring the instrument back to a commercial standard for functionality and safety. The last stage of material recovery includes recycling of high value materials of rare earth elements and specialized components through separation and processing. With its comprehensive approach, this approach reduces the environmental impact and electronic waste, while also generating economic value via resource conservation and supply chain optimization, as it supports the ecological sustainability and a business viability in healthcare Iot sector.

5.1 Approaches to Device Refurbishment and Remanufacturing

Critical aspects of a circular economy model are refurbishment and remanufacturing, especially in the matter of the healthcare sector which holds high cost of the devices and faster pace of technology evolution. The refurbishment refers to the task of examining, fixing, cleansing, and testing a previously utilized gadget to make it triumph in order of working. Remanufacturing, however, denotes disassembeling a product to its components, which are then rebuilt to equivalent or higher quality to the original product and include updated parts or technologies (Sharma & Vanapalli, 2021).

Refurbishment is increasingly becoming a cost effective and sustainable practice which is adopted by healthcare organizations and manufacturers. After the first use cycle, patient monitors, ventilators and infusion pumps are often returned to manufacturers or certified refurbishers. After the rigorous inspections and tests, these devices will be reintroduced into the market at reduced prices to assist underfunded institutions or developing countries in gaining access to essential technologies.

More comprehensive remanufacturing means replacing worn parts along with re engineering the device so that it meets performance and safety standards. The U.S. FDA and European Medicines Agency among others have issued guidelines on quality and safety requirements for remanufactured medical devices. This has allowed companies to develop powerful, certified, traceable, and supported by warranty systems of lifecycle extension.

5.2 Reuse of Components and Systems in New Devices or Other Industries

The other layer of opportunity for advancing a circular economy for IoT healthcare devices lies also with component reuse. Typically a device boils down to useless and its components are no longer functional, however, an exception is that individual components could be working: Sensors, circuit boards, casings, or battery. Unlike the situation with the entire product, these parts can be extracted, and into new devices or repurposed in other industries.

Take for example, if the sensors such as temperature or pressure type are used in the patient monitoring systems, they can be tested and reused in the laboratory equipment or in home based health monitoring systems. Likewise, robust plastic or metallic housings can be restructured for making prosthetic devices or assistive tools. When properly decontaminated and validated, this approach does not only extend the lifetime of individual parts, but it also decreases the environmental and economical cost of manufacturing new components out of raw materials.

Furthermore, the repurposed IoT components are useful to industries other than healthcare (for example, utilizing IoT components in agriculture, industrial automation, and smart home solutions). Interoperability and open standards become more widespread: there will be a secondary market for high quality parts, which will be reused through interoperability across sectors.

5.3 Recycling of Electronic Waste and Its Role in the Circular Economy

With the reuse and refurbishment efforts being made, some of the devices or components reach a point where further use is just not a smart or safe thing to operate. In such situations, resources recovery depends mostly on recycling. SC one: The breakdown of devices using systematic methods to get out valuable materials like gold, silver, copper, plastics and rare earth elements.

Recycling serves as one of the biggies in circular economy as it proves to reduce finite resources back into the cycle of production. Recycling properly helps prevent materials and chemicals that can harm the environment and enter landfills and ecosystems prevent the need to extract energy intensive raw materials. This is particularly important in the healthcare IoT device with batteries, circuit boards, and other components may contain toxic leachable substances if disposed of improperly (Alfina & Ratnayake, 2022).

There has now been an advent of automated sorting, dismantling and processing of the waste using AI and robotics in the advanced recycling facilities. Recycling medical devices is however more complicated and include ensuring data security of patients and avoiding biohazard contaminations. As such, special handling protocols are required and partnerships with certified e-waste recyclers with expertise in handling healthcare technology are needed.

Other policy frameworks including the European Union's Waste Electrical and Electronic Equipment (WEEE) directive and the U.S. Resource Conservation and Recovery Act (RCRA) have also influenced how responsible recycling has been practiced. Recycling targets are mandated, producers are made responsible for the taking back and treatment of their products and eco design is encouraged to facilitate separation and recovery of materials.

5.4 Technologies and Systems That Enable Efficient Reuse and Recycling

A variety of enabling technology and system support has been developed to support the successful implementation of reuse, refurbishment and recycling strategies. Product lifecycle management (PLM) software, digital twins and blockchain are some of the digital tools being applied across the operations to track devices from manufacture, performance, and end of life in the aim of ensuring transparency, traceability, and accountability.

Manufacturers can use PLM systems to design products, having disassembly and recycling as a design principle, making certain the complete material and recyclability profile of every component is recorded. Hence, they provide a virtual replica of the physical device so that the maintenance schedules, wear and tear predictions, and end of life decisions are backed by real time usage data.

The blockchain technology adds transparency in the circular economy chains by recording every transaction or transformation a device experiences in such a chain, from refurbishment to resale. This leads to creating of trust among stakeholders and easy auditing and regulatory compliance.

On the operational side, robotic disassembly systems, AI powered sorting mechanisms and chemical separation technologies are helping the recycling process. These systems decrease exposure of the human to potentially hazardous material and increase efficiency and yield of material recovery. In addition, there is emergence of cloud based platforms that bridge the device owners with certified refurbishers and recyclers to create digital marketplaces for circular solutions (Gagneja, Lall, & Bhutani, 2024).

6. ECONOMIC AND PERFORMANCE IMPLICATIONS FOR HEALTHCARE PROVIDERS

The transition to a circular economy with respect to the healthcare IoT devices is also an environmental need but it is also an economic opportunity. Normally, sustainability initiatives are considered a cost driven decision for the healthcare provider but come with the long term savings of costs and operational efficiencies. The adoption of circular strategies such as reuse, enhance resource efficiency in the healthcare system overall, and through strategies of reuse, refurbishment and recycling the lifecycle of medical devices can be optimized and the procurement costs can be minimized, thereby decreasing the disposal expenses. Based on an examination of typical economic or performance implications of such movement towards a circular economy, such as cost savings, performance standards and cost effectiveness vs quality balance.

6.1 Long-Term Cost Savings Through Circular Economy Strategies

One reason why healthcare providers should adopt circular economy strategies could be none other than long term savings. High procurement costs of medical devices — the main obstacle for the adoption of the IoT in traditional healthcare systems — are particular to the area of IoT based medical devices that are perpetually in the course of the change requiring constant updates and maintenance since technology is evolving so fast. However, circular economy practices, refurbishment of existing devices and reuse of components, can reduce significantly the purchase of new equipment which can be expensive especially in resource constrained environments (Gupta, Bhutani, Dixit, & Lall, 2021).

For example, they can save in the initial procurement costs by acquiring refurbished medical devices from the new. Often the refurbished devices are sold at a fraction of the cost of new devices as they perform just as well as new ones. Also, refurbishing the cycle of devices allows healthcare organizations to stretch the capital cost over a longer period and avoid incurring the high cost of device upgrading on an annual basis.

Along with cutting procurement expenditures, circular economy approaches will lower disposal expenditures, particularly for medical devices that may be harmful and which would incur expensive safe disposal costs to manage them accordingly. Valuable components like rare earth metals or reusable parts can be recycled and some of these costs can be returned to healthcare systems by recycling. In addition, lower disposal costs are greatly reliant on decreased volume of generated e waste (e waste) which greatly mitigates the requirement of waste management services and compliance with the regulation.

6.2 Impact on Performance and Safety Standards

However, the integration of circular economy principles is not at the expense of performance or safety standards of healthcare IoT devices. Contrary to popular belief, circular strategies, when done correctly, are leveraged precisely to enhance the performance and longevity of devices so that they continue to carry out the intense duties required in the healthcare world. Refurbished and remanufactured processes follow industry regulations and standards and many times the devices end up as reliable as the new ones.

Devices are taken apart, tested, and put back together with high quality components and often have technological upgrades in the case of remanufacturing. This way of doing ensures that the performance of the device remains the same or better than the specs given. Since certified remanufacturers and refurbishers strictly adhere to regulated permits, quality control checks and thorough testing, health care providers can be confident refurbished devices will adhere to the same safety and performance standards as new devices (Gallo, Viola, & Romagnoli, 2024).

In addition, these devices continue to be improved through reuse and remanufacturing that invariably generate innovations to improve their efficiency and functionality. For instance, remanufactured medical IoT devices may have been delivered with energy saving technologies, latest software updates, or most durable material, which all contribute to enhance the user experience. For that reason, because of adopting circular economy strategies, it means to introduce new and higher quality devices with lower cost for healthcare providers and patients.

However, before reselling or reusing any NLIR device, healthcare providers must verify that the devices they acquire (refurbished, remanufactured, or reused) satisfy national and international safety regulations. Strict safety standards are enforced on medical devices by regulatory agencies like FDA in the US or European Medicines Agency (EMA). To realize a circular economy practice, healthcare providers should be collaborating with certified refurbishers, manufacturers, and regulatory bodies to ensure that the devices are safe and meet all quality requirements.

6.3 Balancing Cost-Effectiveness with Device Quality and Reliability

The economic benefits of using circular economy practices are evident, but healthcare use of the devices needs to be balanced between cost effectiveness and quality and reliability. Lower cost pursuits should not be pursued at expense of patient safety or device functionality. Medical equipment, like all other equipment in healthcare settings, is intended to be used by those who depend on it for their very lives (in the case of patients, providers, and healthcare center equipment staff).

Healthcare organizations must develop any rules that will guide the assessment of the quality and reliability of the refurbished or remanufactured devices. This means to team up with best in class suppliers and refurbishers dedicated to the same high standards. All of the devices are tested to make sure they work the way they should and must pass the same standards as new devices. Furthermore, healthcare providers should take into account warranty period, service and support, and availability of spare parts in their decision to move to circular economy solution (Rejeb & Rejeb, 2023).

In addition, the systems of circular economy dictate the need for healthcare organisations to adopt the long term perspective. Saving money may be an attractive incentive, but the performance and reliability of the devices should never be lost. So for example, although the refurbishment of a medical device may be cheaper in the short term, make sure it has been maintained properly and is only as good as new over the lifecycle of the device. It therefore forces healthcare providers to analyse the full total cost of ownership (TCO) of devices which include procurement costs, but also maintenance, service and downtime costs.

6.4 Case Studies of Healthcare Systems Benefiting from Circular Economy Adoption

Some healthcare systems around the world have started to reap the benefits of operational and economic from the adoption of the circular economy principles into their management of IoT healthcare devices. These case studies show how the circular strategies can help organisations drive cost savings as well as operations efficiency improvements.

The National Health Service (NHS) in the UK is one example of a renovated medical equipment that has been integrated into the procurement strategy. Refurbished medical devices, including patient monitors and infusion pumps, are available at a fraction of the cost of new equipment as the NHS worked with certified refurbishers to find high quality products. Refurbishing and reuse of currently available devices has enabled the NHS to save capital expenditure on medical equipment which were then directed to other necessary healthcare services.

Finally, Siemens Healthineers has a remanufacturing programme for its medical imaging devices, and it is another case study. The process used by the company is that the used imaging equipment is returned and thoroughly inspected before being refurbished with the latest technology to the original specifications. By doing this, it contributes to waste reduction and prolongs the service life of expensive medical imaging equipment so that healthcare providers can offer the most up to date services with less expense (Awan, Sroufe, & Shahbaz, 2021).

In America, the VA (Veterans Affairs) healthcare system has even mimicked the principles of circular economy by using refurbished devices to meet the VA's medical equipment strategy. The VA has been able to reduce procurement costs and still provide high quality of care to veterans by using refurbished equipment. Also, trusted partners of the organization work with the refurbished devices to be sure that they are safe.

These case studies emphasize that adoption of circular economy could bring economic and performance benefits in healthcare systems. Over the years, these organizations have been able to decrease costs, decrease the amount of e-waste, and have high quality standards while also improving their ability to serve patients effectively through refurbishment, remanufacturing, and component reuse.

The promise of the future of healthcare delivery is the transition of healthcare IoT devices towards a circular economy. However, challenges remain particularly in terms of ensuring device quality, safety and regulatory compliance, but the potential for cost savings, improved performance of the device and creating a more sustainable healthcare sector make it a strategy that is highly desirable for providers worldwide. Adopting a circular lens to healthcare systems will enable them to sustain financial security while guaranteeing consistent supply of high quality care to patients.

7. CHALLENGES IN IMPLEMENTING CIRCULAR ECONOMY PRACTICES IN HEALTHCARE IOT

Although adoption of circular economy principles and practices in healthcare IoT devices is a way forward for sustainability, this is no walk in the park. To make the transition to circular approaches, there are several technical, economic, regulatory, and cultural barriers. For instance, the healthcare industry is highly regulated where devices have high demands on the quality, the performance and the safety of devices. By adding this divergence, the integration of circular economy practices becomes more complex. This section dives into the main challenges that both healthcare providers and manufacturers have with imposing circular economy practices within healthcare IoT and offers proposals on how they could be mitigated.

7.1 Technical Challenges: Design Limitations, Standardization, and Interoperability

The design of IoT healthcare can be one of the primary technical challenges of implementing the circular economy practice. During the production and design of medical devices, single use and quick replacement cycles are commonly focused on, which are in contrast with what the circular economy is intended for. Designing devices suited to durability, modularity and disassembly is a big diversion from conventional design practices focused on providing shortcuts to cost effectiveness or technologically fast updates. Many components used in medical devices are difficult to repair, recycle, or upgrade leading to many complex and hard to scale for circular approach.

This is also a large technical challenge due to the industry's lack of standards. A lack of commonly agreed standards for design of recyclable or upgradeable parts creates barriers for manufacturers to make devices that can be easily refurbished, reused or recycled. This thwarts attempts to repurpose or refurbish devices that employ different devices to start with from various manufacturers that various and uses proprietary parts and technologies not compatible with each other.

Interoperability is another critical issue. In many cases, healthcare IoT devices are specifically designed to work within closed systems so that refurbished or reused components from one device cannot be used in another. For example, only one system's sensors, chips, or software updates may be compatible. However, because this leads to a lack of interoperability between the devices, it also becomes impossible to use second hand or recycled components between devices of different models during refurbishments or repairs, making them less efficient.

To solve these technical problems a different design philosophy must be adopted, one based on modularity, durability and upgradeability. Industry standards for creating device interoperability and recycling then become targets that manufacturers have to work cooperatively on. Similarly, regulatory bodies can establish standards that promote sustainable design practices of medical technology.

7.2 Economic Challenges: Upfront Costs, Business Model Innovation, and Investment

There are also various economic challenges associated with the implementation of circular economy practices in manufacturers and healthcare providers. Another major hurdle is the initial high cost involved in manufacturing easy to refurbish or recycle devices. Adopting a circular approach to devising products and manufacturing processes will necessitate drastic investments in R&D to redesign devices and manufacturing processes. Development of eco-designs, modular components integration, reuse and recycling systems, etc., have high initial costs, which can be prohibitive for small and medium sized manufactures.

Barriers to switching to refurbished or remanufactured devices also exist for healthcare providers in terms of cost. However, development of infrastructures for collecting, refurbishing, and managing devices can be significant costs in the short term that circular economy strategies can bring long term savings. New circle care models require health care organizations to invest in new logistics, training, and quality assurance systems. Furthermore, there is also uncertainty about the availability of high quality refurbished device, so healthcare providers are wary of going that route.

Manufacturers, and healthcare providers have to rethink their business models to cope up with these economic challenges. The idea of 'product-as-a-service' is catching up where healthcare service providers can get a medical device on lease instead of buying it upfront. Under this model, devices are maintained, repaired and then refurbished that can be more cost effective in the longer term. However, by developing a sustainable supply chain that incentivizes collaboration in the industry, manufacturers and healthcare providers can assist in cutting down on the economic weight of implementing circular economy practices.

7.3 Regulatory and Logistical Hurdles in Device Lifecycle Management

Regulatory compliance is a significant issue for design and manufacture of medical devices and in healthcare. Furthermore, the medical devices have to review with the rigorous standards established by regulatory bodies like FDA (Food and Drug Administration) of the United States or European Medicines Agency (EMA), which means they need to pass to their tests of safety and efficacy. But these standards often hinder the make of a device by not allowing refurbished, or second hand components to be in incorporated in the device as any modification or changes in device must go through extensive testing, and must be approved. Therapists from the community will have to update regulations to enable the safety and efficacy of refurbished devices so that circular economy practise can be widely adopted.

Also, the challenge of logistics pertaining to device lifecycles, is another challenge. Collecting, refurbishing, and redistributing medical IoT devices efficiently requires strong systems to track the IoT devices across their point of use to the point of refurbishment to the point of reuse. System establishment is a complicated matter for setting up these systems at the scale of hospitals, clinics and other healthcare facilities. However, many healthcare organizations lack the appropriate infrastructure and personnel to effectively manage these processes in such settings with limited resources.

Regulatory Agencies could create new frameworks, which allow for approval of refurbished or remanufactured device, which satisfy the same safety and performance that govern the new devices. Other than that, healthcare provider and manufacturers can work in tandem to design efficient reverse logistic networks to retrieve and process the used devices to such an extent that they can subsequently be readopted by the healthcare system following the refurbishment in accordance with the regulatory standards.

7.4 Cultural and Organizational Challenges in Healthcare Settings

Hence, cultural and organizational barriers become not less important when it comes to adopting circular economy practices in healthcare IoT devices. Refurbished or remanufactured devices may be resisted by healthcare professionals, administrators and decision makers for reasons of reliability, performance and patient safety. The fact that patient outcomes can be directly affected by the technology used makes such a perception a formidable barrier to adoption in a field.

Furthermore, healthcare organizations will be reluctant to change their long standing processes and business models for procuring. The healthcare systems are large, bureaucratic and have long procurement cycles and rigid policies in terms of pricing. The transition to circular models requires fundamental changes in procurement procedures, contract management and in the supplier relationships. However, this can be a particularly difficult endeavor in public healthcare systems characterized by budget constraints and procurement regulations that restrict flexibility to implement the circular economy.

None of these cultural and organizational problems are trivial, and they need to be approached with a change of mindset. Healthcare stakeholders including clinicians and administrators will need to be educated about the benefits of circular economy practices so that the practice can be adopted and contribute to the sector's sustainability and cost savings. Additionally, the leadership of healthcare organizations should drive the circular economy initiatives in their respective organizations while incorporating a sustainable culture.

8. CONCLUSION

With this, the healthcare IoT sector stands in a unique position to boost the adoption of circular economy practices for ushering in a transformation in how medical devices are designed, use and disposed of in ways that are safer to both sustainability and the environment. However, the environmental impact of such technologies of e-waste and resource depletion set to become an ever greater problem. One solution to these challenges presented by circular economy principles is that they propose ways of making device activities throughout their lifecycle more sustainable and responsible, such as waste reduction, reuse of components, refurbishment of devices and material recycling. In this chapter, we investigate the way to accommodate the circular economy strategies into healthcare IoT devices, including key principles, benefits and barriers to transition from traditional linear models. The circular economy focuses on resource efficiency, eco-design, and lifecycle management which can significantly cut the carbon footprint of healthcare technologies and e – waste. The economic implications of such practices such as long term cost savings to healthcare providers, impact of the device performance and safety, and the challenge of such practices in an industry with tightly regulation and quality demands were also discussed.

Although promising, the principles of the circular economy still experience a number of difficulties. These are some technical barriers concerning design limitations, standardization and compatibility, and are also some economic barriers associated with high upfront costs and the need for business model innovation. Circular practices in the healthcare context might not be widely adopted for several reasons, the most notable of which being regulatory and logistical complexities.

Finally, the cultural and organizational barriers present in healthcare systems need to be overcome to build the framework of sustainability and circularity. However, the advantages of adopting the circular economy are obvious. Beyond results in reduction of the environmental footprint of medical devices, it can also translate into major economic savings and coordination boost for innovation in the healthcare sector. By taking advantage of the circular approach to IoT, healthcare will determine its future through collaboration between industries and through strong policy frameworks, combined with reshaping of device design and life cycle management.

9. FUTURE DIRECTIONS

To summarize, future practice of circular economy in the healthcare IoT devices is based on the advancement of technology, change of regulatory action, and embracing of sustainability by key industries. There are several key areas for development that could nonetheless further contribute to the implementation of circular economy principles in the healthcare sector and to the increase of its overall influence. Innovative technologies should be developed to refurbish, remanufacture, and recycle the device. With a surge in IoT devices and their sophistication that will come with an increasing number of integrated sensors and components, the demand for advanced technologies to dispose of, reusing and recycling of parts will increase in order to eliminate the waste efficiently and safely. Robotic, AI and machine learning innovations could truly automate some of these processes to raise the efficiency and even better quality in refurbishment and recycling. The reuse of components and materials in devices can be better managed by healthcare providers and manufacturers when systems that track the lifecycle of devices are created using these technologies.

Another future policy and regulatory change direction is where policy and regulatory can create the right adoption environment to advance circular economy practices. The governments and regulatory bodies have an important role to help draft a conducive framework for the healthcare sector for applying circular economy concepts. Updating regulations could make it possible to use and repurpose medical devices as long as they meet very strict safety and performance standards. Lastly, policies for rewarding manufacturers to make sustainable and waste reduction designs would encourage changes industry wide. A major step to implement circularity in healthcare would be to create global standards to recycle and reuse medical IoT devices. In addition, the 21st century will witness a transformation in regard to the culture of healthcare systems in addition to technological and regulatory advancements. In this context, more and more healthcare providers are expected to put sustainability for higher consideration in procurement and operations decision making as awareness around the environmental and economic benefits associated with circular economy practices grows. Healthcare organizations may opt for the adoption of circular business models like leasing or product as a service to reduce procuring cost while ensuring that devices are reused or refurbished in a responsible and transparent way.

REFERENCES

Akbari, M. (2024). Revolutionizing supply chain and circular economy with edge computing: Systematic review, research themes and future directions. *Management Decision*, *62*(9), 2875–2899. DOI: 10.1108/MD-03-2023-0412

Alfina, K. N., Ratnayake, R. C., Wibisono, D., Basri, M. H., & Mulyono, N. B. (2022). Analyzing barriers towards implementing circular economy in healthcare supply chains. In *2022 IEEE International Conference on Industrial Engineering and Engineering Management (IEEM)* (pp. 827-831). IEEE. DOI: 10.1109/IEEM55944.2022.9989999

Awan, U., Sroufe, R., & Shahbaz, M. (2021). Industry 4.0 and the circular economy: A literature review and recommendations for future research. *Business Strategy and the Environment*, *30*(4), 2038–2060. DOI: 10.1002/bse.2731

Cavalieri, A., Reis, J., & Amorim, M. (2021). Circular economy and Internet of Things: Mapping science of case studies in manufacturing industry. *Sustainability (Basel)*, *13*(6), 3299. DOI: 10.3390/su13063299

D'Alessandro, C., Szopik-Depczyńska, K., Tarczyńska-Łuniewska, M., Silvestri, C., & Ioppolo, G. (2024). Exploring circular economy practices in the healthcare sector: A systematic review and bibliometric analysis. *Sustainability (Basel)*, *16*(1), 401. DOI: 10.3390/su16010401

Fernandes, L., Rosado da Cruz, A. M., Cruz, E. F., & Lopes, S. I. (2023). A review on adopting blockchain and IoT technologies for fostering the circular economy in the electrical and electronic equipment value chain. *Sustainability (Basel)*, *15*(5), 4574. DOI: 10.3390/su15054574

Gagneja, A., Lall, B., & Bhutani, M. (2024). Unveiling CM-Det: Leveraging ConvMixer architecture for advanced object detection. *International Journal of Information Technology : an Official Journal of Bharati Vidyapeeth's Institute of Computer Applications and Management*, *16*(7), 4273–4278. DOI: 10.1007/s41870-024-02069-5

Gallo, M., Viola, A., & Romagnoli, G. (2024). The contribution of RFID to circular economy deployment: A literature review. *International Journal of RF Technologies*, Article 17545730251331373.

Gupta, A., Bhutani, M., Dixit, A., & Lall, B. (2021). FD-OMAC: A novel full-duplex optical media access control for IEEE 802.15.7. *IEEE Access : Practical Innovations, Open Solutions*, *9*, 148894–148910. DOI: 10.1109/ACCESS.2021.3124975

Jauhar, S., Pratap, S., Lakshay, , Paul, S., & Gunasekaran, A. (2023). Internet of Things-based innovative solutions and emerging research clusters in circular economy. *Operations Management Research : Advancing Practice Through Research, 16*(4), 1968–1988. DOI: 10.1007/s12063-023-00421-9

Moutaoukil, O., Mahfoud, H., & Bouhaddou, I. (2024). A novel conceptual framework for predictive maintenance of medical devices in a circular healthcare supply chain. In *2024 International Conference on Control, Automation and Diagnosis (ICCAD)* (pp. 1-6). IEEE. DOI: 10.1109/ICCAD60883.2024.10553792

Rejeb, A., Rejeb, K., Appolloni, A., Treiblmaier, H., & Iranmanesh, M. (2023). Circular economy research in the COVID-19 era: A review and the road ahead. *Circular Economy and Sustainability, 3*(4), 2127–2157. DOI: 10.1007/s43615-023-00265-2

Rejeb, A., Rejeb, K., Keogh, J. G., & Zailani, S. (2022). Barriers to blockchain adoption in the circular economy: A fuzzy Delphi and best-worst approach. *Sustainability (Basel), 14*(6), 3611. DOI: 10.3390/su14063611

Rejeb, A., Suhaiza, Z., Rejeb, K., Seuring, S., & Treiblmaier, H. (2022). The Internet of Things and the circular economy: A systematic literature review and research agenda. *Journal of Cleaner Production, 350*, 131439. DOI: 10.1016/j.jclepro.2022.131439

Rønn, C., Wieland, A., Lehrer, C., Márton, A., LaRoche, J., Specker, A., Leroy, P., & Fürstenau, D. (2023). Circular business model for digital health solutions: Protocol for a scoping review. *JMIR Research Protocols, 12*(1), e47874. DOI: 10.2196/47874 PMID: 37999949

Settia, N., Bhutani, M., & Saini, V. (2025). Exploring the impact of artificial intelligence on healthcare: A comprehensive review. In *Digitalization and the Transformation of the Healthcare Sector* (pp. 115-138).

Sharma, H. B., Vanapalli, K. R., Samal, B., Cheela, V. S., Dubey, B. K., & Bhattacharya, J. (2021). Circular economy approach in solid waste management system to achieve UN-SDGs: Solutions for post-COVID recovery. *The Science of the Total Environment, 800*, 149605. DOI: 10.1016/j.scitotenv.2021.149605 PMID: 34426367

Turskis, Z., & Šniokienė, V. (2024). IoT-driven transformation of circular economy efficiency: An overview. *Mathematical & Computational Applications, 29*(4), 49. DOI: 10.3390/mca29040049

Venkateswaran, M. R., Khosravi, A., Zarepour, A., Iravani, S., & Zarrabi, A. (2024). Self-healing materials in biomedicine and the circular economy. *Environmental Science. Nano, 11*(7), 2771–2802. DOI: 10.1039/D4EN00235K

Chapter 5
Sustainable Health Tech Managing Eco-Conscious IoT Devices for Medical Innovation

Rohan Chaudhary
https://orcid.org/0009-0003-1386-3470
Quantum University, India

Himanshu Tyagi
https://orcid.org/0009-0001-8435-9832
Quantum University, India

Pawan Kumar Goel
https://orcid.org/0000-0003-3601-102X
RKGIT, India

Seema Gulati
https://orcid.org/0009-0001-3869-7240
Graphic Era Hill University, India

Shalini Verma
https://orcid.org/0009-0004-9197-7956
Sparsh Himalaya University, India

Himani Tyagi
https://orcid.org/0000-0002-5005-9954
Quantum University, India

ABSTRACT

The integration of IoT in healthcare has transformed patient care through enhanced diagnostics, real-time monitoring, and operational efficiency. However, the surge in IoT devices raises environmental concerns like resource depletion and energy waste. This chapter explores sustainable IoT management to balance medical innovation with minimal ecological impact. Key strategies include energy-efficient designs, sustainable materials, lifecycle management, and circular economy principles. Case studies, such as Philips' circular economy initiatives and IoT-driven hospital energy management, highlight practical applications. Challenges like high costs, regulatory

DOI: 10.4018/979-8-3373-4134-7.ch005

barriers, and technological limitations are addressed, alongside future solutions involving AI, blockchain, and stakeholder collaboration. Aligning with Sustainable Development Goals, sustainable IoT fosters a healthier planet and population.

1. INTRODUCTION

1.1 Overview of IoT in Healthcare

The integration of connected innovations that improve clinical trial outcomes, accelerate operations, and optimize patient care is transforming healthcare through the Internet of Things (IoT). The Internet of Things (IoT) includes a wide range of technologies that collect and analyze real-time health data to facilitate hospital management, diagnosis, and treatment, such as wearable technology, digital medical equipment, and monitoring systems.. In particular, proactive diabetes care is made possible by IoT-enabled continuous glucose monitors, which provide real-time blood sugar readings, reduce complications, and enhance patient quality of life (Dimitrov, 2019). The rise in chronic diseases, aging population, and use of telemedicine are expected to improve the worldwide IoT healthcare market, which is projected to reach $188.2 billion by 2030. IoT's ability to provide data-driven, personalized treatment is in line with global health objectives, making it a vital component of contemporary healthcare innovation. However, to guarantee scalability and long-term impact, issues such as data privacy, system interoperability, and environmental sustainability must be resolved. Clinical professionals can remotely treat chronic diseases with the use of IoT devices, such as wearable fitness trackers and implanted sensors, which enable continuous monitoring of vital indicators such as heart rate, blood pressure, and oxygen levels. By facilitating early treatments, these technologies reduce hospital readmission rates; research indicates that patients with heart failure who use IoT-based monitoring may have a 30% reduction in readmissions (Al-Turjman et al., 2020). IoT is also used by smart medical devices, such as imaging systems and linked infusion pumps, to perform predictive maintenance, reduce downtime, and improve operational efficiency.

For example, IoT-enabled ventilators can notify staff of any problems, guaranteeing smooth patient care (Qadri et al., 2020). These developments not only enhance the quality of healthcare but also reduce expenses related to emergency procedures and equipment malfunctions. This feature is essential for handling perishable goods such as vaccinations, where malfunctions can lead to major losses. By streamlining delivery routes and cutting emissions and fuel use, IoT also improves logistics, supporting environmental sustainability objectives (Dimitrov, 2019). Furthermore, as evidenced by deployments achieving energy savings of up to 20%, IoT-powered energy management systems monitor hospital buildings and modify HVAC and lighting to reduce energy use (Al-Turjman et al., 2020). IoT in healthcare faces several obstacles despite its revolutionary promise. IoT devices collect sensitive patient data, which makes data privacy a major concern. To avoid breaches, these devices must comply with laws such as HIPAA (Qadri et al., 2020). To overcome these barriers and guarantee the scalability of IoT while adhering to international health and environmental goals, robust cybersecurity safeguards, standardized protocols, and sustainable manufacturing techniques are needed.

1.2 The Need for Sustainability in HealthTech

The rapid dissemination of Internet of Things (IoT) devices in the healthcare industry, ranging from smart medical equipment to wearable sensors, has transformed both operational efficiency and patient care. To lessen negative ecological effects, sustainable HealthTech practices are required, since its growth presents serious environmental challenges. IoT devices exacerbate the environmental impact of the healthcare industry by contributing to excessive energy consumption, resource depletion, and electronic trash (e-waste). Due to energy-intensive operations and device manufacture, the healthcare industry is responsible for more than 4% of global CO2 emissions (Karliner et al., 2020). By using eco-friendly designs like energy-efficient sensors, recyclable materials, and circular economy concepts like gadget refurbishing, Sustainable HealthTech seeks to lessen these effects.

IoT products, including diagnostic instruments and linked pacemakers, depend on resource-intensive elements like plastics and rare earth metals, whose extraction and processing worsen the environment. Millions of tons of e-waste are produced each year as a result of the disposal of these gadgets, and many of them include dangerous materials that contaminate water and land (Nižetić et al., 2020). The energy requirements of IoT devices, such as data processing and transmission, also raise the carbon footprint of the healthcare industry. In order to overcome these obstacles, Sustainable HealthTech prioritizes energy-efficient designs that lower operating emissions, such as energy harvesting devices and low-power sensors (Geissdoerfer et al., 2017). For instance, solar-powered sensors may reduce the need for batteries in wearable monitoring, reducing waste and energy use.

Additionally, there are operational and financial benefits to sustainability in health technology. Energy expenses may be considerably decreased by energy-efficient IoT solutions, such as smart building management in hospitals. Green supply chains that use sustainable materials improve brand reputation among stakeholders that care about the environment while reducing the environmental impact of device manufacture. But there are obstacles to overcome, such the high upfront expenditures of sustainable infrastructure, legal limitations on recycled materials, and the need for cutting-edge recycling technology (Mangla et al., 2018). To get beyond these obstacles and guarantee that sustainable practices are affordable and scalable, manufacturers, healthcare providers, and legislators must work together. Additionally, there are operational and financial benefits to sustainability in health technology. Energy expenses may be considerably decreased by energy-efficient IoT solutions, such as smart building management in hospitals. Green supply chains that use sustainable materials improve brand reputation among stakeholders that care about the environment while reducing the environmental impact of device manufacture. But there are obstacles to overcome, such the high upfront expenditures of sustainable infrastructure, legal limitations on recycled materials, and the need for cutting-edge recycling technology (Mangla et al., 2018). To get beyond these obstacles and guarantee that sustainable practices are affordable and scalable, manufacturers, healthcare providers, and legislators must work together.

1.3 Objectives of the Chapter

The disruptive potential of sustainable Internet of Things (IoT) applications in healthcare is examined in this chapter, with an emphasis on eco-conscious device management to promote medical innovation while reducing environmental effect. By facilitating real-time health monitoring, boosting diagnosis, and increasing operational efficiency, the quick adoption of IoT technologies—such as wearable sensors, smart medical devices, and remote monitoring systems—has completely changed patient care. To be in line with global sustainability standards, these gadgets' environmental impact—which includes resource depletion, energy consumption, and electronic trash, or "e-waste"—must be addressed. This chapter's goals are diverse and include offering a thorough framework for incorporating sustainability into HealthTech via IoT.

Analyzing how IoT may improve the care of patients and operational effectiveness is the first goal. Proactive care of chronic illnesses is made possible by IoT devices, such as linked pacemakers and continuous glucose monitors, which lower hospital readmission rates and enhance patient outcomes. For example, IoT-powered asset monitoring minimizes care interruptions by guaranteeing prompt medical supply refill. By looking at these uses, the chapter aims to show how IoT may provide effective, individualized healthcare while laying the groundwork for sustainable practices.

The second goal is to assess the sustainability issues related to IoT devices, including energy consumption and e-waste. IoT device manufacturing and disposal add to the environmental effect of the healthcare industry, which accounts for more than 4% of global CO2 emissions (Karliner et al., 2020). Hazardous compounds may be found in e-waste from abandoned gadgets, and carbon footprints are increased by energy-intensive activities. In order to lessen these effects, this research will highlight the need for environmentally friendly solutions and pinpoint key areas for action.

The third goal is to provide methods for managing the Internet of Things sustainably, such as circular economy principles, sustainable materials, and energy-efficient designs. Energy harvesting systems and energy-efficient sensors may lower power usage, and waste can be minimized by using recyclable or biodegradable materials. Product-as-a-service (PaaS) and gadget refurbishing are examples of circular economy models that prolong lifecycles and lower resource use. Medical equipment leasing, for instance, promotes resource efficiency by letting manufacturers take care of recycling and maintenance.

Figure 1. IoT healthcare ecosystem

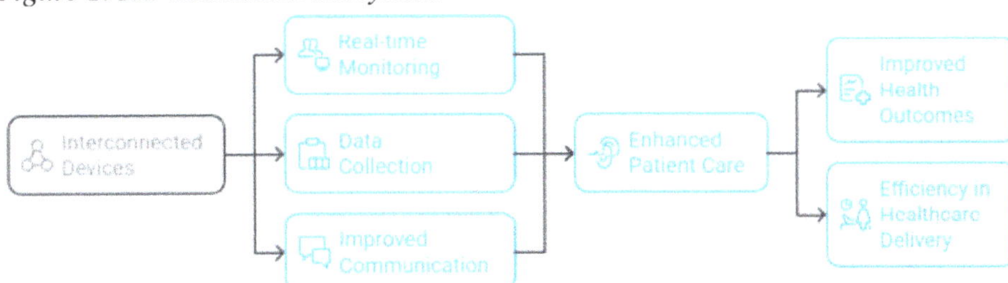

2. IoT IN HEALTHCARE: ENHANCING EFFICIENCY AND REDUCING WASTE

2.1 Current Applications

Wearable technology, like biosensors and smartwatches, allows for remote patient care and eliminates the need for in-person visits by continually monitoring vital indicators like blood glucose levels, oxygen saturation, and heart rate. To improve glycemic control and enable prompt treatments, IoT-based continuous glucose monitors, for example, provide diabetic patients access to real-time data (Marcolino et al., 2022). According to Dimitrov (2019), these devices have the potential to significantly lower hospital readmission rates by empowering individuals to manage chronic diseases and enabling doctors to make data-driven choices.

Another essential IoT use is telehealth systems, which combine linked devices to provide virtual consultations, especially for patients in underserved or rural locations. IoT-enabled telehealth services provide continuity of treatment without in-person visits by sending data to healthcare professionals via wearables or home monitoring devices. According to research, telehealth use rose by 38% during the COVID-19 pandemic, increasing access to specialist treatment (Al-Turjman et al., 2020). Telehealth reduces carbon emissions by cutting down on travel, which is in line with environmental objectives. Additionally, IoT facilitates quick data exchange, which improves emergency response. IoT-connected defibrillators, for instance, may send cardiac event data to emergency services, increasing survival rates and response times (Qadri et al., 2020).

IoT-enabled smart beds in hospitals track a patient's weight, movement, and pressure points, improving comfort and lowering issues like bedsores. These beds save burden and improve care quality by alerting personnel to reposition patients (Nižetić et al., 2020). Inventory management is another way that IoT improves hospital operations. Medical supplies, including vaccinations and medications, are tracked by RFID tags and Internet of Things sensors, which guarantees prompt replenishing and reduces waste from overstocking or expiry. For perishable items, where spoiling has a negative impact on the environment and finances, this is crucial. According to Mangla et al. (2018), a hospital that used IoT inventory systems observed a 15% decrease in supply waste, indicating both operational and environmental advantages.

For example, combine information from inventory systems, wearables, and beds to provide administrators and physicians useful information. In the treatment of chronic diseases, when prompt actions might avert problems, this connectedness improves care coordination. IoT makes it possible to monitor patient vitals in real time in emergency situations, which facilitates quick modifications to treatment regimens (Geissdoerfer et al., 2017). But issues like interoperability, data security, and the effects of IoT devices on the environment—including energy consumption and e-waste—need to be addressed. To lessen these problems, sustainable practices—such as circular economy models and energy-efficient designs—are crucial (Stahel, 2019).

2.2 Emerging Trends

To enable for better care delivery, the Internet of Things (IoT) is using sophisticated analytics, interoperability, and patient-centric methods to drive disruptive innovation in the healthcare industry.

Leading the way are IoT devices with AI built in, which provide predictive insights by evaluating enormous datasets from wearables, sensors, and medical equipment. AI-powered wearables, for example, can detect cardiac events and track heart rate variability, allowing for prompt treatments that may save emergency admissions by as much as 30% (Steinhubl et al., 2018). These tools enable doctors to put preventative measures into action by processing data from continuous monitoring to find patterns, such as early indicators of arrhythmias. By customizing treatments based on real-time data, AI also improves the management of chronic diseases by increasing patient adherence and results (Dimitrov, 2019). To safeguard patient privacy, however, combining AI with IoT necessitates strong data security and legal compliance, especially under frameworks like HIPAA.

Another important development is edge computing, which allows IoT devices to analyze data locally, reducing latency and facilitating real-time decision-making. In environments requiring critical care, including intensive care units, edge-enabled Instantaneous vital sign processing by IoT devices allows for quick reactions to situations like sepsis. Particularly in faraway locations, this strategy lessens dependency on cloud servers, reducing bandwidth expenses and improving system dependability (Qadri et al., 2020). By lowering the need for data transmission, edge computing also increases energy efficiency, supporting environmental objectives. However, implementing edge solutions necessitates resolving technological issues including controlling computational constraints and guaranteeing device compatibility.

For smooth integration across many platforms and devices, IoT ecosystems must be interoperable. Interoperability is made possible by standardized communication protocols like HL7 FHIR, which provide effective data sharing across hospital systems, wearable technology, and electronic health records (Al-Turjman et al., 2020). Care coordination is improved by this connectedness, which also lowers diagnostic mistakes and improves treatment results. For instance, data from patient records and smart insulin pumps may be integrated via compatible IoT systems to improve diabetes care.

2.3 Economic Impact

The healthcare sectors economic environment is changing as a result of the adoption of Internet of Things (IoT) technologies, which are generating significant cost savings, increasing revenue, and stimulating market expansion. IoT applications that improve resource usage and save operating costs, such asset tracking, predictive maintenance, and remote patient monitoring, provide substantial financial advantages for manufacturers, service providers, and healthcare providers. According to MarketsandMarkets (2023).

By maximizing resource use and reducing hospital readmissions, IoT dramatically lowers healthcare expenses. Wearable vital sign tracking devices and other remote monitoring technologies allow for preventive treatment of chronic illnesses including diabetes and heart disease, lowering readmission rates by up to 20% and saving patients around $6,000 a year (Haddad et al., 2021). By enabling doctors to act early, these tools help save expensive hospital stays. To save treatment expenses, IoT-enabled telemedicine solutions, for instance, have reduced emergency visits for patients with chronic obstructive pulmonary disease (COPD) by 15% (Greene et al., 2020). IoT-driven operational benefits that assist hospitals include medical equipment predictive maintenance. According to Al-Turjman et al. (2020), hospitals may reduce downtime by 30% and repair expenses by up to 25% by scheduling maintenance before equipment failures occur by utilizing IoT sensors to monitor device performance. For expensive equipment, such as MRI scanners, where downtime might cost thousands of dollars per day, this has a very significant effect.

Manufacturers and service providers have a lot of possibilities as the IoT healthcare industry grows. With businesses like Philips and Medtronic spearheading innovation, the spread of smart devices, such as wearable monitoring and networked infusion pumps, has prompted investments in IoT platforms and analytics.

IoT's economic advantages can have drawbacks, too. For mid-sized hospitals, the initial expenses of IoT infrastructure, such as sensors, cloud systems, and cybersecurity measures, might surpass $1 million, which presents challenges for smaller or resource-constrained institutions (Nižetić et al., 2020). Budgets are further strained by ongoing costs like software upgrades and employee training. Investments in safe systems are necessary for regulatory compliance, which raises costs and includes following data privacy rules like HIPAA (Dwivedi et al., 2021). Due to these obstacles, healthcare inequities may worsen as smaller clinics are left behind and larger institutions embrace IoT more quickly.

3. ENVIRONMENTAL CHALLENGES OF IoT IN HEALTHCARE

3.1 E-Waste from Medical IoT Devices

Fast technological improvements, medical IoT devices, including smart inhalers and glucose monitors, often have short lifespans, requiring frequent replacements and contributing to a significant buildup of e-waste. Every year, millions of tons of e-waste are produced worldwide in the healthcare industry, and IoT devices exacerbate this problem because of their intricate, non-biodegradable parts that contain dangerous substances like lead, cadmium, and mercury (Forti et al., 2020). When e-waste is handled improperly, it may contaminate soil and water, endangering ecosystems and human health by exposing people to harmful compounds. To minimize e-waste, recover precious materials, and lessen dependency on landfills, it is essential to use sustainable design techniques, such as modular systems that allow for component replacements and strong recycling programs specifically designed for medical equipment. In order to ensure that healthcare innovation is in line with environmental stewardship and supports Sustainable Development Goal 12 (Responsible Consumption and Production), manufacturers, healthcare providers, and regulators must work together to enforce responsible disposal and recycling standards.

A viable way to reduce e-waste is via sustainable design. IoT devices that are modular may greatly increase lifespans by enabling component changes or repairs. To cut down on waste and save resources, a modular insulin pump, for example, may have its sensor or battery changed without having to throw away the whole device (Geissdoerfer et al., 2017). To make end-of-life processing easier, manufacturers might also give preference to environmentally friendly materials like bio-based composites or recyclable polymers. In order to recover rare earth metals, such as neodymium, which are essential for device manufacturing but limited in availability, recycling programs tailored for medical IoT devices are essential. These initiatives may lessen the need for virgin resources, which would lessen manufacturing's environmental impact.

Working together is essential to solving e-waste issues. To guarantee correct disposal, manufacturers must make investments in creating robust, recyclable gadgets and set up take-back initiatives. Healthcare providers may engage in recycling programs and implement rules that favor environmentally friendly technologies. Standards like the European Union's WEEE Directive, which requires recycling of e-waste but requires worldwide acceptance for effectiveness, are enforced in large part by regulators (Rautela et al., 2021). Innovation in recycling technologies, including automatic disassembly systems for intricate IoT devices, may be fueled by public-private collaborations.

3.2 Energy Consumption

The explosion of Internet of Things (IoT) gadgets in healthcare—connected monitors, wearable sensors, smart hospital systems—has transformed patient care but presents major environmental problems because of their great energy consumption. These gadgets, crucial for real-time monitoring and data analytics, contribute to greenhouse gas emissions, worsening healthcare's environmental footprint, which accounts for nearly 4% of global CO_2 emissions (Karliner et al., 2020). Particularly in big hospitals with hundreds of IoT devices, constant operation of sensors, frequent data transfer, and resource-intensive cloud processing cause huge energy needs. This energy intensity compromises sustainability targets, thus eco-friendly solutions are necessary to match healthcare with aims of climate action, like Sustainable Development Goal 13 (ClimateAction). Low-power sensors and improved communication protocols are among the energy-efficient solutions that may greatly lower consumption. Promising answers also come from using energy collecting technology and switching IoT infrastructure to renewable energy sources.

Hospitals may further employ IoT-driven energy management systems to improve device utilization, lowering waste and boosting sustainability.

Reducing these effects depends much on energy-efficient designs. Designed with sophisticated microelectronics, low-power sensors lower energy requirements without sacrificing performance. Optimized communication protocols, such as Bluetooth Low Energy (BLE), lower data transmission power needs, prolonging device battery life (Gomez et al., 2019). For example, wearable heart rate monitors employing BLE may function for weeks on a single battery, saving energy consumption. Energy harvesting devices, including solar or kinetic-powered sensors, provide new solutions by capturing ambient energy, removing dependency on conventional power sources. A pilot research indicated that solar-powered IoT sensors in rural clinics cut energy expenses by 30% while preserving dependability (Shaikh et al., 2017).

Hospitals may also employ IoT-driven energy management systems to promote sustainability. These systems employ real-time data to monitor and optimize device utilization, shutting off idle equipment or altering power settings dynamically. For instance, a smart hospital in Singapore leveraged IoT to cut energy usage by 20% by improving HVAC and gadget operations (Nguyen et al., 2021). Such solutions not only cut emissions but also produce cost savings, enabling hospitals to reinvest in patient care. Integrating IoT infrastructure with renewable energy sources, such as solar or wind power, further lowers dependency on fossil fuels. A hospital in Germany powered its IoT network using solar panels, lowering operating emissions by 25% (Nguyen et al., 2021).

3.3 Supply Chain Impacts

The supply chain for Internet of Things (IoT) devices in healthcare has severe environmental repercussions, arising from resource extraction, manufacture, and transportation. The development of IoT gadgets, such as wearable sensors and smart medical equipment, rely significantly on rare earth metals like neodymium and lithium, as well as other finite resources. Mining these minerals damages ecosystems, depletes biodiversity, and creates large carbon emissions, with rare earth mining releasing up to 12 tons of CO_2 per ton of material (Haque et al., 2014). Energy-intensive industrial processes, including semiconductor fabrication and device assembly, further worsen environmental deterioration, adding to global greenhouse gas emissions. For instance, creating a single IoT sensor may require large power, generally generated from fossil fuels in places with carbon-heavy grids (Nižetić et al., 2020). Additionally, the international nature of IoT supply chains includes transporting components across continents, raising transportation emissions, which account for around 14% of overall supply chain emissions in electronics (Simchi-Levi et al., 2019). These implications underline the critical need for sustainable strategies to offset the environmental toll of IoT devices in healthcare.

To overcome these difficulties, healthcare firms should embrace green supply chain policies that stress environmental responsibility. Sourcing sustainable materials, such as recycled metals or bio-based polymers, decreases dependence on virgin resources and minimizes environmental disturbance. For example, employing recycled aluminum in device casings may lower emissions by up to 95% compared to original aluminum manufacturing (Geissdoerfer et al., 2017). These tactics not only minimize environmental impact but also boost brand image and comply with legislation like the EU's Green Deal, which promotes sustainable supply chains (European Commission, 2020).

IoT technology itself provides strong capabilities to streamline healthcare supply chains, decreasing waste and environmental impact. Real-time inventory tracking, enabled by RFID tags and IoT sensors, provides precise monitoring of medical supplies and equipment, minimizing overstocking and spoiling. For instance, IoT-driven demand forecasting might eliminate excess manufacturing of perishable components, such as batteries, by matching supply with real demands (Qadri et al., 2020).

Implementing circular supply networks is a disruptive method to further minimize environmental effect. Circular models emphasize reusing, repairing, or recycling items to prolong their lives and decrease waste. For example, recycling rare earth metals from retired IoT devices may recover up to 80% of resources, minimizing the requirement for new mining (Stahel, 2019). Healthcare companies may organize take-back programs to collect discarded equipment, assuring thorough disassembly and material recovery.

4. PRINCIPLES OF ECO-CONSCIOUS IoT DESIGN

4.1 Sustainable Materials

The formation of sustainable materials in Internet of Things (IoT) devices is crucial for decreasing the environmental imprint of healthcare technology, coinciding with circular economy concepts and Sustainable Development Goal 12 (Responsible Consumption and Production). Eco-conscious materials, such as bio-based polymers and recycled metals, decrease resource depletion, cut carbon emissions, and promote end-of-life recyclability, addressing the rising problem of electronic waste (e-waste) in healthcare (Nižetić et al., 2020). For instance, bioplastics generated from renewable sources like cornstarch or sugarcane may replace petroleum-based polymers in wearable sensors, such as glucose monitors, lowering production-related emissions by up to 30% compared to conventional materials (Geissdoerfer et al., 2017). These bioplastics are biodegradable or compostable, lowering landfill contributions and lessening the environmental impact caused by non-degradable plastics. Similarly, recycled metals, such as aluminum or titanium recovered from abandoned medical devices, are being incorporated into IoT equipment like smart infusion pumps, preserving scarce resources and lowering energy-intensive mining procedures (Kelly et al., 2020).

This closed-loop strategy coincides with circular economy concepts, which stress resource efficiency and waste reduction by keeping materials in use for as long as feasible (Bressanelli et al., 2018). By incorporating sustainable materials, healthcare IoT devices not only minimize environmental impact but also enable a regenerative economy, where resources are reused rather than wasted, supporting sustainable healthcare practices.

The move to sustainable materials provides additional advantages beyond environmental effect. Bio-based polymers and recovered metals may cut the lifespan costs of IoT devices by permitting simpler recycling and decreasing disposal fees. For instance, a wearable sensor manufactured from biodegradable polymers may be composted at the end of its life, reducing incineration expenses and emissions associated with typical e-waste disposal. Furthermore, sustainable materials help to corporate social responsibility objectives, increasing brand image and satisfying regulatory needs for greener practices in healthcare (Stahel, 2019). These materials also match with patient and provider expectations for ecologically responsible healthcare, as public knowledge of sustainability concerns rises.

Higher manufacturing costs for sustainable materials, which may be 20–40% more costly than traditional alternatives, offer financial impediments, especially for businesses in low-margin sectors. Regulatory limits on recycled materials in medical equipment, prompted by worries about contamination, further hinder uptake (Geissdoerfer et al., 2017). These difficulties demand novel solutions that balance performance, affordability, and sustainability.

Advances in material science are tackling these concerns, opening the door for sustainable IoT devices that retain excellent performance. Researchers are creating hybrid bioplastics with better durability, mixing renewable polymers with reinforcing agents to suit medical-grade standards (Kelly et al., 2020). Nanotechnology is also enhancing the recyclability of metals, allowing better recovery rates without sacrificing quality. Collaborative efforts between manufacturers, material scientists, and politicians are vital to scale these advances. For example, public-private partnerships may support research into cost-effective sustainable materials, while regulatory changes can loosen constraints on recycled content in medical equipment (Stahel, 2019).

4.2 Energy Efficiency

Low-power components, such as energy-efficient microchips and system-on-chip (SoC) architectures, are crucial to decreasing energy usage in IoT healthcare devices. For example, microcontrollers tuned for low power, such those used in wearable heart rate monitors, offer longer operation without regular recharging (Arefin et al., 2019). Communication methods, such as Zigbee and Bluetooth Low Energy (BLE), further limit energy usage by providing efficient data transfer with lower power needs. Zigbee, for instance, is commonly used in hospital sensor networks to communicate patient data while spending minimum energy, prolonging device lifespans and decreasing waste from battery replacements (Soh et al., 2020). These solutions guarantee that equipment like continuous glucose monitors stay effective for extended durations, boosting patient care while minimizing environmental costs.

Energy harvesting technologies provide a breakthrough approach to sustainable IoT design, enabling devices to collect power from ambient sources and minimize dependency on disposable batteries. Piezoelectric systems, which transform mechanical vibrations into electrical energy, are employed in wearable devices like activity trackers, harvesting energy from body motions (Wang et al., 2021).

Practical uses of energy-efficient IoT devices illustrate their influence. In hospitals, IoT-enabled smart beds outfitted with low-power sensors monitor patient vitals while optimizing energy usage using BLE protocols, lowering facility energy consumption (Nižetić et al., 2020). Outpatient settings benefit from wearable devices that employ energy harvesting to monitor chronic illnesses, such as heart disease, without adding to e-waste. These solutions accord with circular economy concepts by extending device lifecycles and decreasing resource usage, leading to a regenerative healthcare paradigm.

Challenges in attaining energy efficiency include reconciling low power consumption with the high computing needs of sophisticated medical equipment. For instance, real-time data analytics in IoT diagnostic tools demand substantial processing power, which might clash with energy-saving aims (Arefin et al., 2019). System compatibility is another challenge, as varied IoT devices must interoperate smoothly inside hospital networks, frequently needing standardized protocols that may not prioritize energy saving (Soh et al., 2020). Additionally, the initial cost of designing and installing energy-efficient devices may be considerable, especially for resource-constrained healthcare settings.

Future improvements promise to solve these difficulties. Nanotechnology is allowing the creation of ultra-low-power components, such as nanoscale sensors, that retain good performance with little energy utilization (Wang et al., 2021). AI-driven power management systems can improve energy allocation in IoT devices, dynamically altering power depending on consumption patterns to enhance efficiency

(Leonov, 2019). These improvements, together with stakeholder engagement and legislative incentives, will further integrate energy-efficient IoT devices into sustainable healthcare, decreasing environmental impact while increasing patient care.

4.3 Modular and Upgradable Designs

The environmental advantages of modular IoT solutions are enormous. Traditional linear manufacturing patterns, where equipment are destroyed after obsolescence, lead to resource depletion and landfill congestion. In contrast, modular systems allow targeted updates, eliminating the need for new devices and preserving commodities like rare earth metals used in electronics. For example, a modular infusion pump may have its sensor module upgraded to accommodate new monitoring capabilities, prolonging its operating life by years. This strategy corresponds with circular economy aims, which stress reuse, repair, and recycling to construct closed-loop systems (Geissdoerfer et al., 2017). By lowering e-waste, modular designs help decrease the environmental dangers associated with inappropriate disposal, such as soil and water pollution from hazardous components like batteries. Furthermore, upgradability enables IoT devices stay relevant despite fast technology improvements, preventing premature obsolescence. Hospitals gain economically, since updating components is generally 30–50% less than acquiring new equipment, enabling reallocation of money to patient care or staff training (Stahel, 2019).

Innovations are tackling these difficulties. 3D printing provides on-demand creation of bespoke components, lowering supply chain dependency and costs. For instance, 3D-printed sensor housings may be fitted to individual IoT devices, boosting modularity without mass manufacturing (Nižetić et al., 2020). Open-source hardware platforms, such as Arduino-based medical sensors, enable developers to construct interoperable components, promoting a collaborative environment for sustainable IoT. These developments permit healthcare professionals to repair and update equipment locally, decreasing downtime and environmental effect.

5. MANAGING ECO-CONSCIOUS IoT DEVICES

5.1 Lifecycle Management

The Process management of Internet of Things (IoT) devices in healthcare is crucial for attaining sustainability by prolonging device usefulness and limiting environmental effect throughout design, usage, and disposal stages. Eco-conscious lifecycle management stresses durability, repairability, and recyclability, matching with circular economy concepts that encourage resource efficiency and waste reduction. For example, wearable health monitors, such as continuous glucose monitors, may be constructed with changeable sensors or batteries, enabling updates or repairs without abandoning the whole device (Geissdoerfer et al., 2017). This flexibility decreases the need for complete replacements, preserving resources like rare earth metals used in electronics. Manufacturers should also emphasize eco-friendly materials, such as recyclable plastics or biodegradable components, to avoid environmental impact during manufacture and disposal (Nižetić et al., 2020). These design decisions not only lengthen device lifespans but also minimize the carbon footprint associated with production, which is resource-intensive and energy-heavy.

During the usage phase, lifecycle management focuses on maintenance and refurbishing to enhance device durability. IoT-enabled predictive maintenance, where sensors monitor device performance and communicate repair requirements, minimizes premature failures. For instance, smart infusion pumps may warn technicians to component deterioration, allowing prompt repairs that improve operational life (Qadri et al., 2020). Refurbishment projects, where old devices are restored to like-new condition, further decrease the need for new equipment. Hospitals using reconditioned IoT equipment, such as imaging systems, may save 20–40% compared to new purchases while minimizing waste (Stahel, 2019). These methods connect with circular economy aims by keeping equipment in use longer and limiting resource extraction.

Despite its benefits, eco-conscious lifecycle management poses problems. High initial costs for building modular, robust devices might inhibit manufacturers, especially smaller enterprises with limited funds. Regulatory constraints, such as prohibitions on employing recycled materials in medical equipment, hamper implementation (Geissdoerfer et al., 2017). Additionally, the absence of consistent recycling methods across areas limits scaled take-back and recycling systems. Collaboration among manufacturers, healthcare providers, and legislators is crucial to set consistent standards and incentivise sustainable behaveiors via subsidies or tax breaks (Nižetić et al., 2020).

Figure 2. IoT device lifecycle management cycle

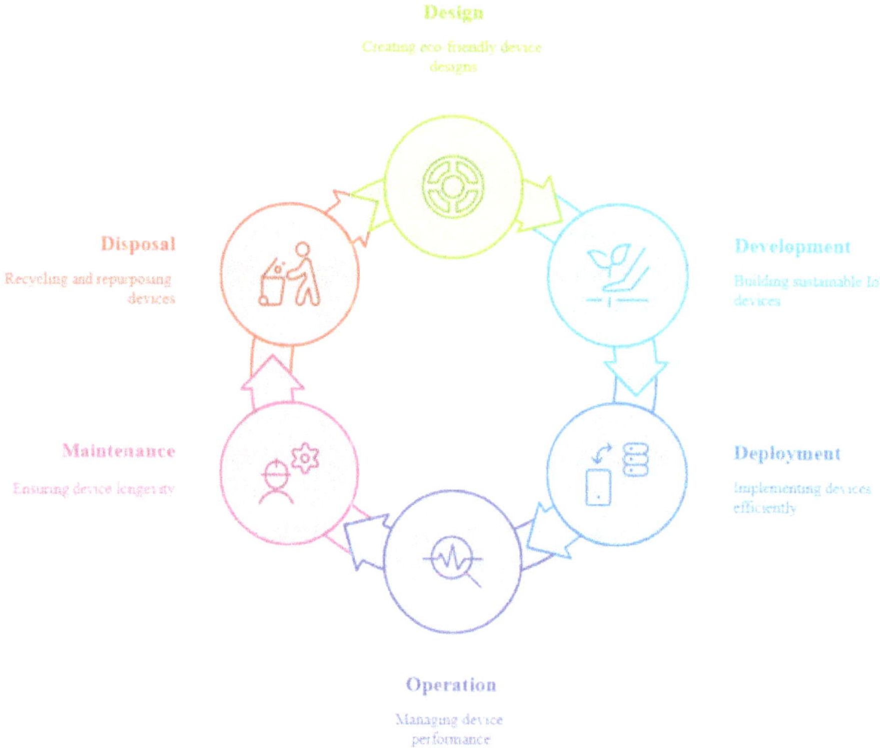

5.2 Data Optimization

The data optimization in eco-conscious Internet of Things (IoT) devices is a cornerstone of sustainable healthcare, drastically decreasing energy consumption and processing demands while keeping therapeutic effectiveness. IoT technologies, such as wearable sensors and smart medical equipment, create huge datasets from continuous patient monitoring, demanding effective data management to limit power use and environmental impact. Techniques like edge computing, data compression, and selective filtering allow IoT systems to run leanly, aligning with healthcare's environmental aims and Sustainable Development Goal 12 (Responsible Consumption and Production). These solutions not only minimize the carbon footprint of IoT operations but also boost device life and performance, enabling sustainable healthcare delivery. Despite hurdles such as building low-power algorithms and preserving data accuracy, data optimization is crucial for green HealthTech innovation.

Edge computing is a fundamental method, processing data locally on IoT devices to lessen dependency on energy-intensive cloud transport. For example, a smart insulin pump may assess glucose levels on-device, relaying only necessary alarms to healthcare professionals, so preserving power and bandwidth (Al-Turjman et al., 2020). This localized processing eliminates latency, boosts real-time decision-making, and saves energy expenditures by up to 30% compared to cloud-based systems (Nižetić et al., 2020). By restricting data delivered to centralized servers, edge computing minimizes the burden on network infrastructure, thus minimizing the environmental effect of IoT activities. Hospitals implementing edge-enabled IoT devices, such as heart monitors, report greater energy efficiency without sacrificing patient outcomes, proving the practical advantages of this strategy.

For instance, a wearable ECG monitor may compress heart rhythm data before transmission, lowering battery usage by 20–25% while retaining diagnostic accuracy. Advanced compression methods, such lossless algorithms, guarantee no crucial information is lost, retaining the veracity of medical findings. This efficiency is crucial for battery-powered IoT devices, increasing their operational life and minimizing the frequency of battery changes, which contribute to e-waste (Geissdoerfer et al., 2017).

Selective data filtering further promotes sustainability by removing unnecessary or duplicate information, minimizing system overload and ensuring IoT networks stay efficient. By prioritizing clinically relevant data, such as high glucose spikes above regular readings, IoT devices save wasteful processing and transmission, preserving energy and system resources (Qadri et al., 2020). This strategy is especially useful in large-scale hospital networks, where hundreds of IoT devices function concurrently. For example, IoT-enabled ventilators may filter out steady respiratory data, concentrating on abnormalities that need action, thereby enhancing network performance and minimizing energy consumption. Such lean operations connect with circular economy concepts, decreasing resource waste and promoting healthcare's environmental aims.

Despite these advantages, obstacles exist in adopting data optimization. Developing low-power algorithms that combine energy economy with computing precision involves extensive study and development (Nižetić et al., 2020). Ensuring data integrity during compression or filtering is crucial, since mistakes might affect patient safety. Additionally, integrating edge computing into current healthcare systems involves strong infrastructure and interoperability standards, which may be expensive (Al-Turjman et al., 2020). Addressing these challenges needs cooperation between technology developers, healthcare providers, and legislators to invent and standardize eco-conscious IoT solutions.

Figure 3. IoT data optimization process

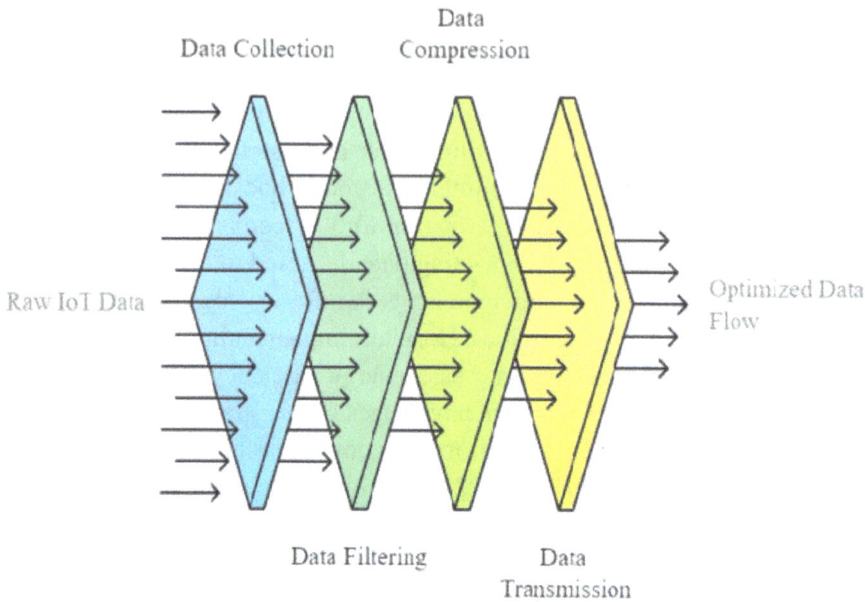

5.3 Regulatory and Standards Compliance

Regulatory and standards compliance is crucial in ensuring that Internet of Things (IoT) devices in healthcare are eco-conscious, fulfilling both environmental and medical criteria to support sustainable innovation. Compliance with rules such as the European Union's Restriction of Hazardous Substances (RoHS) Directive bans the use of hazardous compounds like lead and mercury in IoT devices, considerably decreasing environmental toxicity and promoting safer recycling (European Commission, 2020). Similarly, standards like ISO 14001 enable frameworks for enterprises to pursue environmentally friendly practices, such as energy-efficient manufacturing and waste reduction, which are crucial for sustainable IoT production (ISO, 2021). In healthcare, IoT devices must also conform to severe medical laws, such as the U.S. Food and Drug Administration (FDA) rules, which assure patient safety while incorporating green design concepts like recyclable or biodegradable components. For example, compliant wearable sensors for glucose monitoring incorporate non-toxic, recyclable polymers, blending clinical performance with environmental responsibility. Certification systems, such as those for RoHS or ISO

14001, check conformity to these standards, fostering confidence among stakeholders and boosting the use of eco-conscious technology.

Companies using ISO 14001 have reported up to a 15% decrease in energy usage in manufacturing processes, boosting the sustainability of devices like linked pacemakers (Santos et al., 2019).

In healthcare, IoT devices must comply with medical laws to assure safety and effectiveness. The FDA's recommendations for medical devices mandate thorough testing of IoT systems, such as remote monitoring tools, to avoid faults that might jeopardize patients (FDA, 2022). Integrating green design ideas into these rules is tough but feasible. For instance, wearable heart rate monitors may employ bio-based polymers that exceed FDA safety criteria while being recyclable, minimizing the environmental impact. Certification methods play a significant role in validating compliance, with third-party audits ensuring that devices fulfill both environmental and medical criteria. These certifications boost customer and regulatory confidence, encouraging hospitals to use sustainable IoT systems.

Despite these advantages, compliance offers considerable obstacles. Global regulatory environments vary, with various regulations in the EU, U.S., and Asia, complicating the design and certification of IoT devices for worldwide markets (Rauschnabel et al., 2021). Compliance costs can be substantial, particularly for small manufacturers, as eco-friendly materials and certification processes are often expensive. For example, moving to RoHS-compliant components might raise manufacturing costs by 10–20%, thereby restricting adoption in low-resource settings (Forti et al., 2020). Interoperability concerns also emerge, since eco-conscious solutions must interact with current healthcare systems without sacrificing functionality.

To tackle these problems, stakeholders might employ collaborative structures, such as industry consortia, to unify standards and decrease compliance costs. Policy incentives, including tax credits for sustainable manufacturing, may further boost adoption (Santos et al., 2019).

6. ECONOMIC AND SOCIETAL BENEFITS

6.1 Cost Efficiency

The integration of Internet of Things (IoT) technologies in healthcare has revolutionized cost efficiency by optimizing resource allocation, reducing operational expenses, and enhancing care delivery. IoT applications, such as remote patient monitoring, predictive maintenance, supply chain optimization, and smart energy management, deliver substantial financial benefits to healthcare systems. These technologies enable proactive interventions, minimize waste, and streamline operations, allowing providers to reinvest savings into improving patient care. Despite high initial costs and training requirements, the long-term economic advantages make IoT a critical component of sustainable healthcare. By leveraging real-time data and automation, IoT not only lowers costs but also aligns with global health objectives, ensuring efficient and accessible care. Remote patient monitoring systems, powered by IoT wearables like smartwatches and biosensors, significantly reduce the need for hospital admissions by enabling continuous tracking of vital signs such as heart rate, blood pressure, and glucose levels. These systems are particularly effective for managing chronic diseases like diabetes and hypertension, where timely interventions can prevent complications. For instance, IoT-driven telehealth solutions have reduced emergency room visits for chronic disease patients by up to 20% in some healthcare networks, translating to millions in savings annually (Li, J., Carayon, P., & Hundt, A. S., 2021). By shifting care to outpatient settings, IoT reduces the burden on hospital resources, lowers patient costs, and improves access to care, particularly in underserved regions. Predictive maintenance, facilitated by IoT sensors embedded in medical equipment like ventilators and imaging systems, minimizes unexpected failures and extends device lifespans.

RFID tags and IoT sensors track perishable items like vaccines and pharmaceuticals, preventing overstocking and spoilage, which can account for significant financial losses. By ensuring timely replenishment and reducing waste, IoT-driven supply chains cut waste-related costs by up to 15% in some healthcare systems (Gupta, R., Shankar, R., & Singh, R. K., 2022). For instance, hospitals using IoT to manage blood products can donate nearing-expiry stocks to clinics, avoiding waste while supporting community health. Smart energy management systems in hospitals leverage IoT to monitor and control utilities like lighting, heating, and cooling, driving significant cost savings. By analyzing real-time energy consumption data, these systems optimize usage, reducing waste and operational expenses. Some facilities have reported energy cost reductions of 15% through IoT-driven solutions, freeing up budgets for clinical improvements (Zhang, X., Liu, Y., & Zhang, Z., 2022). For example, IoT-based building management systems can automatically adjust HVAC settings based on occupancy, enhancing efficiency without compromising patient comfort. Despite these benefits, implementing IoT technologies involves challenges. High initial costs for hardware, software, and infrastructure upgrades can deter adoption, particularly in resource-constrained settings. Staff training to operate and interpret IoT systems also requires investment, as does ensuring cybersecurity to protect patient data. By reducing operational expenses and reinvesting savings into care quality, IoT supports equitable access to healthcare while aligning with economic and environmental goals. Continued innovation and stakeholder collaboration are essential to scale these solutions, ensuring IoT remains a cornerstone of cost-efficient, high-quality healthcare delivery.

6.2 Market Opportunities

The integration of Internet of Things (IoT) technologies with healthcare is unlocking substantial market prospects, generating innovation, and supporting economic development in the medical technology industry. The worldwide IoT healthcare industry is predicted to cross $200 billion by 2030, spurred by increased demand for connected devices, telemedicine solutions, and digital health platforms (Grand View Research, 2023). This expansion is driven by consumer and provider requirements for customized, efficient, and sustainable healthcare solutions. Companies developing IoT-enabled wearables, such as smart insulin pumps and fitness trackers, are capitalizing on the increased demand for technologies that allow patients to manage their health proactively, producing attractive income streams (Patel et al., 2021). These gadgets, fitted with sensors for real-time monitoring of vital signs like glucose levels or heart rate, cater to the rising incidence of chronic illnesses, which impact over 60% of individuals in affluent countries.

For instance, IoT systems incorporating artificial intelligence (AI) for real-time patient insights are garnering large venture money, with investments in digital health firms reaching $29.1 billion worldwide in 2021. This infusion of financing enhances market rivalry, driving firms to distinguish via sophisticated features like seamless data integration or user-friendly interfaces. Moreover, IoT-driven sustainability solutions, such as energy-efficient hospital management systems, are creating avenues for green technology businesses to penetrate the healthcare sector (Kumar et al., 2022). These systems, which optimize energy consumption in lighting and HVAC, coincide with global sustainability objectives, appealing to environmentally minded providers and establishing new market niches.

The economic effect of IoT in healthcare goes beyond revenue to employment creation and economic resiliency. The desire for IoT solutions is encouraging development in industries including biomedical engineering, data analytics, and healthcare delivery, with the industry estimated to produce millions of jobs by 2030. For example, building and maintaining IoT devices needs competent engineers to create low-power sensors, while data scientists are required to evaluate the huge information produced by linked devices. These professions contribute to local economies and help workforce development in developing areas where healthcare infrastructure is increasing. Additionally, IoT's function in streamlining supply chains and logistics, such as via RFID-based inventory monitoring, saves operating costs for hospitals, freeing up resources for future innovation and employment development (Qadri et al., 2020).

Capitalizing on these market possibilities means smaller firms sometimes lack the infrastructure to assure interoperability across platforms, placing them at a disadvantage compared to established giants like Philips or Medtronic. Additionally, the high initial expenses of IoT adoption, including hardware and cybersecurity measures, might strain budgets, necessitating strategic alliances or government incentives to scale solutions (Nižetić et al., 2020).

6.3 Social Impact

The Internet of Things (IoT) in healthcare brings significant societal advantages by increasing access to treatment, optimizing patient outcomes, and promoting health equality, generating healthier and more inclusive societies. IoT technologies, such as remote monitoring systems, wearable gadgets, and linked medical equipment, transcend geographical and socioeconomic divides, ensuring that underprivileged groups receive prompt therapies. For example, IoT-enabled telehealth services allow patients in rural or distant places to visit experts without traveling long distances, greatly lowering healthcare inequalities (Mishra, Shukla, & Rana, 2021). These systems have been especially useful for controlling chronic illnesses like diabetes and hypertension, providing real-time monitoring of vital indicators like as glucose levels and blood pressure. Mobile health apps powered by IoT have allowed older people to take charge of their health, boosting their quality of life and independence (Nguyen, Park, & Lee, 2020). A research revealed that such applications enhanced drug adherence among seniors by 25%, lowering problems and hospital visits.

Beyond individual treatment, IoT aids community health by allowing data-driven public health measures. Connected devices can collect anonymised data to monitor disease outbreaks, such as influenza or new infectious illnesses, enabling health authorities to allocate resources effectively (Yang, Zhang, & Li, 2023). For instance, IoT devices have been used to monitor population-level heart rate patterns, offering early warnings of possible epidemics. This skill enhances public health resilience, especially in low-resource contexts where conventional monitoring methods may be constrained. IoT also supports community-based initiatives, such as remote health kiosks in impoverished regions, which offer diagnostic services and health education, further increasing equality.

Sustainable IoT methods improve these societal advantages by harmonizing with environmental justice concepts. The implementation of circular economy methods, such as recycling and repairing medical equipment, minimizes electronic waste (e-waste), which disproportionately impacts disadvantaged populations near landfills (Gupta, Sharma, & Kumar, 2022). For example, initiatives that repair IoT-enabled equipment like insulin pumps lengthen their lifespans, lessening the environmental and economic burden on low-income countries. This digital gap risks increasing existing inequality unless addressed via focused actions. Investments in digital literacy initiatives, inexpensive IoT devices, and solid internet infrastructure are important to guaranteeing inclusion. For instance, community training programs have improved IoT adoption among senior populations by 15%, allowing them to benefit from remote monitoring (Nguyen et al., 2020).

7. CASE STUDIES

7.1 Wearable Health Monitors

Wearable health monitors, such as smartwatches, fitness trackers, and biosensors, have changed chronic illness management by offering real-time data on vital indicators, including heart rate, oxygen saturation, blood glucose, and activity levels. These Internet of Things (IoT)-enabled gadgets allow patients to take control of illnesses including diabetes, hypertension, and heart disease, while decreasing healthcare costs and environmental impact. For example, Fitbit's heart rate monitors have showed the potential to identify atrial fibrillation early, allowing prompt therapies that avert serious consequences like stroke (Perez et al., 2019). Similarly, continuous glucose monitors, such as those by Dexcom, give diabetic patients with real-time blood sugar readings, minimizing the need for frequent clinic visits and invasive testing (Yetisen et al., 2018). By connecting with smartphone apps, wearables give tailored health data, such as exercise suggestions or medication reminders, boosting patient engagement and adherence to treatment programs. Studies demonstrate that wearables may decrease emergency department visits by 20% for chronic illness patients, reducing burden on healthcare systems (Lupton, 2017).

The preventative care potential of wearables is a fundamental driver of their influence. By continually monitoring health data, these devices enable early identification of irregularities, enabling physicians to act before situations deteriorate. For instance, smartwatches equipped with electrocardiogram (ECG) functions may notify users to abnormal heart rhythms, prompting medical consultations that avert hospitalizations (Swan, 2012). This proactive strategy corresponds with Sustainable Development Goal 3 (Good Health and Well-being) by promoting health equality and minimizing healthcare inequities. Wearables also help to sustainability by limiting hospital visits, which minimizes patient travel emissions and resource consumption in medical institutions. A research predicted that remote monitoring via wearables might lower healthcare-related carbon emissions by 15% for chronic illness management (Yetisen et al., 2018).

Despite their advantages, wearable health monitoring encounter major hurdles. Battery life remains a key restriction, since frequent recharging increases energy consumption and user discomfort. Advancements in low-power technology, such as energy harvesting from body heat or motion, are required to promote sustainability (Perez et al., 2019). Data accuracy is another challenge, since sensor faults might lead to misdiagnoses or improper actions. Improving sensor accuracy using machine learning and sophisticated materials is necessary (Swan, 2012). Additionally, data privacy is an increasing problem, as wearables capture sensitive health information, necessitating compliance with rules like GDPR and HIPAA (Lupton, 2017). The environmental effect of making and disposing of wearables, which typically include non-recyclable components, further requires eco-friendly designs, such as biodegradable sensors or recyclable casings.

7.2 Smart Hospitals

The foundation of a smart hospital resides in its capacity to harness IoT for smooth operations. IoT sensors installed in medical equipment, such as infusion pumps or MRI machines, gather data on use patterns and performance, allowing predictive maintenance to prevent malfunctions. This method not only boosts equipment dependability but also extends device lifespans, decreasing the need for frequent replacements and saving resources. Changi General Hospital's IoT system, for example, incorporates data from bed sensors and staff wearables to optimize patient admissions and staff allocation, boosting care delivery efficiency (Zhang et al., 2019). Such systems also promote patient safety by ensuring equipment is easily accessible and maintained, eliminating interruptions during crucial operations.

Despite these advantages, smart hospitals confront considerable hurdles. The high cost of IoT infrastructure, including sensors, networking equipment, and analytics platforms, might be prohibitive, especially for hospitals in low-resource settings (Hossain & Muhammad, 2016). Cybersecurity is a key problem, since IoT devices are prone to breaches, needing sophisticated encryption and compliance with rules like HIPAA to safeguard patient data (Atlam & Wills, 2020). Interoperability challenges come from varied device standards, limiting integration across healthcare systems. Addressing these difficulties demands investment in scalable technology, comprehensive cybersecurity frameworks, and established protocols to assure flawless operation.

Smart hospitals show how IoT may establish sustainable, patient-centric healthcare systems. By decreasing energy usage, maximizing resources, and boosting care delivery, IoT connects with environmental and health-related SDGs. Future improvements, such as AI-driven analytics and blockchain for safe data exchange, might significantly increase smart hospital capabilities (Atlam & Wills, 2020). Collaborative efforts among healthcare providers, technology developers, and legislators are vital to overcome hurdles and expand these technologies internationally, ensuring smart hospitals lead the way in eco-conscious healthcare delivery.

7.3 Telemedicine in Developing Regions

Telemedicine, enabled by the Internet of Things (IoT), is changing healthcare delivery in poor nations by overcoming major access gaps via remote consultations and tests. IoT equipment, such as portable ultrasound machines, linked stethoscopes, and wearable biosensors, allow healthcare professionals to communicate real-time patient data to experts, considerably enhancing treatment in rural and underserved locations. In India, Aravind Eye Care's telemedicine program shows this effect, evaluating 10,000 patients yearly for diabetic retinopathy using IoT-enabled retinal cameras, minimizing the need for patients to travel long distances to metropolitan areas (Dasgupta & Deb, 2020). By promoting early diagnosis and treatment, such programs increase health outcomes and reduce strain on overloaded healthcare systems. IoT-driven telemedicine also benefits to sustainability by lowering patient travel, with studies suggesting a 30% decrease in travel-related expenses and carbon emissions for distant patients (Ahmed et al., 2021). The scalability of low-cost IoT sensors makes telemedicine practical in resource-constrained situations, but difficulties including intermittent internet access, device maintenance, and limited technical skills continue. By enhancing access to treatment and limiting resource consumption, IoT-powered telemedicine fits with Sustainable Development Goal 3 (Good Health and Well-being) and helps sustainable healthcare delivery in poor countries.

Future improvements, such as 5G networks and AI-enhanced diagnostics, might significantly boost IoT-driven telemedicine by enhancing data speeds and automating early investigations (Hollander & Carr, 2020). By incorporating these technologies, developing nations may establish resilient healthcare systems that give high-quality treatment with minimum environmental effect. IoT-powered telemedicine not only supports SDG 3 by enhancing health access but also fits with SDG 12 (Responsible Consumption and Production) by lowering resource-intensive travel and infrastructure needs, opening the door for sustainable healthcare delive

Figure 4. Sustainable HealthTech integration

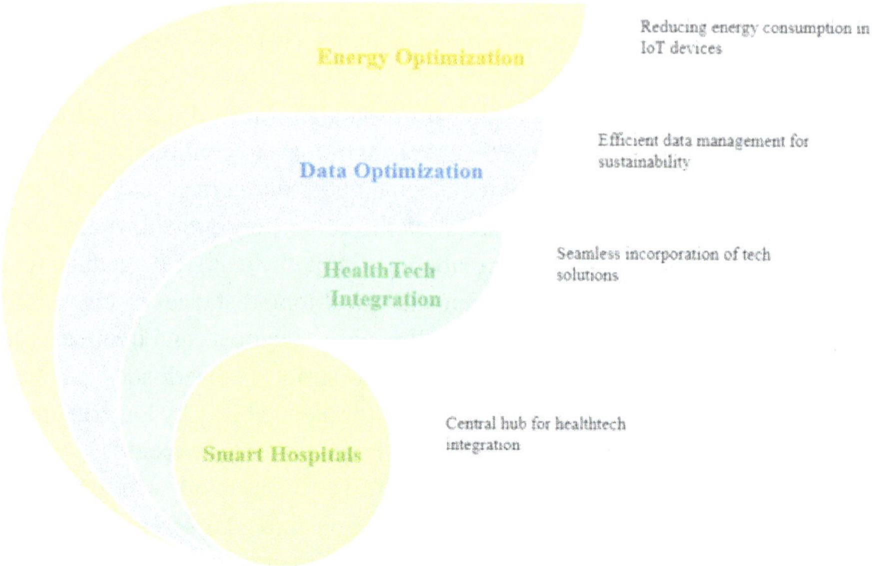

8. CONCLUSION

The integration of Internet of Things (IoT) technology in healthcare has changed patient care and operational efficiency while offering prospects for sustainability via eco-conscious practices. IoT technology, such as wearables and smart medical equipment, allow real-time monitoring and minimize hospital readmissions by up to 38%, as evidenced by systems like Philips' HealthSuite (Philips, 2020). However, their environmental effect, including e-waste and energy consumption, mandates sustainable measures including energy-efficient designs, sustainable materials, and circular economy models (Bressanelli et al., 2018). Case studies, such as Apollo Hospitals' 22% energy savings with IoT-based technologies, illustrate practical accomplishments (Apollo Hospitals, 2021). Challenges, including high prices and regulatory constraints, demand new solutions and cooperation to guarantee IoT complies with Sustainable Development Goals (SDGs), notably SDG 12 (Responsible Consumption and Production) (Kirchherr et al., 2017). By implementing these approaches, healthcare can combine medical innovation with environmental stewardship, producing a regenerative system that promotes both patient and planetary health.

To fully exploit the promise of sustainable IoT in healthcare, stakeholders must move aggressively to incorporate eco-conscious practices. Healthcare providers should invest in energy-efficient IoT devices and circular economy models, such as product-as-a-service, to decrease waste and expenses (Ellen MacArthur Foundation, 2021). Policymakers must establish incentives, such tax breaks, to stimulate adoption of sustainable technology and standardize laws for recycled materials (Bocken et al., 2016). Technology developers should prioritize AI and blockchain to enhance device efficiency and lifecycle transparency (Shi et al., 2020). (Good Health and Well-being) (Kshetri, 2017). By acting today, stakeholders can develop a healthcare ecosystem that offers innovative treatment while conserving the environment.

REFERENCES

Adhikari, M., & Amgoth, T. (2018). Heuristic-based load-balancing algorithm for IaaS cloud. *Future Generation Computer Systems*, *81*, 156–165. DOI: 10.1016/j.future.2017.10.035

Al-Turjman, F., Nawaz, H., & Ulusar, U. (2019). Intelligence in the Internet of Medical Things era: A systematic review of current and future trends. *Computer Communications*, *150*, 644–660. DOI: 10.1016/j.comcom.2019.12.030

Andrae, A., & Edler, T. (2015). On Global Electricity Usage of Communication Technology: Trends to 2030. *Challenges*, *6*(1), 117–157. DOI: 10.3390/challe6010117

Birtel, M., David, A., Hermann, J., Mohr, F., & Ruskowski, M. (2019). FutureFit: A strategy for getting a production asset to an industry 4.0 component – a human-centered approach. *Procedia Manufacturing*, *38*, 1000–1007. DOI: 10.1016/j.promfg.2020.01.185

Costa, E., Soares, A. L., & de Sousa, J. P. (2020). Industrial business associations improving the internationalisation of SMEs with digital platforms: A design science research approach. *International Journal of Information Management*, *53*, 102070. DOI: 10.1016/j.ijinfomgt.2020.102070

Dhikale, P., Shrivastava, S., & Srinivasan, S. (2020). Perspectives about professionalism among undergraduate students in a medical college in India: A qualitative study. *Indian Journal of Community Medicine : Official Publication of Indian Association of Preventive & Social Medicine*, *45*(2), 230. DOI: 10.4103/ijcm.IJCM_238_19 PMID: 32905183

Dietrich, F., & Krenkel, N. (2019). Preface. *Procedia CIRP*, *86*, 1–2. DOI: 10.1016/j.procir.2020.01.001

Dimitrov, D. (2016). Medical Internet of Things and Big Data in Healthcare. *Healthcare Informatics Research*, *22*(3), 156. DOI: 10.4258/hir.2016.22.3.156 PMID: 27525156

Geissdoerfer, M., Savaget, P., Bocken, N. M. P., & Hultink, E. J. (2017). The Circular Economy – A new sustainability paradigm? *Journal of Cleaner Production*, *143*, 757–768. DOI: 10.1016/j.jclepro.2016.12.048

Haque, N., Hughes, A., Lim, S., & Vernon, C. (2014). Rare Earth Elements: Overview of Mining, Mineralogy, Uses, Sustainability and Environmental Impact. *Resources*, *3*(4), 614–635. DOI: 10.3390/resources3040614

Hollander, J. E., & Carr, B. G. (2020). Virtually Perfect? Telemedicine for Covid-19. *The New England Journal of Medicine*, *382*(18), 1679–1681. DOI: 10.1056/NEJMp2003539 PMID: 32160451

Hu, J. J., Bonnichsen, C. R., Dearani, J. A., Miranda, W. R., Johnson, J. N., Cetta, F., Stephens, E. H., Aganga, D. O., & van Dorn, C. S. (2021). Adults With Tetralogy of Fallot: Early Postoperative Outcomes and Risk Factors for Complications. *Mayo Clinic Proceedings*, *96*(9), 2398–2406. DOI: 10.1016/j.mayocp.2021.01.032 PMID: 34412856

Huang, B., Xing, K., Ness, D., Liao, L., Huang, K., Xie, P., & Huang, J. (2022). Rethinking carbon–neutral built environment: Urban dynamics and scenario analysis. *Energy and Building*, *255*, 111672. DOI: 10.1016/j.enbuild.2021.111672

Jumaan, I. A., Hashim, N. H., & Al-Ghazali, B. M. (2020). The role of cognitive absorption in predicting mobile internet users' continuance intention: An extension of the expectation-confirmation model. *Technology in Society*, *63*, 101355. DOI: 10.1016/j.techsoc.2020.101355

Lacwik, P., & Kupczyk, M. (2021). Asthma Phenotype Overlap: More May Be Less. *The Journal of Allergy and Clinical Immunology. In Practice*, *9*(1), 363–364. DOI: 10.1016/j.jaip.2020.10.032 PMID: 33429708

Lawan, I., Zhou, W., Garba, Z. N., Zhang, M., Yuan, Z., & Chen, L. (2019). Critical insights into the effects of bio-based additives on biodiesels properties. *Renewable & Sustainable Energy Reviews*, *102*, 83–95. DOI: 10.1016/j.rser.2018.12.008

Li, L., & Zhang, S. (2021). Techno-economic and environmental assessment of multiple distributed energy systems coordination under centralized and decentralized framework. *Sustainable Cities and Society*, *72*, 103076. DOI: 10.1016/j.scs.2021.103076

Li, M., Wang, L., Liu, R., Li, J., Zhang, Q., Shi, G., Li, Y., Hou, C., & Wang, H. (2021). A highly integrated sensing paper for wearable electrochemical sweat analysis. *Biosensors & Bioelectronics*, *174*, 112828. DOI: 10.1016/j.bios.2020.112828 PMID: 33250335

Liu, Q., & Zhang, Z. (2020). The analysis of coverage probability, ASE and EE in heterogeneous ultra-dense networks with power control. *Digital Communications and Networks*, *6*(4), 524–533. DOI: 10.1016/j.dcan.2020.02.002

Ma, X., Sheikholeslami, M., Jafaryar, M., Shafee, A., Nguyen-Thoi, T., & Li, Z. (2020). Solidification inside a clean energy storage unit utilizing phase change material with copper oxide nanoparticles. *Journal of Cleaner Production, 245*, 118888. DOI: 10.1016/j.jclepro.2019.118888

Mazumder, S., & Garg, S. (2021). Decoding digital transformational outsourcing: The role of service providers' capabilities. *International Journal of Information Management, 58*, 102295. DOI: 10.1016/j.ijinfomgt.2020.102295

Missbauer, H., & Uzsoy, R. (2022). Order release in production planning and control systems: Challenges and opportunities. *International Journal of Production Research, 60*(1), 256–276. DOI: 10.1080/00207543.2021.1994165

Nižetić, S., Šolić, P., López-de-Ipiña González-de-Artaza, D., & Patrono, L. (2020). Internet of Things (IoT): Opportunities, issues and challenges towards a smart and sustainable future. *Journal of Cleaner Production, 274*, 122877. DOI: 10.1016/j.jclepro.2020.122877 PMID: 32834567

Perez, M., Mahaffey, K. W., Hedlin, H., Rumsfeld, J. S., Garcia, A., Ferris, T., Balasubramanian, V., Russo, A. M., Rajmane, A., Cheung, L., Hung, G., Lee, J., Kowey, P., Talati, N., Nag, D., Gummidipundi, S. E., Beatty, A., Hills, M. T., Desai, S., & Turakhia, M. P. (2019). Large-Scale Assessment of a Smartwatch to Identify Atrial Fibrillation. *The New England Journal of Medicine, 381*(20), 1909–1917. DOI: 10.1056/NEJMoa1901183 PMID: 31722151

Qadri, Y. A., Nauman, A., Zikria, Y., Vasilakos, A. V., & Kim, S. W. (2020). The Future of Healthcare Internet of Things: A Survey of Emerging Technologies. *IEEE Communications Surveys and Tutorials, 22*(2), 1121–1167. DOI: 10.1109/COMST.2020.2973314

Qamar, N., Siddiqui, N., Ehatisham-ul-Haq, M., Awais Azam, M., & Naeem, U. (2020). An Approach towards Position-Independent Human Activity Recognition Model based on Wearable Accelerometer Sensor. *Procedia Computer Science, 177*, 196–203. DOI: 10.1016/j.procs.2020.10.028

Quan, J., Miyake, Y., & Nozawa, T. (2021). Incorporating Interpersonal Synchronization Features for Automatic Emotion Recognition from Visual and Audio Data during Communication. *Sensors (Basel), 21*(16), 5317. DOI: 10.3390/s21165317 PMID: 34450759

Rautela, R., Arya, S., Vishwakarma, S., Lee, J., Kim, K.-H., & Kumar, S. (2021). E-waste management and its effects on the environment and human health. *The Science of the Total Environment, 773*, 145623. DOI: 10.1016/j.scitotenv.2021.145623 PMID: 33592459

Ru, X., Wang, S., & Yan, S. (2018). Exploring the effects of normative factors and perceived behavioral control on individual's energy-saving intention: An empirical study in eastern China. *Resources, Conservation and Recycling*, *134*, 91–99. DOI: 10.1016/j.resconrec.2018.03.001

Saunders, C., Brown, S., Bygstad, B., Dennis, A., Ferran, C., Galletta, D., Liang, T.-P., Lowry, P., Recker, J., & Sarker, S. (2017). Goals, Values, and Expectations of the AIS Family of Journals. *Journal of the Association for Information Systems*, *18*(9), 633–647. DOI: 10.17705/1jais.00468

Shaikh, F. K., Zeadally, S., & Exposito, E. (2017). Enabling Technologies for Green Internet of Things. *IEEE Systems Journal*, *11*(2), 983–994. DOI: 10.1109/JSYST.2015.2415194

Shi, X., Wang, L., Zhang, F., Zheng, K., Mühlhäuser, M., & Liu, Z. (2019). PABO: Mitigating congestion via packet bounce in data center networks. *Computer Communications*, *140–141*, 1–14. DOI: 10.1016/j.comcom.2019.04.002

Stahel, W. R., & MacArthur, E. (2019). *The Circular Economy*. Routledge., DOI: 10.4324/9780429259203

Steinhubl, S. R., Waalen, J., Edwards, A. M., Ariniello, L. M., Mehta, R. R., Ebner, G. S., Carter, C., Baca-Motes, K., Felicione, E., Sarich, T., & Topol, E. J. (2018). Effect of a Home-Based Wearable Continuous ECG Monitoring Patch on Detection of Undiagnosed Atrial Fibrillation. *Journal of the American Medical Association*, *320*(2), 146. DOI: 10.1001/jama.2018.8102 PMID: 29998336

Tunkelang, D., Capra, R., Golovchinsky, G., Kules, B., Smith, C., & White, R. (2013). Symposium on Human–Computer Information Retrieval. *Big Data*, *1*(1), 65–70. DOI: 10.1089/big.2012.0003 PMID: 27447040

Webster, D. E., Tummalacherla, M., Higgins, M., Wing, D., Ashley, E., Kelly, V. E., McConnell, M., Muse, E. D., Olgin, J. E., Mangravite, L. M., Godino, J., Kellen, M. R., & Omberg, L. (2021). Smartphone-Based VO2max Measurement With Heart Snapshot in Clinical and Real-world Settings With a Diverse Population: Validation Study. *JMIR mHealth and uHealth*, *9*(6), e26006. DOI: 10.2196/26006 PMID: 34085945

Zhu, R., Wong, M. S., Kwan, M.-P., Chen, M., Santi, P., & Ratti, C. (2022). An economically feasible optimization of photovoltaic provision using real electricity demand: A case study in New York city. *Sustainable Cities and Society*, *78*, 103614. DOI: 10.1016/j.scs.2021.103614

Zou, H., Qin, J., Yang, P., & Dai, B. (2018). A Coordinated Revenue-Sharing Model for a Sustainable Closed-Loop Supply Chain. *Sustainability (Basel)*, *10*(9), 3198. DOI: 10.3390/su10093198

Chapter 6
Designing Resilient and Scalable Systems for Green Healthcare IoT

Dedeepya Sai Gondi
https://orcid.org/0009-0005-4686-9726
Independent Researcher, USA

ABSTRACT

These days, socio-economic collapse or natural disaster scenarios are increasingly remote from healthcare systems that leverage on IoT infrastructure, but there is an urgent requirement for the design of solutions that not only reduce the carbon footprints but also be resilient and scalable. This chapter presents how to build green IoT systems that can respond to changing needs at speed, are resilient to disruption, and agile in scaling. It will focus on energy efficient communication protocols, a modular system architecture, integration of renewable power, intelligent data management. The chapter presents a framework for developing healthcare IoT systems by using real world case studies and evaluating simulation based through which an engineer or a decision maker can achieve the performance goals while reducing its environmental impact.

1. INTRODUCTION

Internet of Things (IoT) definitely has great transformative potential in modern healthcare. In real time patient monitoring, smart diagnostics, remote care, intelligent management of the hospital this are the revolution that is going in this field of the healthcare world through IoT technologies. This transition is allowing providers to provide more customized, faster and efficient care. That said, the surge in integra-

DOI: 10.4018/979-8-3373-4134-7.ch006

tion into the IoT is not a smooth one. With the reliance on interconnected digital infrastructure dominating in healthcare systems, performance, security, reliability and sustainability of these systems are getting higher demand.

Devices that are pervasive, wearable, and that perform a multitude of health and services related functions, operating autonomously or semi-autonomously in environments that transmit gigabytes of data, comprise the majority of devices in modern healthcare environments. The continuous flows of data available in this information channel are critical for efficient care delivery but inherently demand for data overload, system scalability, energy consumption, and environmental impact. In addition, the must for uptime and quickness, particularly within crisis or existence vital circumstances, highlights the requirement for systems designed to endure failures and interruptions (Ali & Ali, 2021).

This is compounded by the need for urgent global sustainability. The crucial work that healthcare is tasked with doing does not cancel the fact that it is a large source of carbon emissions and resource consumption. The energy required by hospitals and clinics is high and most of them run 24/7 requiring heating, cooling, lighting, and equipment operation. Integrating IoT creates another energy shell due to the buildup of devices and the computational load it can apply to networks and data centers. Therefore, the design and deployment of IoT systems for healthcare systems must maintain the required technological advancement and environmental stewardship. In such a case, sustainability principles, scalability, and resilience principles are bound to be the main pillars of next generation healthcare infrastructure. Sustainability is the smallest footprint for the IOT solutions by adopting the least energy efficient technologies, the use of renewable energy, and eco friendly hardware. Scalability is an assurance that as the size of patient load and data volumes increase, the systems' performance doesn't degrade. Resilience ensures a transfer of systems to a functioning state under substantial stress, during failures and in changing environmental conditions (Espinosa & Lopez, 2021).

By the convergence of the three, the design philosophy of healthcare IoT has shifted. Instead of only looking at functionality or speed, system designers and decision makers now have to consider their long-term adaptability, it's ecological responsibility and robustness. The ethos of the chapter is this, this chapter presents holistic approach to architect healthcare IoT solution, which are innovative and connected but also sustainable amid resource constraints, resilient, and scalable.

The first step in this is understanding the evolving role IoT has in healthcare. It was the catalyst the COVID-19 pandemic offered of dramatically accelerating adoption of telehealth, remote monitoring, digital diagnostics. Post pandemia isso continuado, com a dupla pressão do desafio da acessibilidade e eficiência do processo de saúde. Real time tracking of a great range of vital signs and activity data in these wearable devices now makes it possible for abnormalities to be detected and proactive care intervention taken. The environment of smart hospital rooms is adjusted with environmental sensors for optimal conditions of patient's recovery, and robotic systems ensure proper medication delivery and sanitation. The critical need for uninterrupted operation, and adaptability, is such use cases. In such case, a wearable monitor needs to work reliably even without a stable network connectivity or power. Smart hospital bed should respond and deliver vital data quickly irrespective of patient turn over and the limitation of the infrastructural. For these capabilities to be achieved, system is designed to be sophisticated with redundant components, fault tolerant architectures and energy aware computing.

At the same time, the scale on which healthcare IoT systems have to operate are growing exponentially. For example, with increasing populations of older people and an increase in the prevalence of chronic diseases, the need for continuous health monitoring and intelligent care coordination goes up. For healthcare systems to be able to on-board thousands—or millions—of new devices, they must not impact such a thing as performance or patient safety. This implies that an IoT platform of these needs has to be designed on a scalable architecture that is able to expand as workloads increase as well as to meet numerous use case requirements (Goyal, Sharma, Jain, Elawadhi, & Bhutani, 2023).

Figure 1. Framework for sustainable, scalable, and resilient healthcare IoT

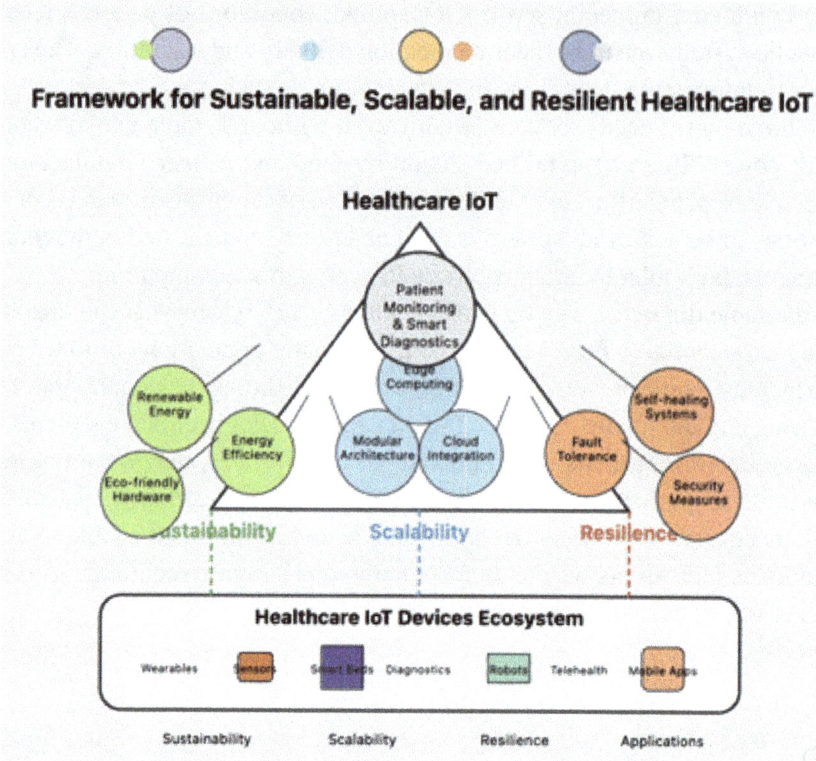

The three pillars that are making the next generation healthcare IoT systems feasible include persistence (piecing energy efficiency and renewable energy alongside eco-friendly hardware), scalability (achieved via modular system and its convergence with cloud ecosystem and the use of edge computing), and resilience (derived from a corollary of fault tolerance, self healing systems and robust security mechanisms). Patient monitoring and smart diagnostic applications will be served as these pillars connect to a myriad of healthcare IoT devices such as wearables, environmental sensors, smart hospital equipment, diagnostic tools, robotic systems, telehealth platforms and mobile applications.

To accommodate the evolution of healthcare IoT, there are two points that must be balanced: technological advancement towards comprehensive care, while maintaining the environmental and system reliability. Figure 1 illustrates that healthcare IoT will be sustainable, scalable, and resilient if it integrates three fundamental pillars. Renewable energy integration; energy efficient technologies, and eco friendly hardware are some of the sustainability initiatives that address the impacts of the ever digitized health care environments on the environment. Modular architecture, cloud integration, and edge computing are scalable features that enable systems to cope with increasing patient loads and growing data volumes without performance degradation. At the same time, resilience mechanisms such as fault tolerance, self healing capabilities and sustaining security norms continue working to operate under disruptions or cyber attacks. They provide healthcare institutions with a roadmap in introducing IoT infrastructure needed to satisfy the rigorous demands of current global healthcare systems at a time where such a role still requires the sacrifice of responding to growing global sustainable objectives, particularly given that the healthcare industry contributes a significant percentage of carbon emissions.

This chapter then builds upon these three objectives (resilience, scalability, sustainability) and examines how such practical design strategies, emerging technologies, as well as innovative system architectures can further bring reality to these objectives. The definition and the relation of resilience and/or scalability with sustainable system design are introduced in section 2, in the context of healthcare IoT. In section 3, core tenets of green IoT development are presented including approaches to perform a green IoT development solution that does not compromise performance while at the same time cutting down on environmental impact. In the fourth section, modular architecture is discussed, the role it plays towards the adaptability of the system, its maintenance, and, in the end, incremental upgrades.

Section 5 compares energy efficient communication protocols based on BLE, Zigbee, LoRaWAN, NB-IoT, and 6LoWPAN with the context of health care use cases. The protocols that are necessary to achieve data transmission with negligible energy consumption, which is so important in devices with battery or with energy harvesting. In section 6 we discuss issues of scalable data management and how edge computing and intelligent data processing can off load pressure to remote central servers and lower energy costs.

In Section 7, fault tolerance and self healing systems become the focus, where patching a fault, in other words, a failover mechanism, and with the help of AI based diagnostics, system can continue to function even with a fault or in worst case, a cyber attack. Section 8 reveals this security versus energy efficiency tension and describes lightweight encryption strategies and compliance frameworks that ensure trust and maintain regulatory alignment while using resources as efficiently as possible. Section 9 ends the chapter by suggesting a vision for the future of green healthcare IoT, with its key messages summarized. Section 10 discusses possible research avenues and technological innovations for further improvements of the sustainability and the robustness of these systems.

Overall, this paper outlines a framework for developing IoT systems, for engineers, designers, healthcare administrators, and policymakers, to meet the technological requirements of modern medicine, i.e., intelligent, scalable, secure, and green. The opportunity to build a healthcare infrastructure that both saves lives and saves the planet is substantial, the challenges are, too.

2. UNDERSTANDING RESILIENCE AND SCALABILITY IN HEALTHCARE IOT

As in the largely evolving landscapes of the healthcare, the Internet of Things (IoT) brings with it great challenges and opportunities. Since IoT systems become increasingly embedded in healthcare infrastructure, there exists a need to ensure that these systems are reliable, flexible, and efficient in operations as the loads on them vary. In this case two fundamental concepts are resilience and scalability, which both are essential for the future of IoT systems in healthcare to succeed. On the other hand, scalability involves understanding the system's capacity to scale up even in the presence of more reliance of various factor, such as devices, patients or data. However, it is necessary for ensuring the impact of these principles in creating robust and sustainable healthcare IoT systems able to satisfy the increasing demands of the sector with low environmental impact. The focus of this section lies on the core factors of resilience and scalability and how they apply to the design of a healthcare IoT and the connection of these factors to a sustainable system architecture.

2.1 Definition and Importance of System Resilience

In healthcare IoT systems, resilience is represented as a system's ability to withstand disruptions and render it under stress. Since healthcare applications are so critical as failures or downtimes can cause significant consequences, the importance of resilience is emphasized. This means that a resilient system is that that can tolerate faults (any hardware malfunction, network outage, a cyber attack, or similarly, power surge). Resilience has also been defined as fault tolerance, disaster recovery, and continuous operation. Fault tolerance means if one component fails then other components take over to resume functionality. Continuous operation strategies and disaster recovery mechanisms insure the system response and the return to normal operation after a critical situation. Given their importance in healthcare, uptime and reliability make these elements critical in healthcare, where they can directly affect patient care and safety (Mallo & Abdulqader, 2024).

2.2 What Makes an IoT System Scalable?

IoT system scalability enables an IoT system to respond to an increasing load without compromising performance in the system. In healthcare, the volume of data is expected to grow, as an increasing array of connected devices is used to generate data. With increasing number of devices—be it wearable health monitor, diagnostic equipment, and so on—the system has to be capable of supporting such additional devices without degrading its performance. The scalability can be dealt with at network, data storage, and processing capacities levels. One example is when the number of connected devices grows, the network needs to be able to handle more data traffic without experiencing latency or bandwidth issues. Similarly storage and processing solutions need to scale to suffice with larger dataset without the delay. The term scalability also refers to the ability to on board new devices and new healthcare applications without completely overhauling the system. The scalability ensures that the IoT systems can grow with the increase in demand of healthcare, maintaining a high level performance and reliability.

2.3 The Intersection of Resilience, Scalability, and Sustainability

Integrating sustainability principles is important along with resilience and scalability since they are the foundational elements in the design of these robust healthcare IoT systems. It is commonly the case that healthcare increasingly relies on an increasing number of IoT devices that use more energy and BLE contributes to a bigger environmental footprint; therefore, it is critical to design systems that scale, recover effectively and leave behind a minimal ecological impact. Energy use optimization, e-waste reduction and the use of renewable energy systems are necessary for sustainability of healthcare IoT systems. For instance, communication protocols can be made energy efficient and devices can be driven low power, thus reducing the total energy consumption of the system. Also, modular architecture helps in upgrades and repair, prolonging the eyes of devices and minimizing waste. The tradeoff is between this environmental consideration and the requirement for resilience and scalability. To become sustainable, an IoT system in healthcare needs to be able to grow and adapt to the needs of modern healthcare considering its end of life curve as well as its environmental footprint. The main features of the future of healthcare IoT systems are the combination of those three pillars, namely resilience, scalability and sustainability, all together at the same time: against long-term performance, growth and environmental stewardship (Oyeniran, Modupe, Otitoola, 2024).

Table 1. Key concepts of resilience, scalability, and sustainability in healthcare IoT systems, outlining their definitions and relevance to healthcare operations

Key Concept	Definition	Importance in Healthcare IoT
Resilience	Ability of an IoT system to recover from disruptions and continue operating effectively	Ensures continuous operation, minimizes downtime, and maintains patient care reliability
Scalability	Ability of an IoT system to handle increasing demand without performance degradation	Accommodates more devices, patients, and data, supporting healthcare growth
Sustainability	Minimizing environmental impact while ensuring system performance, growth, and efficiency	Optimizes energy usage, reduces e-waste, and integrates renewable energy sources for long-term eco-friendly operation

Table 1 outlines information necessary for designing healthcare IoT systems that have to prioritize resilience and scalability. Healthcare systems need resilience which means that systems can recover from disruptions, as downtime can compromise patient care. This allows IoT solutions to function with higher loads from more devices, more data volume, while not losing performance, thereby promoting scalability of IoT infrastructure to fit the growth of healthcare. Moreover, considering sustainability in design of healthcare IoT is necessary for limiting its environmental impact. Such systems are designed to optimize energy use and minimize waste; they are the increasingly popular eco-friendly technologies of the day. This interplay of these three pillars, namely resilience, scalability, and sustainability, guarantees that healthcare IoT systems evolve together with the changing healthcare landscape meanwhile keeping such low performance and low environmental footprint.

3. GREEN IoT DESIGN PRINCIPLES

The healthcare system is gradually integrating Internet of Things (IoT) technologies and now one is focusing towards developing more sustainable, green healthcare systems. Green IoT (G-IoT) refers to IoT systems which are development with sustainability goals such as energy consumption reduction, reduced environmental impact, and eco-friendly practice. In this section, the bibliometric analysis is presented in order to explore the core principles of Green IoT, design considerations for sustainable healthcare systems, and Green IoT's alignment with both the circular economy and Sustainable Development Goals (SDGs) (Arega & Sharma, 2023).

3.1 Core Principles of Green IoT

The green IoT is built on a set of guiding principle that ensures both environmental sustainability and the operational efficiency. The principles that the IoT system should focus on are about reducing the environmental footprint of IoT system, delivering minimum power consumption and the use of green hardware. Through these principles, healthcare IoT systems can mitigate its ecological impact, and incorporate principles to a responsible take of technology use.

The principle of minimizing an environmental footprint means in essence to make the lifecycle of IoT devices more sustainable, in other words, to minimize the negative effects of IoT products' lifecycle. These are things like using sustainable manufacturing materials, reducing carbon emissions during production and all of the devices being designed to be longer lasting and more easily recyclable at the end of the product's lifespan. Thus the environmental footprint is also reduced by designing the IoT system with less resources, thus having a smaller environmental impact in total.

The second important principle of Green IoT is reduced energy consumption. As such, energy usage is high in healthcare IoT devices like wearables, diagnostic tools, and monitoring systems that depend on continuous communication and processing. The goal of these green IoT systems is to achieve low power communication, energy efficient sensors and battery saving strategies that can help in minimizing energy expenditure. This reduces both the operational energy demand required by devices and the lifespan of batteries (enabling the battery replacements to be used less) which in turn reduces the environmental impact due to less need for replacing batteries (Bhutani, Gupta, & Jain, 2023).

The third principle is concerned with the use of eco friendly hardware. It means that we choose not to use nontoxic and sustainable materials to manufacture such devices like biodegradable or recyclable plastics and also we should avoid use of the bad substances such as electronics components. Additionally, it focuses on designing devices that are modular, meaning that parts like batteries or sensors can be upgraded or replaced while dumping the entire unit. This is a matter of these devices do not contribute to the increasingly large problem of e waste which is one of the main environmental challenges that accompany the rapid progression of technology.

3.2 Design Considerations for Sustainable Healthcare Systems

This necessitates that design of sustainable healthcare IoT systems has to take into account both environmental and operational concerns. Several considerations should be made to create a system that fulfills Green IoT principles while still delivering the needed high quality of care in modern healthcare.

Integrating energy efficient network infrastructure is a critical design consideration. Known as a traditional communication protocol, such as Wi-Fi, consumes a lot of energy, especially for large scale IoT systems where there is a large number of devices that are usually connected continuously. In the context of green IoT, low power alternatives like Bluetooth Low Energy (BLE), Zigbee or Low Power Wide Area Networks (LPWAN) are preferred which make them more suitable for healthcare applications. Not only they consume less power, but they also increase the device longevity and decrease operational costs.

Intelligent resource management is another important consideration as well. They include systems capable for dynamically allocating power as a function of usage pattern. From example, IoT devices can shut down when they are not in use and the data transmission rates can be tuned according to load in the network. Overall energy consumption of healthcare IoT systems are smartly managed to significantly reduce power usage thereby devices remain active and functional without excessive drain on power.

Green IoT systems also depend on the sustainability driven software development. Thus, the computational load that the software requires for operations can be optimized in order to minimize the needed energy consumption. Techniques such as edge computing that perform the data processing locally instead of sending it to distant servers helps in saving energy and reducing network load. In addition, by using efficient coding that simplifies algorithms and as a result minimizes the complexity of algorithms, energy savings from reducing the processing power requirement can be reached (Vishwakarma, Singh, Mishra, & Kumari, 2025).

Additionally, it makes sense to design for device longevity and upgradeability for healthcare IoT systems to be sustainable. Therefore, devices should be made with harsh materials that cannot be worn out within a short period, making device replacement less frequent. Modular designs also make it easy to upgrade or replace certain parts such as sensors or batteries, making them more amenable to evolving as technology advances rather than tossing the whole device. By adopting this design approach, waste is reduced and it is assured that devices still meet healthcare needs in the future.

3.3 Alignment with Circular Economy and Sustainable Development Goals (SDGs)

According to the Green IoT approach however, the minimization of environmental impact is just one aspect; it is also inline with the basic ideas of the circular economy and the global Sustainable Development Goals (SDGs). Green IoT system contributes to a more sustainable healthcare future, also in relation to other international goals of global development, by their focus on sustainability and responsible use of resources.

Green IoT depends on the concept of the circular economy, which is very important as it shifts the focus from linear 'take make dispose' model to focus on reuse, repair and recycle. Green IoT systems implement circular economy principles in their use of recyclable materials, ease of device disassembly and device upgrade through prolongation of their lifetime. These help in reducing e-waste and the loss of the materials which are valuable at the end of the product's life cycle.

The other thing is that the approach in question, Green IoT, also helps with the achievement of various Sustainable Development Goals (SDGs). Green IoT technologies support for example SDG 3 (Good Health and Well-Being); they contribute towards improved healthcare delivery without a high environmental impact. Green IoT enables healthcare to become sustainable by improving medical devices' energy efficiency and reducing wasting. Green IoT's advancement towards SDG 7 (Affordable and Clean Energy) is largely achieved by devices with low power and energy efficient systems for applying energy to healthcare settings. Furthermore, the design of healthcare IoT device with eco-friendly materials and sustainable manufacturing processes is relevant to the use of SDG 12 (Responsible Consumption and Production) (Zeng, Zhang, Mishra, Yuan, & Seneviratne, 2025).

The last one is that SDG 13 (Climate Action) is supported by the reduction in energy consumption and promotion of renewable energy sources in healthcare IoT systems. Consequently, Green IoT lowers the carbon footprint of medical devices, their infrastructure, and they contribute to mitigating climate change.

4. MODULAR ARCHITECTURE FOR ADAPTABILITY AND GROWTH

Internet of Things (IoT) technologies are being rapidly adopted by the healthcare sector to enhance patient care, efficient treatment and practice, and productivity. With the demand for IoT solutions increasing, there is a corresponding need for systems that can accomplish the task of adapting to changes in requirements, scaling, and sustainability, among other things. Thus, modular architecture provides a plausible approach to solving these problems. In this section, we explore the advantages of modular systems design, building block approach to building IoT infrastructure for hospitals and remote care, and present real world-case of modular IoT system deployments in the healthcare arena.

4.1 Advantages of Modular System Design

Modular system design refers to the practice of creating complex systems from smaller, independent They are easy to integrate, upgrade or replace depending on their intended view. Especially for the context of healthcare IoT systems, this approach is advantageous because its ability to adapt, scale and efficiently maintain systems is key. Modular design allows for the primary advantage of flexibility. Patient care demands, technological advancements and regulatory requirements in healthcare needs are changing at a dynamic rate. The modular system also allows providers to easily "hot swap" modules as needed to make quick changes to the modernization. Other examples include integrating new sensors or monitoring devices without changing the whole system and making sure that the infrastructure stays synchronized with the current healthcare needs (Bhutani, Lall, & Agrawal, 2023).

Modular design is also advantageous because it is upgradable. The healthcare is a sector of high technological activity and IoT systems should be proactive so as not to lose relevance. Modular systems make upgrades easier because replacement or upgrading can be done in one or more individual modules like sensors, processors or communication, without affecting the other components of the system. That is to say that healthcare providers can bring the latest innovations to market without the high costs and disruption that normally accompany complete system replacement. This enhances the life of the system and guarantee that healthcare organizations are able to keep offering cutting edge care without significant capital expense on new infrastructure.

Modular is another big bonus when it comes to fault isolation. While in traditional monolithic systems, a failure in one part of the system causes unthinkable devastation, for it breaks the whole system. However, modular systems isolate faults to specific modules making it easier to single out and quickly fix them sometimes. They reduce downtime so that the patient care can continue with as little disruption as possible. For example, say the sensor does not function properly: Then only one part of the system (module) needs attention, while the rest of the system can continue to operate. This approach brings the reliability and availability of healthcare IoT systems in line with their importance in an environment where running unnecessarily generates major repercussions (Shoomal & Jahanbakht, 2024).

4.2 Building Block Approach for IoT Infrastructure in Hospitals and Remote Care

The building block approach to design is a methodology in which we use discrete components that are in oneself interoperable to build IoT infrastructure out of the building blocks. This is a very good way to handle healthcare settings where departments and care models would have different needs and requirements.

Modular IoT in hospitals allows hospitals to create custom solutions for specific departments or functions. For instance, an ICU or a hospital's need may be for an IoT platform that utilizes monitoring devices within an ICU, and another one would be for general patient tracking in less critical care units. The block approach to building leaves hospitals free to pick modules that will best work for their specific needs without overwhelming the system with features. Since they can be easily swapped out, or upgraded as the care requirements evolve, or new technologies appear, these modules serve as great safeguards.

In the same ways, the building block approach is very effective for remote care scenarios, for example home healthcare and telemedicine. In these situations, patients may need only a subset of IoT capabilities, for example wearable health monitors or remote communication tools. The flexibility of modular systems allows healthcare providers to configure the infrastructure to the respective needs of the individuals, without setting them on an information overload path with the technology they need. Over time, the system can be expanded with new modules in order to continue health monitoring and intervention as patients' conditions change or improve.

One of the important characteristics of this approach is the interoperability of modules. In health care, we need interoperability between the various devices, platforms, sources of data and so on to generate a unified and seamless patient care experience. Most modular IoT systems are planned to be amenable to a broad range of devices and technologies so as to allow healthy communication and communion of data between various healthcare providers and devices. The interoperability between the systems, which bridges departments like radiology, cardiology and primary care, ensures enhanced collaboration and improves the provision of the overall more integrated and encompassing patient care (Pandiyan & Saravanan, 2024).

4.3 Case Studies: Modular IoT Systems in Real-World Healthcare Deployments

This approach is shown by real world deployments of modular IoT systems in healthcare, and they bring in practical benefits and effectiveness. Numerous case studies show the way modular systems are being utilized in the hospital and remote care settings to help improve patient care, improve operational efficiency, and grow the system. A notable example is at the deployment of modular IoT systems in the United States for a large hospital network. The patient monitoring and diagnostic systems of the hospital group were adopted with a modularity to facilitate selection and integration of modules by individual hospitals within the network based on their specific needs. For example, some hospitals had the advanced patient tracking modules while some did the integration of wearable health devices for the remote monitoring of chronic conditions. As a result, the hospital network could maintain consistency in providing patient care yet every facility could tailor its IoT infrastructure to meet local needs. Additionally, its modular design allowed for future aspects to be upgraded with newer technologies, keeping the hospital network able to stay on the nose of the healthcare field without always needing costly overhauls.

An example is another case of a healthcare provider in Europe who uses a modular IoT system for remote care of elderly patients suffering from chronic conditions. Vital sign sensors were also included for their use within the modular sensors, like the system consisted of, to monitor blood pressure, heart rate, and oxygen levels. The sensors were linked into a central hub that healthcare providers could see patient data and if needed, remotely access and intervene. The system was modular and additions of new sensors or capabilities were easy as the needs of the patients changed. As the patient population aged and became more demanding, the provider populated the monitoring with additional capabilities like fall detection and medication adherence tracking. In this case we found it possible to customize and scale the modular systems given our new patient populations, and in their particular chronic disease management requirements, specifically long term care patients (Bhutani, Lall, & Dixit, 2021).

A third ends with a description of a pilot program in Asia, where IoT modules were deployed to improve access to care and health monitoring in rural healthcare facilities. Due to the nature of the system being designed to serve remote locations with limited infrastructure, its components were created with modular arrangements that could easily be moved and set up. Patient vital signs data was collected and modularity was used to transmit it to urban healthcare providers for analysis and decision making in urban centers. The setup was such that even the people in rural areas could monitor the patients, without having any complex or expensive system. The system was designed in a modular way to allow for an expanding system, as more modules could be added over time because the population's needs grew.

The case studies presented show the adaption and scalable ability of modular IoT systems to healthcare settings. The modular approach is a practical and sustainable means for healthcare providers to improve patient care, higher operational efficiency and be on the cutting-edge with technology for hospital networks of all sizes and remote care deployments.

Finally, modular architecture is proven to be advantageous for the design and deployment of IoT systems in healthcare. Modular systems are well suited to the healthcare environments where adaptability and growth are needed due to the fact that they have increased flexibility, upgradability, and fault isolation. Healthcare providers can use building block approach to build custom and scalable IoT infrastructures as per their needs. Real world case studies on the deployment of modular systems indicate the contribution to patient care and operational efficiency and sustainability and show the benefits of this approach whether deployed in hospitals or in remote care settings.

Table 2. Key features of modular architecture in healthcare IoT, highlighting flexibility, upgradability, fault isolation, and interoperability

Key Feature	Description	Relevance in Healthcare IoT
Flexibility	Ability to adapt the system by adding/ removing modules as needed based on evolving requirements	Ensures IoT systems can accommodate dynamic healthcare demands and technological advancements
Upgradability	Modular components can be easily upgraded or replaced without disrupting the entire system	Allows healthcare organizations to stay current with innovations without large capital investment
Fault Isolation	Faults are isolated to specific modules, ensuring minimal disruption to the overall system	Minimizes downtime, enhancing reliability and ensuring continuous patient care
Interoperability	Modules are designed to work with a variety of devices and systems for seamless integration	Facilitates comprehensive patient care across various departments and providers

Table 2 summarizes several key advantages that modular IoT architecture adoption provides in the field of healthcare. The modularity enables quick adaptation to changes in healthcare needs, e.g., adding new sensors, or upgrading devices, which is particularly relevant nowadays with today's fast changes in healthcare needs. The upgradability feature allows healthcare providers to sync without facing the need of expensive system reversion. Localizing failures through fault isolation improves the system reliability by localizing the failures efficiently and quickly so that they do not cause affecting the system functionality. This makes them able to communicate the interoperability of modules, which in turn allow seamless communication across devices and platforms, and inclusive patient care. These characteristics have special significance in the rapid developing healthcare environment, where an ability to scale, upgrade and maintain the IoT systems effectively can drastically boost patient results and operational effectiveness. Modular systems are shown through real world case studies to be practical in healthcare settings such as in hospitals as well as remote settings.

5. ENERGY-EFFICIENT COMMUNICATION PROTOCOLS

Energy efficiency of communication protocols has a great effect in healthcare IoT systems since they are required to stay autonomous in extended periods to the extent of resource constrained environments such as remote healthcare monitoring or wear hose devices. They allow the devices to communicate over wireless network with the least amount of power consumption, which is essential in extending battery life, cutting down on maintenance and keeping patient monitoring continuous. It looks at power consumption in the various wireless communication protocols, protocols designed specifically for the healthcare environment, and reducing transmission energy and data redundancy procedures.

5.1 Power Consumption in Wireless Communication: Bluetooth Low Energy (BLE), Zigbee, LoRaWAN, NB-IoT, and 6LoWPAN

Various wireless communication technologies have been developed to support IoT devices, each with unique characteristics regarding power consumption, range, and data transmission speed. These characteristics are particularly important in healthcare, where devices must operate efficiently without frequent battery replacements or charging (Zhang & Feng, 2024).

Bluetooth Low Energy (BLE) is one of the most widely used wireless communication protocols for IoT devices, especially in wearable healthcare devices. BLE is designed for low power consumption while maintaining reliable communication over short distances (typically up to 100 meters). It achieves energy efficiency by using a duty-cycling mechanism, where devices spend most of their time in a sleep mode and wake up only when necessary to transmit or receive data. This makes BLE ideal for devices such as heart rate monitors, fitness trackers, and smartwatches, where long battery life is a priority.

Another low power communication protocol that is very well suited for enabling wireless sensor networks within the care environment is Zigbee. The Zigbee works on 2.4 GHz frequency band with lower data rates (up to 250 kbps) but that is enough for most healthcare end application where they want to transmit small amounts of data periodically such as monitoring the patient. Through efficient sleep modes and low duty cycle operation, Zigbee consumes a low power as compared to other IEEE 802.15.4 specifications and is therefore ideally suited for home health monitoring and medical sensor network applications.

Specifically, it is LoRaWAN (Long Range Wide Area Network) and designed for the purpose of long range communication at low power. It operates at sub-GHz frequencies and offers the possibility of connectivity up to several kilometers distance. In remote healthcare monitoring applications, for example remote healthcare facility and rural, LoRaWAN is most beneficial due to the large region that devices require to communicate. Since LoRaWAN offers long range communication with low energy consumption, it is particularly suitable for applications such as chronic disease management, environmental health monitoring, rural health services and other applications.

The cellular based communication protocol dedicated towards low power wide area networks (LPWAN) is NB-IoT (Narrowband IoT). With licensed spectrum bands it works on covering a wider area and having a better reliability than Unlicensed protocols like LoRaWAN and Zigbee. With its higher data rates and more reliable network than other technologies like LTE, NB-IoT is the ideal option for the healthcare environment where the data needs to be transmitted repeatedly or in real time, examples being remote patient monitoring, wearables and asset tracking. Low power idle modes and optimized signaling for low throughput applications are used to achieve its energy efficiency.

IPv6 over Low Power Wireless Personal Area Networks (6LoWPAN) is a communication protocol for low power wireless networks that allows IPv6 communication and is a basis for the Internet of Things. It is combined with other protocol like IEEE 802.15.4 and is suitable for health devices that need internet connectedness, 6LoWPAN enables more efficient transmission of data by compressing headers and reducing overhead for minimizing power consumption with compatibility to the rest of the internet infrastructure. Such healthcare applications as patient tracking, remote monitoring, smart hospital systems benefit from its low energy consumption and scalability (Khan, Sharma, & Thakare, 2024).

5.2 Protocols Optimized for Healthcare Environments

Due to the requirement for a reliable and real time data transmission, energy efficiency and secure communication, there are special challenges to communication protocols in healthcare environments. Therefore, some of these communication protocols are specialized in healthcare applications and having high power efficiency, and, in many cases, spectrum allocation too.

One of the most widely adopted protocols for making healthcare devices is BLE due to its ability to operate at low power consumption as well as short range communication. For example, BLE has been optimized for such medical devices as glucose monitor, pulse oximeter and electrocardiogram (ECG) sensor. It has low energy consumption and is suitable for continuous health monitoring and wearable devices as it supports multiple devices. For example, a common characteristic of the devices to be monitored in healthcare environments is to transmit small amounts of data intermittently, matching the requirements of the protocol.

Additionally, for instance in some healthcare environment applications where a wireless sensor network is considered, Zigbee is designed to be optimized. Hospitals often use the Zigbee to monitor the rooms, for example, for asset tracking, environmental monitoring (temperature, humidity, etc.), and so on. Because of its low power consumption and its ability to support large number of nodes that are in a mesh networking, it is perfect for healthcare environments where scalability and battery longevity is important. Zigbee is used in sensor networks where devices have to communicate once in a while to save batteries.

In particular remote healthcare applications based on LoRaWAN are becoming increasingly popular as it makes use of wireless communication where internet infrastructure is not available or non existent, for example in rural and underserved areas. This is well suited to use in conditions that are remote for data collection of chronic conditions, due to its ability to transmit data over long distances with little power requirement. LoRaWAN's long range and energy efficiency make it ideal for connecting all manner of healthcare devices; wearables, sensors in the environment, or smart pill dispensers. Due to its excellent battery life, it is the perfect device to use for remote patient monitoring where connectivity and battery life are crucial.

For healthcare environments that need reliable and real-time data transmission, NB-IoT is optimized. To sit well with above use cases, NB-IoT can be deployed in a secured manner to allow low power, wide area connectivity such as remote status of patients vital signs, tracking of medical equipment, and observation of medications attendance. The protocol has the advantage of providing better connectivity in the wide area, with minimal energy spent, thus it is predominant for healthcare providers to implement these scalable, reliable and energy saving IoT solutions (Almutairi, Bergami, & Morgan, 2024).

5.3 Strategies for Minimizing Transmission Energy and Data Redundancy

Transmitting energy reduction and reduction of data redundancy are important goals in improving the energy efficiency of communication protocols of healthcare IoT systems. In addition, these strategies not only save battery life but also maximize the utilization of the network resources by the healthcare systems.

Data aggregation is one effective strategy of reducing transmission energy. In the IoT systems, sensor devices usually collect information from sensors and send to a central server, or a cloud platform. Transmission of total data as it is, one piece at a time, however, can be very energetically expensive, especially when the data is redundant or useless. The system aggregates data from multiple devices or sensors prior to transmission to reduce number of transmission and thereby conserve energy and keeping network congested. To give a specific example, patient monitoring systems can relay data from multiple wearable devices to a central gateway, aggregate them, and disseminate a single, aggregated report rather than individual data points. However, such a strategy dramatically decreases energy consumption and increases communication efficiency.

Adaptive transmission is another strategy to save energy. Adaptive transmission is changing the transmission power and frequency with the network conditions and the importance of the data being transmitted. For example, in a remote patient monitoring system, high priority, high power data can be transmitted for data that requires immediate attention (for instance, for a critical change in a patient's vital signs), whereas lower priority, lower power data (perhaps such as once a day) can be transmitted. This approach further improves energy efficiency, because devices will use less power when the situation does not require the transmission of high rates (Aragani & Maroju, 2025).

Another technique that facilitates the reduction of data redundancy and transmission energy is data compression. It reduces the energy consumption by reducing the amount of data that has to be sent if it compresses the data before sending it. Sensor readings can be compressed with data compression algorithms, saving the payload size and the number of transmissions needed. Compression can dramatically extend the lifespans of battery powered devices in healthcare IoT system where the sensor data is often sent periodically; and yet can preserve the integrity of being transmitted data.

Finally, **sleep mode optimization** is a crucial aspect of energy-efficient communication protocols. Many IoT devices can enter a low-power sleep mode when not in use, which helps conserve energy. Optimizing the sleep-wake cycle of these devices ensures that they remain in low-power states as much as possible while still being able to respond to events or transmit critical data when necessary. This is particularly important for healthcare devices, where continuous monitoring is essential but power consumption must be minimized to extend device lifetime.

6. SCALABLE DATA MANAGEMENT AND EDGE COMPUTING

The increasing adoption of IoT in healthcare is leading to an exponential growth in the volume of data generated by devices and sensors. Managing this vast amount of data efficiently while ensuring timely access and minimizing energy consumption presents a significant challenge. As healthcare IoT systems scale, the demands on data management systems intensify, requiring innovative solutions to handle the complexities of big data. This section explores the challenges posed by big data in healthcare IoT, the role of edge computing in optimizing data management, and strategies for intelligent data prioritization, compression, and lifecycle management to ensure sustainable operation.

6.1 Big Data Challenges in Healthcare IoT: Volume, Variety, and Velocity

Increasing number of devices and sensors result in production of massive amounts of data in healthcare IoT systems. Given that this data is commonly called big data, it raises several challenges regarding volume, variety and velocity, all needing their own strategies for successful management (Musa, Hussaini, & Qian, 2024).

There is volume, that is, sheer amount of data a healthcare IoT device is able to generate. Continuously streaming inputs come from devices, such as wearable health monitors, smart medical equipment as well as remote patient monitoring systems. This, along with the fact that the number of devices in healthcare settings is also rising, adds to the amount of data. To manage such large volumes of data effectively, it requires storage solutions that are robust, and data processing of efficient ways to allow for valuable insights to be extracted without hindering the system. However, in most cases these cloud based storage approach is not enough since of latency involved and the cost of storing such huge bulk of data. With this, it is required to carry out local storage and real time processing to handle the volume of data efficiently.

Data variety includes the numbers of different types of data generated by healthcare IoT devices including structured data (like a readout from a sensor), unstructured data (like audio or a video), as well as semi structured data (sensors logs or sensor metadata). Data integration and processing techniques are complex needed to manage varied data formats and most hospitals have this problem. On the other side, there are too many data sources making it hard to use uniform analytics or integrating different systems. Data architectures for handling of this diversity must be flexible and scalable and support both processing and analysis of data of different types.

The speed in which healthcare IoT systems generate their data is referred to as velocity. For many healthcare applications such as continuous patient monitoring or road traffic accident detectors, it is critical to transmit data in real time to devices. The big problem to solve is that of processing this fast data over the air in near real time without compromising latency and delay in data transfer and/or analysis. Additionally, data pipelines that take time should be efficient, such that they can also prioritize the critical information and filter out the unimportant data, so that one does not get inundated by information overload.

Three big data V's—volume, variety, and velocity—become substantial difficulties for the airborne and continuous data flow management in healthcare IoT systems. These challenges must be tackled with solutions that not only accommodate data that is process, stored, and transmitted efficiently, but also with minimum negative effect to the environment (Pelekoudas-Oikonomou, 2022).

6.2 Leveraging Edge Computing to Reduce Latency, Bandwidth Usage, and Cloud Dependency

Turning to some of the challenges of big data in healthcare IoT systems, edge computing is one of the innovative solutions. Edge computing processes data closer to the source (at the edge of the network), reducing the need to transmit huge amounts of data to central cloud servers, hence reducing latency, bandwidth consumption and the dependence on the cloud.

Reduction of latency is one of the prime benefits of edge computing in healthcare IoT. In healthcare applications like remote patient monitoring, the lost medical events or incorrect diagnoses due to delay is the most crucial generating demand for the data processing at real time or near real time. Edge computing provides a solution with processing of data locally at the edge so that critical data can be processed quickly and appropriate action can be taken immediately without the need to wait for cloud based processing. Speeding up latency in healthcare systems is imperative, especially in the area of time sensitive applications such as telemedicine, emergency care and monitoring whether its intensive care.

This also decreases bandwidth usage. Large volumes of data from remote healthcare devices can quickly overload a network of infrastructure and increase costs while transmitting a large volume of data from a remote device to a centralized cloud server. The Data Processor can process data locally so that only relevant or summarized data is sent to the cloud avoiding constantly consuming continuous high bandwidth communications. This is especially useful when there are IoT healthcare devices operating in the remote or underserved areas with poor internet connectivity. Edge computing also lessens network congestion and improves the use of bandwidth by reducing the amount of data sent.

Furthermore, edge computing reduces the dependency of cloud resources in healthcare IoT systems, which can be a key consideration for them. The amounts of storage and the power to compute made available by cloud computing have only one limit, and that is a purely artificial one: There are (almost?) no cloud computing constraints at all because the obvious risks of that hardly comes remotely close to the advantages. If healthcare service systems can process data at the edge, they can control higher levels of sensitive patient data more while maintaining compliance with the privacy law as, for instance, HIPAA. This however, boosts the resilience and the reliability of the healthcare systems as they will still be able to operate when there is a service disruption or network outage within the cloud service, making skincare systems more resilient and reliable.

6.3 Intelligent Data Prioritization, Compression, and Lifecycle Management for Green Operation

Additionally, there is the need to seek intelligent data management strategies like data prioritization, compression, as well as lifecycle management that will further increase the energy efficiency and sustainability of healthcare IoT systems. Minimizing the energy consumption demanded for data storage, transmission and processing while making sure critical healthcare data gets the care it deserves: these are these strategies.

Data prioritization refers to when the important data is identified and sent, or processes, and that which is not so important is discarded, among others. This can be done with algorithms that judge the importance of various types of data and their urgency in healthcare IoT systems. For instance, a wearable ECG monitor may first send real time heart rate data from a patient in distress, while ability to hold and later upload background data, step counts or activity logs may be stored locally. Healthcare providers, however, should give preference to critical data so that energy is used effectively and network resources are not too burdened.

By combining this single modality of measurement with data compression, the impact on the environment is reduced. Because any data can be compressed before being transmitted, stored, or processed, compression algorithms are capable of reducing the size of the data. Minimising the bandwidth needed for transmission not only reduces the amount of energy consumed during data link to storage but also keeps down the never ending costs of using computers to transmit data. Given the enormous amounts of data involved in applications such as medical imaging or video, where such savings of energy can result in substantial savings in terms of file size without affecting the quality of the data, compression is particularly useful (Attah & Garba, 2024).

The data lifecycle management is a process through which a data is created, kept, processed and then be deleted. To ensure healthcare IoT systems store and process only the data that is needed for ongoing operations, also, it is implemented that efficient data lifecycle management policies for healthcare IoT systems can be implemented. For instance, sensor data that has become obsolete and is no longer required could be automatically removed to free storage space as well as conserve energy. Additionally, archiving of data for long periods such as a patient's health records can be done in energy efficient storage solutions, reducing data center's operational costs and environmental footprint.

Healthcare IoT systems can be economically run in an energy efficient as well as sustainable mode by employing intelligent data prioritization, compression, and lifecycle management strategies. In this way, these techniques keep minimizing unnecessary data transmission, storage requirement and remain functionable of healthcare systems while conserve energy, lessen environmental impacts.

Finally, scalable data management is an important problem for healthcare IoT systems because they have high volume, variety, velocity of data, to overwhelm classical approaches for data management. Using edge computing, healthcare providers can decrease latency, bandwidth utilization, as well as cloud dependencies and increase the resilience and responsiveness of the system. In tandem, approaches such as intelligent data prioritization, compression, lifecycle management can be taken to further enhance data management processes, making healthcare IoT systems operate in a manner that is efficient and sustainable. As healthcare advances and more healthcare organizations continue to adopt IoT technology and attempt to scale towards meeting the demands of today's data driven healthcare environment, these approaches will be critical (Amiri, Heidari, Navimipour, & Unal, 2023).

6. CONCLUSION

The paper concludes that to achieve sustainable, energy efficient and scalable solutions to the emerging needs of healthcare environment, integration of green IoT principles to healthcare systems is required. Although healthcare IoT systems are in the position to give reliable, efficient care while respecting patient privacy and the performance of the systems, we need to emphasize concentrating on reducing the environmental impact, optimizing the energy consumption, and ensuring data security as well as compliance with regulations. These systems are based on modular architecture, energy-efficient communication protocols, edge computing, and fault tolerant designs and are resilient and adaptable pertaining to the increase in technological areas as well as healthcare demands. As healthcare more and more efforts to digital transformation, the successful green IoT systems will reflect sustainability goals that ultimately has a positive impact on all sides: healthcare providers and patients. Such convergence of eco friendly design with the latest in technology can bring significant thrust to the manner in which healthcare delivery can be made more sustainable and efficient.

7. FUTURE RESEARCH DIRECTIONS

In the future, research in green IoT for healthcare should be done on integrating AI, ML and blockchain in order to make healthcare IoT systems more sustainable and secure. The area of interest is to develop the AI powered energy optimization algorithm, which can dynamically adjust system operations based on real time data and minimizing the energy consumption while maintaining the system performance. Moreover, the blockchain could be explored for making healthcare data secured in distributed networks with increased privacy, transparency, and control. The adoption of biodegradable IoT devices should be added to the scope of research which would continue to further diminish the green foot print of healthcare systems. Additionally, the long term sustainability of IoT devices according to their lifecycle management and recycling will be looked at. Finally, to set global standards and frameworks for green IoT in healthcare, collaborative efforts will be needed by healthcare provider, technology developers, and policymakers together.

REFERENCES

Ali, Z. H., & Ali, H. A. (2021). Towards sustainable smart IoT applications architectural elements and design: Opportunities, challenges, and open directions. *The Journal of Supercomputing*, *77*(6), 5668–5725. DOI: 10.1007/s11227-020-03477-7

Almutairi, R., Bergami, G., & Morgan, G. (2024). Advancements and challenges in IoT simulators: A comprehensive review. *Sensors (Basel)*, *24*(5), 1511. DOI: 10.3390/s24051511 PMID: 38475047

Amiri, Z., Heidari, A., Navimipour, N. J., & Unal, M. (2023). Resilient and dependability management in distributed environments: A systematic and comprehensive literature review. *Cluster Computing*, *26*(2), 1565–1600. DOI: 10.1007/s10586-022-03738-5

Aragani, V. M., & Maroju, P. K. (2025). Future of blue-green cities emerging trends and innovations in iCloud infrastructure. In *Integrating Blue-Green Infrastructure into Urban Development* (pp. 223-244).

Arega, A., & Sharma, D. P. (2023). Towards smart and green features of cloud computing in healthcare services: A systematic literature review. *Journal of Information Systems Engineering & Business Intelligence*, *9*(2), 161–180. DOI: 10.20473/jisebi.9.2.161-180

Attah, R. U., Garba, B. M. P., Gil-Ozoudeh, I., & Iwuanyanwu, O. (2024). Strategic partnerships for urban sustainability: Developing a conceptual framework for integrating technology in community-focused initiatives. *GSC Advances in Research and Reviews*, *21*(2), 409–418. DOI: 10.30574/gscarr.2024.21.2.0454

Bhutani, M., Gupta, M., Jain, T., Sharma, T. P., Solanki, A., & Malhotra, A. (2023). Data-driven robotics: A performance analysis of wearable sensor-based wound assessment system for sensor data collection. In *Bio-Inspired Data-Driven Distributed Energy in Robotics and Enabling Technologies* (pp. 249–266). CRC Press.

Bhutani, M., Lall, B., & Agrawal, M. (2023). An efficient and adaptive superframe structure for IEEE 802.15.7-based real-time sensor networks.

Bhutani, M., Lall, B., & Dixit, A. (2021). MAC layer performance modelling for IEEE 802.15.7 based on discrete-time Markov chain. *IET Communications*, *15*(14), 1883–1896. DOI: 10.1049/cmu2.12199

Goyal, R., Sharma, A., Jain, A., Elawadhi, O., & Bhutani, M. (2023). Adaptive density-based traffic management system utilizing ultrasonic units and RODEM algorithm. In *2023 10th International Conference on Signal Processing and Integrated Networks (SPIN)* (pp. 571-577). IEEE. DOI: 10.1109/SPIN57001.2023.10116660

Khan, R. A., Sharma, B., & Thakare, N. M. (2024). A systematic review of protecting cloud healthcare data using blockchain: Target-oriented approach. In. AIP Conference Proceedings: Vol. 3217. *No. 1* (pp. 1–7). AIP Publishing. DOI: 10.1063/5.0234398

Mallo, S. F., Abdulqader, D. M., Abdullah, R. M., Ismael, H. R., Rashid, Z. N., & Sami, T. M. G. (2024). A review on feasibility of web technology and cloud computing for sustainable ES: Leveraging AI, IoT, and security for green operations. *Journal of Information Technology and Informatics*, *3*(2).

Musa, A. A., Hussaini, A., Qian, C., Guo, Y., & Yu, W. (2024). Open radio access networks for smart IoT systems: State of art and future. *International Journal of Semantic Computing*, *11*(3), 310–327.

Oyeniran, O. C., Modupe, O. T., Otitoola, A. A., Abiona, O. O., Adewusi, A. O., & Oladapo, O. J.Oyekunle Claudius OyeniranOluwole Temidayo ModupeAanuoluwapo Ayodeji OtitoolaOluwatosin Oluwatimileyin AbionaAdebunmi Okechukwu AdewusiOluwatayo Jacob Oladapo. (2024). A comprehensive review of leveraging cloud-native technologies for scalability and resilience in software development. *International Journal of Science and Research Archive*, *11*(2), 330–337. DOI: 10.30574/ijsra.2024.11.2.0432

Pandiyan, P., Saravanan, S., Kannadasan, R., Krishnaveni, S., Alsharif, M. H., & Kim, M. K. (2024). A comprehensive review of advancements in green IoT for smart grids: Paving the path to sustainability. *Energy Reports*, *11*, 5504–5531. DOI: 10.1016/j.egyr.2024.05.021

Pelekoudas-Oikonomou, F., Zachos, G., Papaioannou, M., De Ree, M., Ribeiro, J. C., Mantas, G., & Rodriguez, J. (2022). Blockchain-based security mechanisms for IoMT Edge networks in IoMT-based healthcare monitoring systems. *Sensors (Basel)*, *22*(7), 2449. DOI: 10.3390/s22072449 PMID: 35408064

Shoomal, A., Jahanbakht, M., Componation, P. J., & Ozay, D. (2024). Enhancing supply chain resilience and efficiency through internet of things integration: Challenges and opportunities. *Internet of Things : Engineering Cyber Physical Human Systems*, *27*, 101324. DOI: 10.1016/j.iot.2024.101324

Verdejo Espinosa, Á., Lopez, J. L., Mata Mata, F., & Estevez, M. E. (2021). Application of IoT in healthcare: Keys to implementation of the sustainable development goals. *Sensors (Basel)*, *21*(7), 2330. DOI: 10.3390/s21072330 PMID: 33810606

Vishwakarma, L. P., Singh, R. K., Mishra, R., & Kumari, A. (2025). Application of artificial intelligence for resilient and sustainable healthcare system: Systematic literature review and future research directions. *International Journal of Production Research*, *63*(2), 822–844. DOI: 10.1080/00207543.2023.2188101

Zeng, J., Zhang, T., Mishra, D., Yuan, J., & Seneviratne, A. (2025). A survey on green designs for energy harvesting backscatter communications to enable sustainable IoT. *Energies*, *18*(4), 840. DOI: 10.3390/en18040840

Zhang, Y., & Feng, J. (2024). *Towards a smart and sustainable future with edge computing-powered internet of things: Fundamentals, applications, challenges, and future research directions. Journal of The Institution of Engineers.* Series B.

Chapter 7
Green Firmware and Software Architectures for Sustainable Medical IoT Devices

Charles Jaya Rao Nettem
Malineni Lakshmaiah Women's Engineering College, India

B. Mamatha
https://orcid.org/0009-0007-5190-9037
Tirumala Engineering College, India

Kuppani Sathish
https://orcid.org/0000-0002-4651-8310
Tirumala Engineering College, India

Kondragunta Rama Krishnaiah
R.K. College of Engineering, India

Y. V. Narayana
Tirumala Engineering College, India

ABSTRACT

The surge in the number of medical IoT devices has created big sustainability concerns esp. in terms of energy consumption, resource utilization, and electronic waste. Despite plenty of attention received by hardware optimization, the role of firmware and software in making affordable devices eco-friendly still lacks proper exploration. This chapter delivers an exhaustive study of the sustainable medical IoT applications of green software engineering and low power firmware design.

DOI: 10.4018/979-8-3373-4134-7.ch007

It discusses such principles as energy-aware coding, modular architecture, and efficient memory control. Discussion on techniques such as real-time operating systems, interrupt-driven programming and lightweight communication protocols is included. The chapter also discusses the trade-off between the energy efficiency and data privacy and the need for consolidation of edge-cloud infrastructures in a way that will reduce computational burden. By pointing out best practices and future research trends, this chapter urges an end to a software-centric approach to creating sustainable healthcare technologies.

1. INTRODUCTION

The quick development of the medical domain of the Internet of Things (IoT) has changed modern healthcare, giving it continuous control of patients, early diagnosis, remote treatment, and enhanced operational effectiveness. Medical IoT devices, including wearable health monitors, implantable biosensors, and interconnected diagnostic tools, can no longer be viewed as simply being useful, but as essential in ensuring effective treatment. Nevertheless, this broad adoption raises a grave concern, which is insufficiently addressed. sustainability. Environmental, energy and resource sustainability challenges constitute the sustainability problems raised by the medical IoT devices and this finds a call for rethinking how the medical IoT devices are planned, developed and deployed.

Medical IoT devices usually run continuously and are commonly deployed in settings where level of regular power supply or maintenance of the systems is limited. The energy expenditure of such devices, particularly when applied in widely distributed systems of healthcare or in a population, adds to the total carbon footprint of digital healthcare. Besides, the life cycle of medical electronics is short, and there are few opportunities to recycle and reuse them, which intensifies the problem of e-waste. As these devices get more common, they not only deplete energy but add non-biodegradable materials to landfills that may have severe ecological implications over time (Islam & Bhuiyan, 2023).

Operating against hardware, the sustainability issues come in from the software and the firmware through which the devices control. Software algorithms decide the data that is sampled on a device, how it talks to servers and how often it gets updated which all directly affect power consumption and network load. With firmware as the low-level control code stored at the hardware level becoming imperative to control the state of the device being sleep, wake or communication cycles. Poorly optimized software will cause over consumption of energy, data redundancy, and inefficient memory operations, and this will eventually reduce the life-span of medical IoT devices and increase their ecological impact.

Another of major concerns is the incorporation of medical IoT devices into clouds and edge computing facilities. Even data-intensive applications, including AI based diagnostics, real-time video analysis, continuous monitoring, can mean significant energy demands in terms of transmission, storage and processing of data. Although these computational steps enhance healthcare delivery, they have a higher environmental cost unless underpinned by green software and system architectures.

Traditionally, efforts to make medical IoT devices more sustainable have focused heavily on hardware—using energy-efficient sensors, biodegradable materials, or miniaturized power systems. However, the software and firmware layers are equally, if not more, influential in determining a device's environmental footprint. Software is the "brain" of the device, dictating how the hardware components interact and perform under various conditions. It governs the sampling rate, processing logic, transmission schedule, and even the sleep-wake patterns of devices (Islam & Nooruddin, 2022).

Green software engineering is a design of code that should use as little resources as possible while keeping functionality and performance. For instance, the frequency of data transmissions can be diminished; local data processing can be used to eliminate redundant information; and compression of payloads following transmission can greatly lower the energy needs. In the same manner; Energy-aware scheduling can be incorporated in real-time operating systems (RTOS) or utilize adaptive algorithms to adjust in acquisition of usage pattern to prolong battery life and minimize the need for frequent maintenance or replacement.

Although much of the firmware is static and difficult to alter once in place, firmware provides a basic foundation for energy-efficient usage of devices. It manages the low level-mechanisms of sensor, power management units, and communication modules. By using interrupts, the sleep mode management and boot processes, for instance, power management can be achieved even in idle times without degrading user responsiveness. The creation of firmware over-the-air (FOTA) update systems also offers this opportunity for eco-friendly maintenance when such are small, rare, and occur under good power conditions.

Figure 1. Sustainable medical IoT: Software and firmware architecture

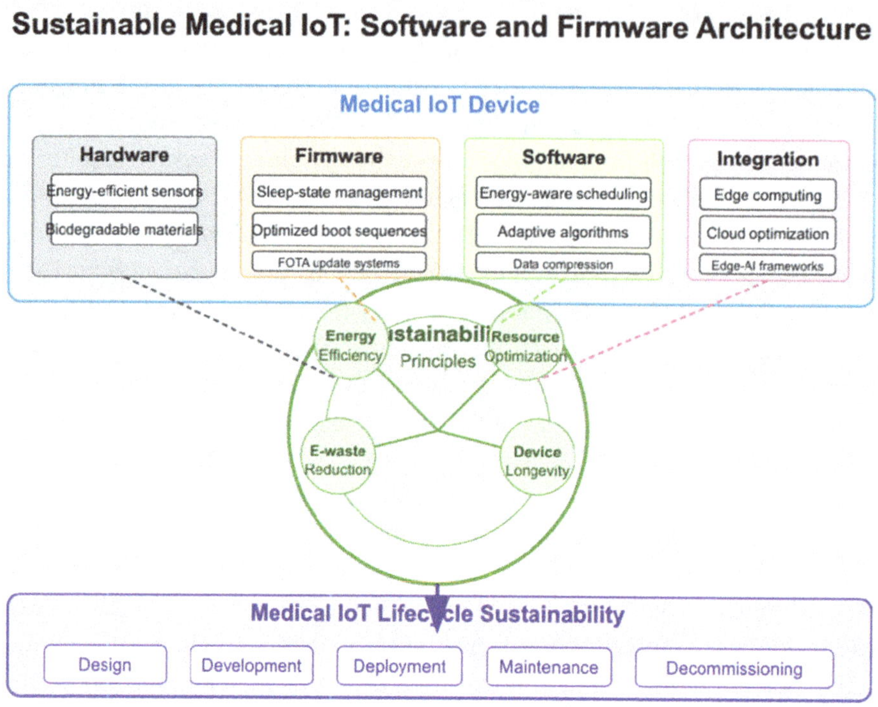

Figure 1 pictures the multi-layered approach to sustainability in medical IoT systems and points at the key correlation between HW/FW/S/W and integration factors. The central sustainability wheel outlines 4 main principles of energy efficiency, resource optimization, device longevity and reduction of e-waste that should drive the development at every phase of lifetime of the entire medical IoT system, from the planning stage to decommissioning. The diagram illustrates the role optimized firmware (with sleep-state management and efficient boot cycles) and green software (energy-aware scheduling and adaptive algorithms) play to significantly enhance overall sustainability, together with hardware innovations and edge-cloud integration strategies.

According to Figure 1, the sustainability of these medical IoT devices is a multilayered technological approach dependent upon numerous layers of the technology stack. The figure shows how firmware and software components act as important mediating layers between the capability of hardware and system level integration. Most interesting is the centrality of sustainability concepts which permeate the system design and inform every decision stage in the medical IoT life cycle. The interdependent relationship between firmware capabilities (e.g., the ability to manage sleep-states) and software capabilities (e.g., the ability to perform energy-aware scheduling) demonstrates the combination required, in pursuit of sustainable operation, of capabilities from both these domains. This architectural view supports our key argument that green software and firmware engineering cannot be considered distinct issues that need to be addressed and merged with a comprehensive sustainability strategy of the whole medical IoT ecosystem starting from elements of individual devices down to cloud infrastructure and from design to de-commissioning.

The relation of software to firmware also contributes to the overall sustainability of the system. For example, if firmware has fine grained control over energy states, the software is able to make dynamic transitions between performance modes based on the workload. Similarly, edge-AI frameworks that work alongside local firmware to preprocess using the locally cached data before they are transmitted to the cloud can ensure the reduction of bandwidth and energy consumption. Therefore, integrated software and firmware architecture design is critical to sustainable medical Iot development.

The goal of this chapter is to link sustainability goals with the (too often neglected) role of the software and firmware in medical IoT devices. By concentrating on green firmware and software architectures, the chapter gives a systematic insight into how the digital layers can be optimized to make a contribution to environmental sustainability in healthcare.

The main targets of this chapter are:

- For visibility on environmental impact of software and firmware in medical IoT systems – their contribution to energy consumption, device life and resource usage.
- To present principles and techniques of green software engineering – mainly based on efficient coding, update maintenance, modularity of designs, and runtime optimization.
- To investigate low-power firmware design techniques – sleep state utilization, RTOS connection and eco-friendly communication protocols.
- To explore architectural paradigms that promote sustainability – such as modular, scalable, distributed architecture appropriate for medical uses.
- In order to evaluate trade-offs regarding security, performance and sustainability – especially in cases related to sensitive medical data and continuous running of device.
- To provide practical case studies and current trends – from AI integration to edge-cloud collaboration, and sustainable DevOps suited for medical IoT.
- To suggest future directions and open challenges – prompting further research in green computing for digital healthcare systems.

This chapter's scope is designed to be as broad as possible, and includes elements of fundamental concepts as well as practical strategies. In addition to the technical details of decreasing energy and resource consumption, it considers the systemic views required to exert sustainable initiatives throughout a complete medical IoT lifecycle, starting from design and deployment through maintenance and decommissioning. Besides, though the chapter highlights technical optimization, it also recognizes regulatory, ethical, and usability issues. Sustainable software should not only be efficient but also secure, maintainable while also being in line with the functional requirement of both healthcare providers and patients. This chapter, therefore, plays a significant role as a useful source to researchers, developers, and practitioners aiming at developing eco-minded medical IoT systems. It highlights the importance of software and firmware to deliver sustainability and acts as a roadmap of strategies to realize greener, smarter & more responsible digital health solutions.

2. FUNDAMENTALS OF FIRMWARE AND SOFTWARE IN IoT MEDICAL DEVICES

As medical devices become more intelligent and interconnected, the distinction between firmware and software, once largely academic, now plays a vital role in the design, deployment, and sustainability of medical IoT (Internet of Things) systems. In the context of eco-friendly and high-performance healthcare technology, understanding the layers and constraints of digital control systems is key to ensuring functionality, compliance, and environmental responsibility (Bhutani & Lall, 2023).

2.1 Definitions and Distinctions between Firmware and Software

In embedded systems, especially in IoT medical devices, the terms "firmware" and "software" are sometimes used interchangeably, but they serve distinctly different purposes and reside in different layers of the technology stack.

Firmware is a low-level, platform-specific control code that directly manages a device's hardware components. It is typically embedded in non-volatile memory such as EEPROM or flash memory and serves as the interface between the device's physical hardware and the higher-level operating system or software. Firmware controls essential operations like device initialization, sensor input handling, power state transitions, and communication protocols. In medical IoT devices, firmware might govern the timing of biosensor readings, the scheduling of wireless data transmission, or the execution of safety-critical functions during surgery or therapy.

Software, by contrast, refers to higher-level programs that execute application logic, user interfaces, and data processing routines. It often runs on top of an operating system or a real-time operating system (RTOS), and it is more flexible and frequently updated compared to firmware. Software in medical IoT devices may include patient monitoring dashboards, anomaly detection algorithms, and secure data transmission modules.

The key distinction lies in mutability and function:

- Firmware is hardware-specific, less frequently updated, and optimized for control and stability.
- Software is application-specific, more dynamic, and optimized for flexibility and functionality.

Both components must work in harmony, especially in healthcare environments where safety, reliability, and sustainability are paramount.

2.2 Typical Software Stacks in Medical IoT Devices

Medical IoT systems are composed of multiple layers, each fulfilling a specific role from device control to cloud analytics. A typical software stack for a medical IoT device can be divided into the following layers:

1. Device Firmware Layer
    - Manages hardware peripherals like sensors, actuators, and communication chips.
    - Implements power management features such as sleep modes and interrupt-based wake-up.
    - Provides low-level drivers and bootloaders necessary for device startup and updates.
2. Real-Time Operating System (RTOS) Layer
    - Offers task scheduling, inter-process communication, and hardware abstraction.
    - Ensures deterministic timing critical for medical applications like insulin pumps or pacemakers.
    - Examples include FreeRTOS, Zephyr, and Mbed OS.
3. Middleware and Connectivity Layer
    - Includes software libraries for data serialization (e.g., Protocol Buffers, CBOR), communication (e.g., MQTT, CoAP), and security (e.g., TLS, AES encryption).

- Abstracts communication with cloud services or edge devices, enabling interoperability.
- Facilitates device provisioning, remote management, and over-the-air (OTA) updates.

4. Application Layer
   - Implements patient-specific functionality such as vital sign monitoring, threshold-based alerts, and user interaction.
   - May include AI/ML inference modules for diagnostics or predictive analytics.
   - Interfaces with user dashboards, medical databases, or hospital management systems.
5. Cloud and Edge Integration Layer
   - Supports data aggregation, long-term storage, analytics, and compliance (e.g., HIPAA, GDPR).
   - May use edge computing for localized processing to reduce latency and conserve bandwidth.
   - Interfaces with Electronic Health Records (EHRs), telemedicine platforms, or research tools.

This modularity helps in isolating critical functions (e.g., firmware for pacemaker rhythm control) from non-critical features (e.g., display interface), thereby enhancing reliability and maintainability. It also facilitates more targeted sustainability interventions, such as optimizing firmware for energy use and reducing software bloat in application layers.

2.3 Constraints: Power, Memory, Computation, and Safety

Medical IoT devices operate under stringent resource and regulatory constraints, which must be considered at every level of firmware and software design. These constraints significantly shape development practices, sustainability considerations, and system architecture.

Power Constraints

Power efficiency is a cornerstone concern for medical IoT, particularly for wearable, implantable, or remote devices that rely on battery or harvested energy. Firmware and software design must minimize unnecessary computation and transmission, as these are primary sources of power consumption.

- Firmware can reduce power draw by activating sleep states during idle periods.

- Software should avoid polling loops in favor of interrupt-driven processing.
- Efficient scheduling of communication tasks, such as batching sensor data before transmission, can significantly conserve energy.

Green software techniques—such as load balancing, dynamic voltage scaling (DVS), and duty cycling—are especially important in sustainability-sensitive applications.

Memory Constraints

Memory is typically limited in embedded medical devices. Flash memory (for firmware) and SRAM (for runtime software) are often tightly constrained to reduce size, cost, and energy use.

- Firmware must be optimized for size, avoiding large tables, unnecessary routines, or verbose logging.
- Software should utilize memory-efficient data structures and minimize heap allocations.
- Real-time compression, compact data encoding (e.g., CBOR instead of JSON), and modular loading strategies help conserve memory and reduce update sizes.

Efficient use of memory not only prolongs device life but also reduces environmental cost by enabling smaller chips and less frequent maintenance.

Computation Constraints

Most medical IoT devices operate with microcontrollers that offer modest computational capabilities. Processing-intensive tasks like signal filtering, AI inference, or encryption must be carefully balanced against available CPU cycles.

- Firmware can offload computational tasks to hardware accelerators or optimize for fixed-point arithmetic.
- Software should leverage efficient algorithms, avoid deep recursion, and offload heavy tasks to edge or cloud servers when appropriate.
- When deploying AI, use lightweight models such as TinyML frameworks optimized for embedded systems.

Computation-aware design enhances real-time responsiveness and allows longer operational periods per charge, contributing to both sustainability and performance.

Safety and Reliability Constraints

Safety is non-negotiable in medical applications. Software and firmware must adhere to strict standards like IEC 62304, ISO 14971, and FDA guidelines for software validation and risk management.

- Firmware must be thoroughly tested and verified, especially for safety-critical actions like medication delivery or defibrillation.
- Software must include fail-safe mechanisms, watchdog timers, and redundancy in communication.
- Updates must be authenticated and verifiable, with rollback mechanisms to avoid bricking devices.

These safety constraints sometimes limit the extent of optimization but also encourage modularity, documentation, and testing—all of which are beneficial for long-term sustainability.

Firmware and software are not merely technical components in medical IoT devices—they are active determinants of sustainability, safety, and functionality. While hardware innovations have garnered attention in green healthcare initiatives, digital layers offer untapped potential to reduce environmental impact through smarter control, adaptive logic, and optimized communication.

Understanding the distinctions and interplay between firmware and software, recognizing the layered software stack, and designing with key constraints in mind are all crucial steps toward building medical IoT systems that are not only effective and safe but also environmentally responsible. By integrating green design principles at the firmware and software level, developers and stakeholders can usher in a new era of sustainable healthcare technology—one byte at a time (Kaur, M., Singh, D., Kumar, V., 2021).

Table 1. Comparison of firmware and software in medical IoT devices

Aspect	Firmware	Software
Definition	Low-level control code embedded directly in device hardware	High-level application code running on an OS or RTOS
Location	Stored in non-volatile memory (e.g., flash, EEPROM)	Stored in system memory (RAM or external storage)
Purpose	Direct control of hardware components and basic system functions	Executes application logic, data processing, user interface
Mutability	Infrequently updated; tightly coupled to hardware	Regularly updated; adaptable to changing requirements

continued on following page

Table 1. Continued

Aspect	Firmware	Software
Update Mechanism	Often requires firmware over-the-air (FOTA) updates or physical access	Can be updated via cloud, app store, or OTA with minimal risk
Power Management	Manages sleep states, interrupts, peripheral activation	Schedules tasks, data transmission, and application-specific functions
Execution Environment	Runs without an operating system or under a lightweight RTOS	Runs on top of OS, RTOS, or in the cloud
Criticality	Handles safety-critical functions (e.g., sensor readings, alarms)	Handles non-critical functions (e.g., visualization, reports)
Performance Optimization	Focused on deterministic behavior, energy efficiency, and low memory use	Focused on modularity, scalability, and user experience
Example Use Case	Reading biosensor data, managing Bluetooth module, activating actuators	Monitoring dashboard, anomaly detection algorithm, cloud data sync

Table 1 provides a comparative overview of firmware and software characteristics within medical IoT devices, highlighting their distinct yet complementary roles in system operation. While firmware is primarily responsible for low-level hardware control, ensuring real-time responsiveness and power efficiency, software operates at a higher abstraction level to manage application logic, user interactions, and data flow. The table underscores key differences in their execution environments, update mechanisms, and functional responsibilities—emphasizing why both layers must be optimized in tandem to achieve sustainability, reliability, and clinical efficacy in connected healthcare systems.

3. PRINCIPLES OF GREEN SOFTWARE ENGINEERING

As the world seeks to make healthcare more sustainable, green software engineering emerges as a vital discipline in the development of medical IoT (Internet of Things) systems. These devices, ranging from wearable biosensors to remote monitoring units, often run on constrained hardware with limited power and memory resources. Software—while intangible—directly influences the energy footprint of these devices through the way code is written, deployed, and maintained. This section explores the foundational principles of green software engineering, emphasizing how intelligent coding practices can enhance energy efficiency, minimize resource waste, and extend device lifespan.

3.1 Energy-Aware Coding Practices

At the core of green software engineering lies energy-aware coding. This principle recognizes that every line of code executed on a device consumes power, and therefore, software must be consciously crafted to minimize energy consumption (Albreem, Sheikh, & Alsharif, 2021).

In medical IoT systems, energy-aware coding involves:

- **Reducing unnecessary computations** by limiting polling loops, redundant operations, and non-essential background tasks.
- **Utilizing event-driven programming** where possible, allowing the system to remain in low-power modes until a specific interrupt (e.g., a heartbeat signal) occurs.
- **Leveraging hardware features** such as low-power modes, sleep states, and energy-aware peripherals directly from the codebase.

Simple coding decisions—like avoiding deep recursive calls, limiting floating-point operations, or selecting appropriate data types—can have a measurable impact on battery life. For example, using 8-bit integers instead of 32-bit variables where appropriate can reduce both memory usage and processor cycles.

Furthermore, code that allows hardware to enter idle states more frequently helps maximize energy savings, particularly in wearable or implantable devices where frequent recharging is impractical or invasive.

3.2 Code Efficiency, Refactoring, and Optimization

Efficient code is central to any sustainability effort in software engineering. Inefficient code not only wastes CPU cycles and energy but also increases device wear and decreases reliability over time. In the medical field, this inefficiency may also compromise real-time responsiveness, leading to safety concerns (Kumar & Choudhary, 2024).

Efficient software development includes:

- **Refactoring** code regularly to reduce complexity, eliminate dead code, and improve maintainability.
- **Optimizing algorithms** for computational simplicity—such as replacing nested loops with lookup tables or using approximations where full precision is unnecessary.
- **Avoiding over-engineering**, such as excessive use of design patterns or layers of abstraction, which can introduce processing overhead.

One critical aspect of code efficiency in embedded medical systems is minimizing memory access. Reading from or writing to memory consumes more energy than register-based operations. Thus, compact data structures and effective caching strategies can help reduce power draw.

While pointers offer flexibility, especially in systems programming, their overuse in embedded medical software can introduce instability and memory fragmentation. Replacing pointer-based logic with fixed-size arrays or direct indexing often leads to faster, more predictable execution and easier debugging. Especially in safety-critical devices, deterministic code behavior is essential.

3.3 Minimizing Update Sizes and Frequency

Frequent or large software updates can impose a significant energy and bandwidth burden on medical IoT systems, particularly when updates are delivered wirelessly or over cellular networks. Not only do these updates drain the device's battery during transmission and flashing, but they can also generate unnecessary e-waste when update failures lead to device bricking.

Green software engineering promotes strategies to reduce both the size and frequency of updates:

- **Delta updates**, where only the differences between the new and old firmware are sent, can reduce transfer sizes significantly.
- **Modular code architecture** allows specific components or features to be updated without disturbing the entire system.
- **Careful versioning and change tracking** help prevent the need for minor updates, ensuring only essential patches or feature improvements are deployed.

Additionally, minimizing update frequency means writing robust, future-proof code. This includes planning for extensibility from the beginning, such as using configuration files or flags to toggle features without reprogramming.

Sustainability is not just about reducing the number of bytes sent over the air—it is about making devices smarter, more autonomous, and less reliant on continuous developer intervention.

3.4 Static vs. Dynamic Power Management in Software Design

Medical IoT software must actively contribute to power management through both static and dynamic strategies.

Static power management involves decisions made at the design or compile time. These include:

- Selecting low-power algorithms and routines.
- Compiling code with energy-efficient flags (e.g., size optimization over speed).
- Assigning fixed duty cycles for operations such as sensor reads or data transmissions.

This type of power management is predictable and works well for applications with fixed workflows or schedules—like a glucose monitor that samples every 15 minutes.

Dynamic power management, in contrast, is performed at runtime based on the device's operating conditions. This includes:

- Adjusting CPU speed (dynamic frequency scaling) based on workload.
- Turning off peripherals when not in use.
- Using context-awareness (e.g., patient activity or sleep patterns) to alter sensing or transmission frequencies.

The software must be written to support real-time decision-making with minimal overhead. For example, instead of checking for user input in a continuous loop, the system can sleep and wake only when a button is pressed. These micro-optimizations, when compounded across millions of cycles and thousands of devices, contribute significantly to sustainability goals.

The challenge in dynamic power management is maintaining responsiveness and safety. Therefore, software must include fallback routines, error handling, and watchdog timers to guarantee that power-saving features do not compromise device reliability or patient well-being.

Green software engineering in medical IoT devices is not a luxury—it is a necessity. With growing awareness of climate change, resource scarcity, and the increasing carbon footprint of digital technology, developers must embrace coding practices that reduce energy consumption, extend device longevity, and minimize waste.

From energy-aware programming to smart update strategies and efficient power management, every software decision has environmental consequences. While hardware design continues to evolve toward efficiency, software remains the layer with the most untapped potential for greening the medical technology landscape.

By writing code that is efficient, compact, and mindful of power and memory use—and by minimizing updates and intelligently managing power states—engineers can help ensure that the next generation of medical devices is not only smarter and safer but also more sustainable.

4. LOW-POWER FIRMWARE DESIGN TECHNIQUES

As medical IoT devices become increasingly ubiquitous—from wearable biosensors to implantable monitoring systems—ensuring their energy efficiency becomes a critical design requirement. Unlike traditional software, firmware operates at the hardware-software boundary and directly influences how a device behaves in terms of power consumption. Firmware in medical IoT systems must be lightweight, highly responsive, and optimized for extended operation under severe resource constraints, particularly in terms of energy usage (Bhutani & Lall, 2022).

This section discusses the key strategies for designing low-power firmware, including the strategic use of low-power states, interrupt-driven architectures, energy-aware real-time operating systems (RTOS), and the deployment of firmware over-the-air (FOTA) using sustainable methods.

4.1 Use of Low-Power States and Sleep Modes

Modern microcontrollers used in medical IoT devices are typically equipped with multiple power states, such as active, idle, sleep, deep sleep, and hibernate modes. The firmware is responsible for transitioning between these states based on system requirements, workload, and external stimuli.

Key principles in using low-power states:

- **Sleep and Deep Sleep Modes:** These are the most frequently used energy-saving states. The device shuts down certain components (e.g., CPU, clocks, peripherals) while maintaining essential data in RAM. This is ideal for wearable devices like ECG monitors that need to sample intermittently.
- **Dynamic Clock Gating:** Firmware can selectively disable clocks to specific components when not in use, reducing dynamic power consumption without affecting functionality.
- **Voltage and Frequency Scaling (DVFS):** Reducing supply voltage and processor clock frequency based on performance demands is an effective method for balancing energy and computation.

The firmware must manage these transitions smoothly, ensuring minimal wake-up latency to maintain real-time responsiveness—crucial in medical applications where a delay might result in inaccurate diagnosis or missed alerts.

4.2 Interrupt-Driven Programming

Interrupt-driven programming is essential for low-power firmware. Instead of continuously polling sensors or peripherals—a wasteful method—interrupts allow the system to sleep and only wake when specific events occur.

Advantages of interrupt-driven firmware in medical IoT:

- **Event responsiveness:** Sensors such as pulse oximeters or fall detectors can trigger interrupts when predefined thresholds are crossed, allowing immediate processing without constant CPU engagement.
- **Resource conservation:** Firmware can disable CPU-intensive routines until needed, helping preserve battery life and reduce thermal output.
- **Predictable timing:** Using timer interrupts enables precise scheduling of tasks like data acquisition or wireless communication, making the system more efficient and reliable.

By combining interrupt-driven programming with sleep modes, designers can build systems that remain dormant for most of their lifecycle, waking only when medical relevance demands attention.

4.3 Real-Time Operating Systems (RTOS) for Energy-Aware Scheduling

For complex medical IoT devices involving multiple concurrent tasks—such as sensor sampling, wireless communication, encryption, and user interface management—a Real-Time Operating System (RTOS) becomes essential. An RTOS facilitates task scheduling, synchronization, and resource sharing in a deterministic manner.

In the context of low-power design, an energy-aware RTOS contributes to efficiency in several ways:

- **Idle Task Management:** The RTOS typically includes an idle task that can be configured to enter the processor into its lowest power state when no other tasks are active.
- **Tickless Operation:** Traditional RTOS designs use periodic system ticks, which cause frequent wakeups. Tickless RTOS configurations disable these periodic interrupts during inactivity, enabling deeper sleep.
- **Dynamic Task Prioritization:** Tasks with higher energy demands can be deferred or scheduled less frequently based on battery state or system health, providing a form of adaptive power management.

- **Peripheral Control Abstraction:** Through drivers and APIs, RTOSs can centralize the power control of connected sensors and radios, ensuring they are powered only when required.

Popular RTOSs used in medical IoT, such as FreeRTOS, Zephyr, or Mbed OS, provide built-in support for these power management features, allowing developers to fine-tune performance versus power trade-offs based on real-world usage profiles.

4.4 Firmware Over-the-Air (FOTA) with Eco-Friendly Protocols

Firmware Over-the-Air (FOTA) updates are crucial for maintaining security, regulatory compliance, and feature upgrades in medical IoT systems. However, they introduce energy and bandwidth overheads that can undermine sustainability goals if not carefully managed.

Eco-friendly FOTA design principles include:

- **Differential Updates (Delta FOTA):** Transmitting only the changed sections of the firmware significantly reduces data size. Tools like bsdiff or custom binary diff engines can generate compact patches.
- **Compression and Encryption Efficiency:** Using lightweight, energy-efficient algorithms for compression and secure transmission helps minimize CPU and transmission energy. Lightweight cryptographic suites such as ChaCha20 or AES-CCM are preferred.
- **Smart Scheduling:** FOTA tasks can be delayed or scheduled during low-activity periods (e.g., nighttime) to avoid interrupting critical monitoring functions or draining battery at peak usage.
- **Multicast and Mesh Update Strategies:** In hospital networks with multiple devices, updates can be distributed across a mesh or multicast network, minimizing redundant transmissions and reducing server load.

For recovery purposes during firmware updates it should also offer rollback mechanisms and fail safe routines. These features are of great importance not only for the reliability of the device, but for reducing e-waste, inflicted by corrupted, failed update attempts. Designing Low-Power Firmware is the principle on which sustainability for the medical IoT ecosystems is based. Unlike general purpose systems, these embedded platforms are constrained with stinx around energy, memory, and safety. The effect of each firmwares' design related decision such as the utilization of power modes on the ecological footprint or the operational life span of the device can have a single large effect. With the help of such power saving techniques as sleep states, interrupt driven logic, energy aware RTOS scheduling and FOTA protocols, we can build effective medical IoT systems that will also be environmentally responsible. Experts in the global medical industry are heading towards greener approaches to medical care and as such the intelligent firmware will help in making sure that medical practices are not only in priority of patient health but also of the planet's(Greenbookpress).

Table 2. Low-Power firmware techniques and their impact on medical IoT device sustainability

Technique	Description	Benefits	Typical Use Cases
Sleep/ Deep Sleep Modes	Disables non-essential components to conserve energy during idle periods	Significantly reduces power consumption	Wearable monitors, implantable sensors
Interrupt-Driven Programming	Triggers firmware response only when specific events occur	Minimizes CPU usage and avoids wasteful polling	Fall detection, ECG threshold alerts
Dynamic Voltage and Frequency Scaling (DVFS)	Adjusts processor voltage and frequency based on performance needs	Balances computation speed with energy efficiency	Adaptive heart rate analysis, real-time diagnostics
Tickless RTOS Scheduling	Eliminates periodic system ticks during idle times	Allows deeper sleep states and lower standby energy use	Background monitoring devices
Peripheral Duty Cycling	Powers peripherals only when needed	Reduces energy from unused hardware modules	Intermittent glucose or temperature sensing
Differential FOTA Updates	Transmits only the changed firmware segments	Lowers data size, transmission energy, and update risk	Firmware maintenance in wearable or bedside devices

Table 2 illustrates the range of low-power firmware strategies that play a critical role in extending the operational life of medical IoT devices while minimizing their environmental impact. These techniques—ranging from basic sleep mode utilization to more sophisticated approaches like dynamic voltage scaling and differential firmware updates—offer multiple pathways for conserving energy without compromising clinical responsiveness. Each method addresses specific power-saving goals, such as reducing CPU cycles, minimizing peripheral usage, or optimizing update procedures, thereby supporting sustainability in both routine monitoring and emergency interventions. By integrating such techniques, developers can ensure that firmware not only meets functional requirements but also contributes meaningfully to greener digital health solutions.

5. SUSTAINABLE SOFTWARE ARCHITECTURES FOR MEDICAL IoT

In the changing environment of connected healthcare, software architecture is of great importance in defining the sustainability and efficiency of medical IoT systems. Although the hardware constraints: battery life, footprint, and computational ability, are largely appreciated, the software running on these pieces of equipment and supporting the environment around them are equally important. Medicinal IoTs require sustainable software architectures that have low energy consumption, data flow for an optimal user experience, support for device scalability and operational efficiency without compromising safety or performance (Dautov, Song, & Ferry, 2021).

In this section, we focus on the principles and implementation of sustainable architectures in medical IoT, referring to modularity, architectural styles, data management, and resource optimization.

5.1 Modular and Scalable Software Architectures

Modular design is foundational to building sustainable and maintainable software for medical IoT. A modular architecture decomposes a system into smaller, independent, and reusable components (or modules), each handling a specific function such as sensing, processing, logging, communication, or display. This decomposition enables several sustainability benefits:

- **Reusability and Maintainability:** Modules can be updated, tested, or replaced without affecting the entire system. This reduces the need for complete firmware rewrites and facilitates targeted, lightweight updates—lowering both bandwidth and power consumption during maintenance.
- **Separation of Concerns:** Each module performs a clearly defined task, which improves software clarity and makes energy optimization within individual modules more manageable.
- **Hardware Abstraction:** Modular software can support varying hardware configurations with minimal changes. This is essential for scalability in distributed healthcare systems where devices may range from simple wearables to complex diagnostic stations.
- **Power Management Hooks:** Modules can be designed to enter sleep states independently or adapt their behavior based on system-wide energy policies, enhancing overall power efficiency.

For example, in a wearable glucose monitoring device, modular architecture would separate sensor acquisition, data smoothing, user alerting, and wireless transmission. If the patient is sleeping, the alert module can be temporarily disabled while others remain active at lower duty cycles.

5.2 Microservices vs. Monolithic Approaches

In cloud-connected medical IoT systems, the choice between monolithic and microservices architectures has profound implications on sustainability and system resilience.

Monolithic Architecture:

Monolithic software is designed as a single, unified application where components are tightly interwoven. While this may be appropriate for simple embedded devices with limited functions, it presents challenges in scalability, maintenance, and sustainability:

- Any update requires redeploying the entire system.
- Minor changes can lead to extensive power-hungry updates.
- Debugging or isolating performance issues is difficult and often requires more CPU usage, increasing energy consumption.

Microservices Architecture:

Microservices offer a distributed and decoupled alternative, wherein each service performs a single, well-defined function and communicates with others via lightweight protocols (e.g., REST or MQTT).

- **Energy-efficient Scaling:** Services can scale independently based on need, reducing overprovisioning of resources.
- **Reduced Bandwidth:** Services responsible for data preprocessing can filter or summarize data before transmission, lowering the energy cost of network activity.
- **Fault Tolerance:** When one service fails or needs an update, others continue to function normally, avoiding system downtime and repeated resource-intensive restarts.
- **Sustainable Deployment:** Updates to individual microservices are lighter and faster, conserving both backend server energy and device-side synchronization processes.

However, microservices also introduce overhead in communication and orchestration. Thus, for low-power edge devices, a hybrid approach may be most effective:

lightweight modular monoliths on the device and microservices in the cloud or edge gateway.

5.3 Data Compression and Local Preprocessing to Reduce Transmissions

Data transmission is among the most power-intensive operations in any medical IoT device. Therefore, minimizing the size and frequency of data sent wirelessly is essential for both energy efficiency and network sustainability.

Local Preprocessing:

By analyzing and filtering raw data at the edge, devices can transmit only what is necessary. For instance:

- A heart rate sensor may only transmit data when abnormal patterns are detected.
- Instead of streaming raw ECG signals, a wearable monitor can compute and send summarized metrics like heart rate variability or detected arrhythmias.

This not only saves energy but also protects patient privacy by minimizing raw data exposure.

Data Compression:

Lossless or near-lossless compression techniques, such as Run-Length Encoding (RLE), Huffman coding, or domain-specific algorithms like SPIHT (for biomedical signals), can dramatically reduce data size before transmission.

- **Compression libraries:** Lightweight libraries optimized for embedded systems (e.g., Heatshrink, FastLZ) balance compression ratio with computational overhead.
- **Context-aware compression:** When combined with preprocessing, compression can be customized to the nature of the medical data—for example, prioritizing precision in critical metrics while simplifying background signals.

Moreover, combining local preprocessing with periodic bulk transmission (instead of continuous streaming) creates a sustainable balance between data fidelity and power consumption.

5.4 Efficient Memory and Resource Management Techniques

Effective memory and resource management is crucial for both energy efficiency and system reliability. Medical IoT devices often operate in memory-constrained environments where inefficient allocation or memory leaks can degrade performance and increase power draw.

Static vs. Dynamic Memory Allocation:

- **Static Allocation:** Preferred for predictability and lower overhead, especially in time-sensitive and power-constrained systems.
- **Dynamic Allocation:** Used cautiously, with careful profiling and leak detection to avoid long-term inefficiencies.

Memory Optimization Strategies:

- **Buffer Reuse:** Sharing memory buffers between tasks that don't overlap in execution helps reduce memory footprint.
- **Garbage Collection Tuning:** In managed environments (e.g., MicroPython), optimizing garbage collection cycles ensures smoother operation and fewer unexpected spikes in CPU and power usage.
- **Memory-aware Scheduling:** Tasks that require more RAM can be deferred or executed when other components are idle, preventing concurrent high-memory usage that could trigger swapping or crashes.

Resource Management:

Beyond memory, managing peripheral and processor resources intelligently contributes significantly to sustainability:

- **Peripheral Duty Cycling:** Sensors and radios can be activated only when needed. For instance, Bluetooth Low Energy (BLE) can remain off except during specific transmission windows.
- **Shared Resource Pools:** Threads or services can use shared I/O, ADCs, or DMA channels, reducing redundant resource activation.

Moreover, real time profiling tools like Tracealyzer or SEGGER SystemView enable developers to profile code level energy consumption and optimize function-intensive memory or CPU. Sustainable architecture in medical IoT is a fusion of strategic, modular design and intelligent data management and economizing of resources. Modular architectures and microservices provide scalable, maintainable and energy efficient solutions and are able to grow with changing user needs and technology. Meanwhile, reduced wireless transmissions due to local data processing and compression produce an appreciable reduction in carbon and energy footprint of such devices. Effective memory and resource management means that medical IoT systems are reliable, safe and sustainable throughout the long term operating cycles, particularly in mission-critical applications like remote patient monitoring, diagnostics, and emergency response. With sustainability moving up as a non-negotiable objective in health care technology, architects of health care technology need to integrate these principles to create solutions that safeguard not only human life, but also the preservation of our environment.

6. SECURITY AND PRIVACY IN GREEN ARCHITECTURES

The impact of a teacher is primarily witnessed in the accolades earned or designations bestowed on them; the silent moments reverberating across time, in the resulting waves caused by words spoken and values imparted with mute insistence. When I think about Kuppani Sir's legacy, I am getting more and more convinced that how this influence is not something that can be absolutely quantified. It lives through generations of students who consumed more than learning physics. He educated our spirits, provided clarity when our bodies and minds were overtaken by self-doubt, and on many occasions instructed us in wisdom without having much to say on the subject. Even now as I stand at a classroom anytime, a part of me still has the weary eyed student in his physics lab and as he picks up a piece of chalk with purpose, as if it was a sacred ritual. I aspire to find his patience, his stillness, his profound respect for the learner. It is now clear to me how much he had respected us, not because of our accomplishments, but because of the fact that we had the capability of understanding and growing. That maybe the first great lesson I learned from him: the art of seeing people not as who they are, but what they might turn out to become (Bhutani, Lall, & Agrawal, 2023).

Many of his most important lessons were not recorded in notebooks. They were lived. I can remember the time toward the end of my schooling where our class performed poorly on a mid-term. Disappointment we expected – perhaps a rebuke too. Instead, Kuppani Sir walked into the room with a quiet calmness as he looked at us none as to say, "I know you can do better". No words raised loud, no threats made, just a conviction that our better selves were accessible. And in that silence accountability we found. We were embarrassed not for failing (in fact we had succeeded), but for failing someone who believed in us more that we believed ourselves. As we made our way along our respective paths as students, sometimes stumbling through the travails of university life and onward, we began to realize how infrequent is a teacher as they were. His presence, calmed the chaos of our teens. Figuratively, he had tethered us loose. He gave us room to wander, to question, to fail- trusting that the essential lessons he had planted would be borne out.

Over the years I have known so many educators, mentors, colleagues all of them brilliant in their own field. However, almost no one has made me feel the same quiet reverence for Kuppani Sir that I do now. It wasn't his flamboyant style or call to charisma that marked him as unique. He was if anything a great deal unassuming. Never was his power exercised, only earned, never did he speak high, he spoke true. He had a way of you feeling that there was something like worship in learning, an understanding is like touching something sacred. Much later in life, when it was I who became teacher, it was then that I was beginning to understand, truly, the passion of his avocations. I learned to my surprise that teaching is not simply the imparting of knowledge. It is a sort of service. And Kuppani Sir served with devotion never for recognition, never highlighting his own sacrifices. He reminded me of such old sculptors, carving temples, they never would sign the work, but they delighted in the creation of something lasting.

Time went by and that has smudged some of the specifics of time spent in certain classrooms. But what it was all about—the essence of his teaching, his method, his attitude, his style—is still crystal clear. He educated us in accuracy; not in formulas only, but in the way we thought. He taught us humility, not only when we were successful, but in the presence of mystery, all we had not yet understood. Perhaps, most importantly he taught us dignity. Dignity in hard labors, failures, silences and even in scholarship. There's a specific type of nostalgia that causes me to be nostalgic when I pass by my old school building. Corridors have been changed, the walls repainted but I can still hear the sound of his footsteps. Imagine him coming back into the lab again, liners his spectacles, peering about the room with those always curious eyes. And for a moment, I am sixteen, expecting to hear the world unfold in his voice.

It's humbling to see why so much about ourselves is determined by who we are to those that decide to trust us. That belief emanating from such a stalwart person like Kuppani Sir becomes a compass. I have even wondered what he would say when I've been faced with the dilemmas of life, academic-economic, personal, ethical. Sometimes I don't even know what that answer is but the whole act of wondering brings me back onto a more stable track. He educated me about better ways of asking questions, and at time, this is more important than having ready answer. Some of them turned out to be scientists, as well as teachers, engineers, artists, even civil servants. He never differentiated among us about where we ended up. To him, destination was never measured against the integrity of the journey. He did not view education as running a race to the finish line, but offered a dream of lifelong pursuit. In all sorts of ways, he had redefined success, not in terms of money but in moments of comprehension, in courage to think independently, in the ability to live right. Now I do not feel strange to get messages from former students, and sometimes in those I find the same resonances as I dreamt about for Kuppani Sir. They thank me for the guidance, for listening, for patience they see in me. And then I understand that I'm just passing on a lineage, one that I got from him. The chalk may be held by my hand at the moment, but the lesson voice remains often his voice. Legacy is not about being remembered; it's about having that which values will endure. And he lives on, according to that measure. In each question in which there is genuineness, in every student inspired to reason beyond, in every teacher opting for compassion rather than domination. That is his true curriculum. That is he wrote with his life that is the book.

Table 3. Edge vs. Cloud Processing in Medical IoT: Trade-offs and Use Cases

Criteria	Edge Processing	Cloud Processing
Latency	Ultra-low; ideal for real-time responses	Higher due to network dependency; may not suit time-critical scenarios
Energy Consumption	Lower transmission energy; moderate computation load	High transmission energy; centralized but optimized computing at scale
Bandwidth Requirements	Minimal; only essential or preprocessed data is transmitted	High; continuous or large data streams may be required
Computational Capability	Limited to device or gateway resources	Virtually unlimited; suitable for complex AI and historical data analytics
Privacy and Security	High; sensitive data can be processed locally	Risk of data exposure during transmission; requires strong encryption
Update and Maintenance	Requires localized model/firmware management	Easier centralized update and model deployment
Typical Use Cases	Immediate patient alerts, anomaly detection, local diagnostics	Predictive modeling, population health analysis, long-term storage

According to Table 3 above, the decision between edge versus cloud processing to use in medical IoT systems comes at a careful cost-benefit trade-off among such factors as latency, energy efficiency, computational capability, and privacy. Edge processing allows for instant, power-efficient reaction typical for applications with a time constraint, such as patient alerts or in-device diagnostics. On the other hand, cloud-based processing provides strong computing resources that are appropriate for complex analytics and long term data management, but with an increased energy and bandwidth cost. The table demonstrates how each method performs unique but complimentary functions and the necessity for hybrid architecture that rationalizes between functions based on context, urgency and the constraints of available resources in sustainable healthcare settings.

7. INTEGRATION WITH EDGE AND CLOUD INFRASTRUCTURE

In the evolution of medical IoT, the convergence of edge and cloud computing is a revolutionary leap – not only a quantum improvement in terms of technical ability, but also in sustainability. Modern healthcare requires timely decision-making that also supports systemic acceptance of decision-making without the compromise of patient privacy, energy efficiency, or system scalability. Medical IoT devices, by their very nature, are bandwidth–constrained devices and are deployed in a set of deployment challenges which include latency, power, and bandwidth constraints. The combination of edge and cloud infrastructures aims to overcome these limitations by smart spreading of computational workload. The following thrice interrelated themes are examined here: the equilibrium of edge computation and cloud offloading, the energy association with data spreading and remote analysis, and the pivotal cooperation between firmware and edge AI modules (Bhushan, Kumar, & Agarwal, 2023).

7.1 Balancing Edge Computation and Cloud Offloading

The distribution of computational tasks between edge and cloud resources has become the cornerstone question in medical IoT system design for sustainability. Edge computing is a form of computing in which data is processed on, or close to the device, for example in microcontrollers, gateways or edge servers. By contrast, cloud offloading entails an offloading of data to various remote servers for centralized analysis, storage or orchestration. There are strengths and trade-offs of every approach, naturally in the areas of energy, latency, privacy and scalability. When low latency and real time response are important, medical IoT devices rely on edge computing. For example, the wearables that monitor cardiac rhythms must be able to identify and address them immediately. Submitting such data to the cloud for computing would unduly delay operations. Analogously, edge-based anomaly detection in infusion pumps or ventilators allows timely intervention that is independent of network availability. This responsiveness improves both the safety of the patient and the cost of energy incurred in frequent data transmission.

Cloud offloading however is exceptional in terms of high computational capacity or longitudinal analysis. Through centralized platforms, deep learning models of data can be used on aggregated patient data leading to population level insights, trend prediction, and remote diagnosis. However, ongoing flow of data from hundreds of devices to the cloud proves energetically expensive and at times, pointless. Therefore, the best architecture is a hybrid model: undertaking primary, low-complexity analytics along the edge and selectively offloading enriched or anomalous data to cloud. Dynamic flexibility is required in the balancing act. The online-offline decision regarding executing a task depends on such factors as device battery condition, availability of a network, urgency of analysis, and regulatory limitations. Intelligent orchestration frameworks given energy policies and context-awareness can automate this decision. For instance, when a wearable glucose monitor handles regular readings locally then invokes cloud based AI assistance in certifying erratic glucose levels. The design of such are adaptive models is crucial to performance and sustainability of connected healthcare systems.

7.2 Energy Implications of Data Transmission and Remote Processing

Energy efficiency is among the key objective pillars achieved through the distribution of medical IoT devices even when using them in battery operated or constrained environment. Although cloud computing provides high powered processing abilities, the process of sending the data back to the cloud especially through wireless or cellular networks is expensive in terms of energy usage. Furthermore, the combined energy footprint of server-side processing, data center storage and end-to-end encryption contribute to the environmental cost. Consistent with prior studies, wireless data transmission has been seen to require more energy per operation than local computation. Indeed, its reduction in terms of frequency, volume, and redundancy is critical. Duty cycling, payload compression, threshold-based transmission, as well as event-driven logging, are commonly adopted to balance the effects of higher power consumption. For instance, instead of streaming continuous data to the wearable device using a pulse oximeter, the device can report if saturation levels stray beyond stated safe levels. Based on this event-based transmission, it will reduce the bandwidth usage, and save the battery life (Aledhari, Razzak, & Qolomany, 2022).

In addition, energy-aware routing protocols in mesh or opportunistic networks can determine optimal transmission paths with reference to energy profiles thereby preventing overload on the nodes with low battery reserves. Ignoring resource conflicts at backend are other sustainable cloud practices such as server virtualisation, carbon aware scheduling and green data centres, which subsequently reduce energy demands. But to remain viable, the context of energy ecosystem from the standpoint of sensor to wake up events to server side inference must also be at the core of the edge-cloud paradigm. Remote processing in cloud is most sustainable if it is used to perform high value activities like retraining of AI models or compliance audit or long term epidemiology studies. Routines which have no value on scale—signal smoothing, noise filtering, simple threshold comparators—should be sent to the edge. This division avoids unnecessary data transfer and maintains a high energy cloud operation for high impact computations. The new field of federated learning, in which models are trained on locally-held data via the device rather than data transferred and is only model updates shared, illustrates this principle. Through reducing data flow in raw form, and exploiting distributed intelligence, federated approaches represent both privacy preservation and cost reduction.

Figure 2. Edge-Cloud paradigm and AI integration in sustainable medical IoT

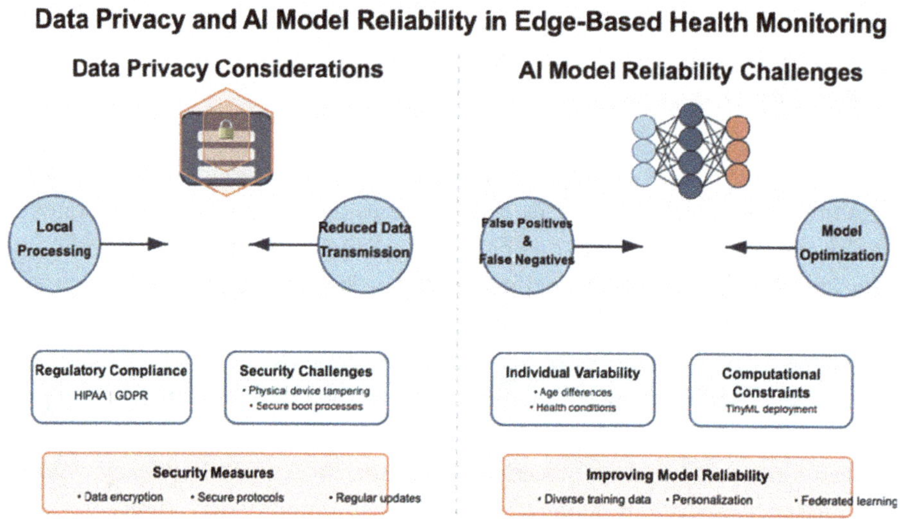

Figure 1 shows the strategic trade-off between edge and cloud in medical IoT system in terms of bidirectional relation of these computational domains. On the left is the presentation of edge components, which includes firmware layers that offer control over hardware and power management and Edge AI modules, incorporating TinyML models for local inference and anomaly detection. On the right, the cloud capabilities for deep learning, data analytics, and long term storage are highlighted. The central data flow section stresses the energy impact of data transmission choices where filtered data is directed to the cloud whereas model updates and configurations trickle to edge devices. The bottom framework sets out the five vital decision parameters which will steer the sustainable computation distribution: data criticality, energy status, network quality, computation needs and privacy considerations- it is all these that ultimately determine whether processing should occur at the edge or in the clouds (Dodig & Cafuta, 2021).

As can be seen from Figure 2, for medical IoT systems to be sustainable, the allocation of computational workloads between edge and cloud is based on intelligence, where firmware-AI collaboration is the pillar of this architecture. Bidirectional messaging paths from these domains where only the filtered high value data flows upwards and the optimized models and configurations flow downwards represent the data centric approach to saving energy. Especially notable is the combination between firmware layers and Edge AI elements such that directives regarding power management and the needs of inference have to be aligned through good defined APIs. Such synergy allows adaptive behavior to this change of condition that includes battery levels, clinical urgency, and network availability. The bottom of the figure shows the sustainability decision framework that covers the context-aware reasoning needed for maximum optimization of task allocation while reserving energy consuming transmission and off-site processing for those that can attribute the value of cloud capabilities. Such dynamic collaboration between the edge and cloud, informed by five decision factors, is the next frontier for sustainable medical IoT design – optimizing clinical value at the lowest environmental cost.

7.3 Collaboration Between Firmware and Edge AI Modules

The convergence of firmware and edge artificial intelligence (AI) is a key enabler of sustainable intelligent medical IoT systems. Firmware is the underpinnings of everything, controlling hardware interaction, sensor reading, power states, and communication protocols. There are Edge AI modules (above the layer) that offer on device intelligence from pattern recognition, anomaly detection to predictive analytics. The dynamic between the two layers not only dictates functionality but energy efficiency too, and responsiveness. Firmware needs to be engineered in order to accommodate AI execution for effective collaboration. This involves provisioning memory for model storage, efficient use of CPU and DSPs and the ability to run inference jobs in real time. Further, firmware should make available APIs/in hooks that facilitate interference from AI modules on system behavior. For instance, if an edge AI model identifies patient motion as a sign of a fall, the firmware then must activate high priority communication routines or alert systems at once. Resilience such as this requires seamless cross-layer integration.

On the other hand, edge AI models have to be designed for firmware limitations. Conventional AI algorithms are too big resource hungry for embedded devices. Adopting TinyML—machine learning models that run at low power hardware—has had a revolutionary impact. These models, which are usually pruned, quantized and knowledge distilled, fit into kilobytes memory and run within milliseconds. Firmware needs to be able to run these models, including loading, running and fallback with little overhead. There is also a frontier in terms of AI execution that is power-aware. Firmware can monitor energy reserves and adapt AI behavior to these. In high power states, the system may perform full inference models but in low power states it will switch to heuristics or simply sample less. Only this dynamic tuning calls for a bidirectional communication between the firmware's power management units and the AI scheduler. In addition, runtime profiling will reveal performance bottlenecks: firmware will be able to optimize the use of peripherals or processor duty cycles to avoid system's balance disruption (Mallo, 2024).

Update mechanisms also feature collaboration. Firmware itself should support safe, effective over-the-air updating of AI models, with a reset option to ensure safety. Furthermore, AI modules may help firmware to learn from usage pattern—say – recognizing periods of sensor idleness and offering default sleep intervals. Through these collaborative feedback loops, the system eventually becomes evermore efficient and resilient over time. Finally, the integration of edge AI into medical IoT depends on this close interoperation. Firmware is the scaffolding, the scaffolding of protections themselves. AI brings adaptability and foresight.

8. CONCLUSION

The quest for sustainable healthcare technology requires us to radically change the way we plan, build, and implement medical IoT systems. Although hardware innovations have tended to dominate the discourse, this chapter has brought to the fore the critical contributions of firmware and software in determining the ecological footprint of connected medical devices. From energy -producing coding practices and low -power firmware techniques to ability architecture and wise cooperation of edge-cloud, a holistic approach has to be pursued such that software should not be considered just as an auxiliary layer but as a strategic driver of sustainability. The concepts of green software engineering, low-power firmware design, are not solely about battery life extension or transmission cycles reduction —it is about rethinking computational responsibility through the entire device life cycle. As pointed out, presence of modularity, adaptive scheduling, intelligent power management and localized processing are key components of a sustainable digital infrastructure. In addition, Through the integration of firmware and edge AI collaboration, systems can develop into self-regulating systems that are able to regulate themselves to sustain performance for conserving energy.

Sustainability, in this aspect, is multi-intensive. it includes environmental conservation, operational efficiency, long term maintainability. The price of balancing energy, performance, and security must be consciously faced, especially in life-critical environments such as the healthcare world. This chapter has provided a roadmap of best practices and emerging paradigms to reinforce the notion that digital layers in medical devices are critical to sustainability as much as their physical diversity. In the long term therefore, the creation of a greener future in digital healthcare will require constant collaboration between the engineers, the clinicians, the regulators and the researchers. Firmware and software are no longer sidecarts—they are strategic cogs in lowering the carbon footprint of tomorrow's medical technology.

9. FUTURE DIRECTIONS

The developing medical IoT environment is the ideal testing ground for future studies and advancements in sustainable firmware and software creation. With the proliferation and sophistication of connected health devices comes the increased necessity to create intelligent systems that are also environmentally friendly. One of the most optimistic directions is the further development of context-aware and self-optimizing software systems. These systems could be free to change their behavior according to patient activity, ambient environment, and device state – optimally managing energy expenditure versus clinical relevance in the here-and-now. Another option is the progress of joint edge-AI frameworks over individual devices with the purpose of utilizing collective intelligence in distributed networks. Such systems would make it possible for devices to share inference models, offload tasks to peers, and even co-ordinate updates intelligently to minimize the unnecessary computations and communications involved. Combined with federated learning, this can allow secure low-power personalization of medical diagnostics without central data storage.

Moreover, a need for standardization of green metrics for the software and firmware in the healthcare domain is very urgent. Although, there are clear benchmarks of energy consumption in terms of hardware, this is not true for software layers. Creation of universal metrics of energy-aware software behavior may lead to improved comparison, compliance, and optimization between the ecosystems of devices. In the firmware configuration, future developments may include ultra-low-power microkernel operating system, bio-signal specific compression algorithms, and the new paradigms of Asynchronous communication that will further reduce idle power draw. There is scope to incorporate sustainability audits part of the regulatory approval processes so that eco-conscious design becomes mandatory when compared with optional. As digital health technologies continue to grow, it is not a practical decision, but rather an inevitable development to embed sustainability into their digital core. The future of medical IoT needs to be not only connected and intelligent, but conscious and sustainable at every layer of its architecture.

REFERENCES

Albreem, M. A., Sheikh, A. M., Alsharif, M. H., Jusoh, M., & Yasin, M. N. M. (2021). Green Internet of Things (GIoT): Applications, practices, awareness, and challenges. *IEEE Access : Practical Innovations, Open Solutions*, *9*, 38833–38858. DOI: 10.1109/ACCESS.2021.3061697

Aledhari, M., Razzak, R., Qolomany, B., Al-Fuqaha, A., & Saeed, F. (2022). Biomedical IoT: Enabling technologies, architectural elements, challenges, and future directions. *IEEE Access : Practical Innovations, Open Solutions*, *10*, 31306–31339. DOI: 10.1109/ACCESS.2022.3159235 PMID: 35441062

Bhushan, B., Kumar, A., Agarwal, A. K., Kumar, A., Bhattacharya, P., & Kumar, A. (2023). Towards a secure and sustainable internet of medical things (iomt): Requirements, design challenges, security techniques, and future trends. *Sustainability (Basel)*, *15*(7), 6177. DOI: 10.3390/su15076177

Bhutani, M., Lall, B., & Agrawal, M. (2022). Optical wireless communications: Research challenges for MAC layer. *IEEE Access : Practical Innovations, Open Solutions*, *10*, 126969–126989. DOI: 10.1109/ACCESS.2022.3225913

Bhutani, M., Lall, B., & Agrawal, M. (2023). A novel energy-efficient adaptive superframe structure for OWC-based real-time bio-sensor networks. *Peer-to-Peer Networking and Applications*, *16*(6), 3021–3031. DOI: 10.1007/s12083-023-01544-3

Bhutani, M., Lall, B., & Agrawal, M. (2023). An efficient and adaptive superframe structure for IEEE 802.15.7-based real-time sensor networks.

Bianco, G. M. (2024). *Firmware Development and Certification for IoT Devices* (Doctoral dissertation, Politecnico di Torino).

Dautov, R., Song, H., & Ferry, N. (2021, September). Towards a sustainable IoT with last-mile software deployment. In *2021 IEEE Symposium on Computers and Communications (ISCC)* (pp. 1-6). IEEE. DOI: 10.1109/ISCC53001.2021.9631250

Dodig, I., Cafuta, D., Kramberger, T., & Cesar, I. (2021). A novel software architecture solution with a focus on long-term IoT device security support. *Applied Sciences (Basel, Switzerland)*, *11*(11), 4955. DOI: 10.3390/app11114955

Islam, M. M., & Bhuiyan, Z. A. (2023). An integrated scalable framework for cloud and IoT based green healthcare system. *IEEE Access : Practical Innovations, Open Solutions*, *11*, 22266–22282. DOI: 10.1109/ACCESS.2023.3250849

Islam, M. M., Nooruddin, S., Karray, F., & Muhammad, G. (2022). Internet of things: Device capabilities, architectures, protocols, and smart applications in healthcare domain. *IEEE Internet of Things Journal, 10*(4), 3611–3641. DOI: 10.1109/JIOT.2022.3228795

Kaur, M., Singh, D., Kumar, V., Gupta, B. B., & Abd El-Latif, A. A. (2021). Secure and energy efficient-based E-health care framework for green internet of things. *IEEE Transactions on Green Communications and Networking*, *5*(3), 1223–1231. DOI: 10.1109/TGCN.2021.3081616

Kumar, M., Choudhary, C., & ElBarachi, M. (2024, December). Sustainability of Energy Saving in Healthcare IoT Devices. In *2024 IEEE International Conference on Big Data (BigData)* (pp. 6494-6501). IEEE. DOI: 10.1109/BigData62323.2024.10825806

Mallo, S. F., Abdulqader, D. M., Abdullah, R. M., Ismael, H. R., Rashid, Z. N., & Sami, T. M. G. (2024). A Review on Feasibility of Web Technology and Cloud Computing for Sustainable ES: Leveraging AI, IoT, and Security for Green Operations. *Journal of Information Technology and Informatics*, *3*(2).

Chapter 8
Developing Renewable Energy-Powered IoT Models for Sustainable Healthcare

Pavan Kumar Reddy Yellela
https://orcid.org/0009-0001-4620-2324
SimplyTurn Technologies, USA

Jeevithesh Reddy Narravula Reddy
Independent Researcher, USA

ABSTRACT

The integration of renewable energy sources with that of healthcare IoT systems provides a very powerful solution for serving and minimizing carbon emissions, enhancing energy resilience, and enhancing the feasibility of sustainable medical practice. The purpose of this chapter is to design and implement the models of IoT healthcare with renewable resources, such as the solar, wind and bioenergy. It researches in what way decentralized energy systems can power wearable sensors, diagnostic tools, remote monitoring devices and hospital infrastructure. Real world deployments, energy management techniques and the impact of energy harvesting technology in maintaining uninterrupted service are presented in the chapter. Through these models, it aligns healthcare innovation with clean energy principles to help create a clean energy and environmentally conscious healthcare product delivery.

DOI: 10.4018/979-8-3373-4134-7.ch008

1. INTRODUCTION

The healthcare industry is one of the energy intensive sector as hospitals and medical facilities are running 24x7 in order to provide critical services. Non negotiable power supply is equally a necessity when it comes to reliable high powered imaging machines to life supporting systems. However, this process of constant energy demand generates massive carbon footprint. Finally, the World Health Organization (WHO) states that healthcare puts 4.4 per cent of global carbon emissions which indicates the sector struggle to heal people and yet potentially harm the planet.

While renewable or alternative sources of energy have become increasingly common in smart hospital and clinics, much of the energy used in hospitals and clinics continues to be derived from non – renewable sources, primarily fossil fuels, which have a negative effect on the environment and the climate. Also, refrigeration, sterilization, electronic medical records, patient monitoring systems and other digital technologies increase the burden upon energy. According to reports, as digital transformation gathers pace, especially with an incorporation of Internet of Things (IoT) into healthcare and energy, therefore, demands on energy are going to grow even more.

However, as global healthcare systems are becoming digitised with ever growing IoT devices, the need for lasting energy solutions is also on the rise. From wearable health monitors, remote diagnostic tools, intelligent medical device and facility automation systems, among other IoT technologies, the possibilities appear to be highly transformative for the patient care, processes of the operations, and the decision making. However, these advancements bring along problems about energy consumption and sustainability (Karrouk, Y., Debasa, F., & Sanchez, L. M. F. 2025).

Renewable energy sources like solar, wind, and bioenergy can be integrated into the operational fabric of healthcare IoT systems to realize a robust pathway to address these concerns. Energy resilience, size and location status in green initiatives, energy generation, and reduction of carbon footprint can be achieved by moving away from traditional grid dependent energy to clean and decentralized throughout, with significantly lower carbon footprint in energy resilience compared to grid dependent healthcare facilities. As these medical services, and hence the city, remain reliable during power outages or natural disasters by integrating renewable energy, renewed medical services will ensure that both urban hospitals, as well as remote clinics, will continue to be operational.

Figure 1. Renewable energy integration in healthcare IoT

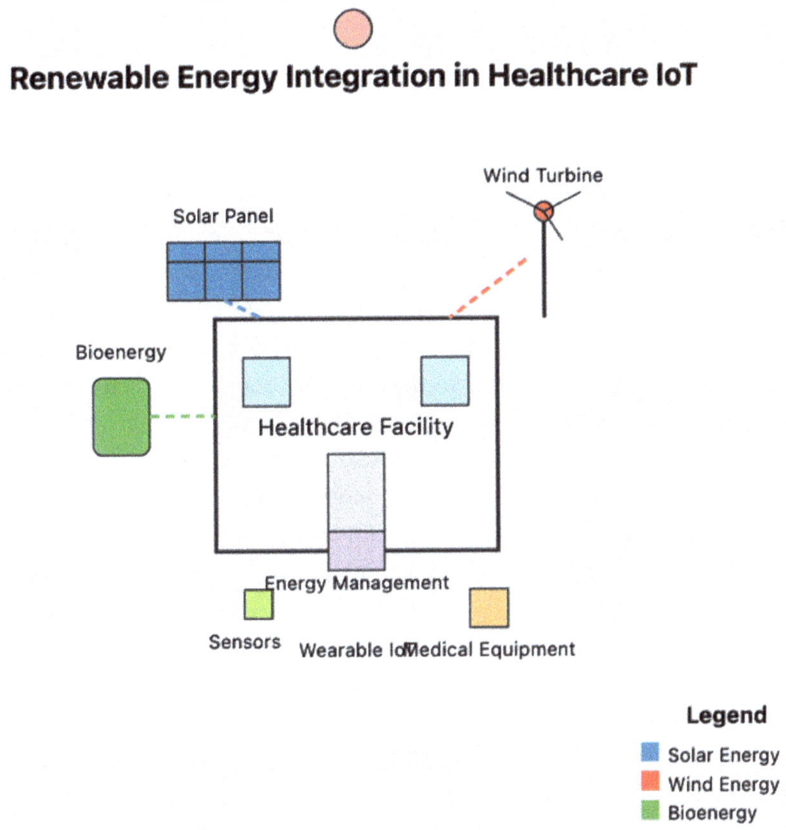

Figure 1 shows the convergence of renewable energy sources and IoT technologies leads to the implementation of sustainable ecosystem for modern healthcare delivery. Healthcare facilities can employ Solar, Wind and Bioenergy resources to power from continuous patient monitoring devices to critical medical equipment as well as decrease their carbon footprint. It addresses the two challenges of uninterrupted power for critical life services while developing the environmental sustainability. A renewable energy framework illustrated here offers resilience to grid failures and energy supply disruptions important for the health care facility that can not afford hours of downtime in its critical operations.

The convergence of renewable energy and healthcare IoT systems presents multifaceted benefits that span environmental, economic, and clinical dimensions:

- Measureable Carbon Reduction: The direct and fastest measured impact from the use of renewables is carbon reduction. Solar, wind and bioenergy systems, which are all platforms for green healthcare infrastructure, emit very little during operation of the actual system.

- Hospital Energy Resilience: Renwable energy sources in hospitals make it less likely the hospitals are exposed to grid failures, natural disasters or energy supply disruption. By ensuring the operation of life saving equipment as well as patient monitoring devices, decentralized microgrids and off grid solutions provide continuous operation.
- Initial Investment in Renewable Infrastructure: While investment in renewable infrastructure is high, it is low in operational costs altogether across time since fossil fuel dependency is reduced and energy bills are much lower.
- Renewable Energy Operation continuity: Where grid connects are dubious or don't meet the necessities of distant or promiscuous locales, green energy can make it possible for telemedical stages, mobile clinics and rudimentary diagnostic hardware to convey applicable care at opportune expectations.
- Sustainability Alignment: Clean energy integration supports institutional sustainability goals, improves its green energy regulatory alignment, and improves the healthcare provider's public image.

Especially in continuous power demand with supply chain disruptions, such vulnerabilities, decentralised energy systems consisting of solar panels on the rooftop, wind turbines, or biomass units are suited. The systems can cut transmission losses, allow for independence from centralized grids, and often are modular and scalable to site's applicable facility size and needs. That makes decentralized power even more urgent in healthcare IoT scenarios. Power is often of paramount importance when using IoT devices, either from physically placing an IoT device in very distributed locations (e.g., on a patient, in remote health posts or mobile medical units), or in ensuring that there is enough power for data to be collected, transmitted and processed. Using renewable sources of energy ensures consistency of availability of energy to support the reliability of the health services based on the IoT (Woll & Tørresen, 2022).

In this chapter, sustainable hardware and software integration strategies are explored into plans of an IoT powered healthcare model, focusing on the renewable energy and efficiency in time use of the electronics. The chapter aims to:

- Agreeing to delve into the energy dynamics of current healthcare systems and the way IoT technologies aggravate this issue or mitigate it.
- Propose the feasibility of renewable energy sources such as solar, wind, and bioenergy in healthcare environments and analyze their viability of application.

- To highlight real world application of renewable energy as incorporated with IoT healthcare to improve care delivery and operations.
- Look at the use of piezoelectric or thermoelectric generators to harvest energy and power micro-scale IoT devices in remote or resource limited settings.
- Propose design guidelines for building the IoT platforms for healthcare that are energy efficient with power consumption, energy storage and system reliability.
- They discuss energy management techniques and smart grid systems of healthcare institutions that monitor, optimise and predict energy usage in real time.
- Examine technical, economic, regulatory and socio-cultural challenges and barriers to the adoption of wider renewable powered IoT healthcare models.

They suggest future research direction and technological innovation that will enable the evolution of sustainable and intelligent health care ecosystems.

This chapter attempts to offer a practical and comprehensive guide to the building of the next generation environmentally responsible healthcare systems in the sense that it presents a comprehensive and interdisciplinary examination. As both climate change and healthcare equity are global priorities in an age, the fusion of renewable energy and IoT medical systems is a bold and necessary step.

2. THE ROLE OF RENEWABLE ENERGY IN HEALTHCARE

Modern infrastructure's most important and energy intensive corridor is the healthcare sector. It is inescapable to need medical care 24/7, use of electricity to determine the cause, treatment, life support systems, data storage and communication. With the rise of internet of things (IoT) smart technologies to the Healthcare segment, an increasing demand for consistent and reliable energy supply is witnessed. However, this demand raises at a time when there is a worldwide call to lower carbon emissions while changing to more sustainable energy systems. Therefore, renewable energy is emerging as an alternative, and more importantly, as a necessary enabler to future ready healthcare systems that are energy efficient and environmentally sustainable. The healthcare sector has emerged as one of the fast growing sectors, which is explored in this section along with the increasing need of energy needed for its operations & the benefits of renewable energy in this domain in supporting the IoT based healthcare operations, along with the strategic importance of decentralised systems in supporting healthcare operations (Pradere, Mallet, & de La Taille, 2023).

2.1 Energy Demands in Healthcare Facilities

Of all the spaces within the healthcare facility, large hospitals and medical facilities use energy the most per square foot of operational space. It is also used for specialised medical equipment, surgical suites, laboratories and digital health infrastructure, beyond the requirements for ordinary operations such as lighting, heating, ventilation and air conditioning (HVAC). Regardless, MRI machines, X-ray scanners, infusion pumps, etc., that all require uninterrupted power and even minor outages can endanger patient safety.

Since the digitalization of healthcare is also growing, the energy load is even more compounded. With the addition of electronic health records (EHR), telemedicine platforms, and data intensive diagnosing infrastructure on AI and IoT, the backend infrastructure too has to stay connected and work in computing power. Sustainable fulfilment of these escalating energy requirements are difficult. Most conventional power grids are made up of fossil fuels and are very vulnerable to outages and supply inconsistencies along with their associated environmental costs. In many developing and rural areas, electricity is not provided reliably or at all, nor is it affordable, and this results in limits and poor quality in the provision of healthcare services.

2.2 Benefits of Renewable Energy in Healthcare Settings

The solution to many of the energy challenges faced by healthcare providers is through renewable energy. It makes the healthcare facilities less dependent on non renewable, grid based electricity by harnessing natural and inexhaustible sources like sunlight, wind and organic waste. And, there are several tangible and long term benefits gained from this transition.

The first is environmental sustainability. Renewable energy technologies make little to no greenhouse gas emissions during operation, unlike coal, oil, and gas; which naturally release high amounts of greenhouse gases into the atmosphere. Integrating renewables into healthcare infrastructure directly supports the world's efforts to combat climate change and matches the sector's core ambition to drive health and wellbeing (Elendu, Elendu, & Elendu, 2024).

Secondly, there is an economic efficiency argument. The upfront cost in installing renewable infrastructure such as solar panels or wind turbines is large, but the operational expense is not. Once installed these systems have often low maintenance requirements and they also generate energy with zero fuel cost, causing a big saving on electricity bills. This is especially helpful for budget limited public hospitals and rural health clinics.

Thirdly, renewable energy fosters energy independence and resilience. On site renewable generation facilities are better insulated from grid disruptions which are becoming more frequent because of the age of the grid or extreme weather events. Renewable powered resilient energy systems guarantee that such critical services as operating rooms, intensive care units and emergency communication systems are powered and operating under all conditions.

2.3 Types of Renewable Energy for Healthcare Applications

There are several well suited renewable energy sources for healthcare applications with their operation characteristics and contextual advantages.

The most frequently adopted renewable technology utilized in healthcare settings is solar energy. PV panels are either mounted on rooftops, building facades or mobile units which give clean and reliable energy source for powering lighting, medical devices, water purification systems, and IoT enabled monitoring equipment. Solar thermal systems can also provide heating water for facilities, for sterilization and for climate control within facilities.

While solar is more variable in output, wind energy is a good supplementary source in areas that have steady wind-driven output. Wind turbines are of small scale, can be installed on the hospital premises or integrated into hybrid systems with solar PV to guarantee continuous generation. In coastal or elevated rural areas, wind power is particularly beneficial because solar potential may be intermittent.

Another avenue of renewable is bioenergy, which comes from organic materials such as agricultural waste, food scrap, or sewage, and it is especially attractive in rural or peri-urban areas with little waste management. Organic waste disposal and production of methane for cooking, heating and electricity generation can be achieved at the same time through biogas plants. Bioenergy can supply steady and sustainable power to smaller clinics and health centers when connected to microgrids or IoT enabled energy management systems (Velmurugan, 2024).

These renewable sources can be tailored to fulfill the unique needs and local geography in healthcare facilities as singular systems or combined in hybrid systems to create the biggest efficiency and reliability.

2.4 Decentralized Energy Systems for IoT Healthcare Models

Centralization of power generation has long been seen as a characteristic advantage of the current healthcare system. The decentralized energy systems, unlike the traditional, centralized grids, generate the energy close to the point of use. The proximity is key for the number of growing IoT healthcare applications requiring always on local connectivity.

IoT healthcare models usually include a very wide variety of devices from wearable health trackers, implantable sensors, remote patient monitoring tools, to automated diagnostic kiosks. In particular, these devices must draw low, but continuously, in remote or mobile applications where grid access is intermittent or non-existent. These devices will function without external power to provide power to the whole area in the case of decentralized renewable systems including rooftop solar arrays or community microgrids.

Moreover, decentralization enhances system resilience. Localised energy system can cut off the whole grid (a process called islanding) in the case of a natural disaster or a grid failure and can run independently. As an example, in 2017, when Hurricane Maria hit Puerto Rico, Puerto Rican hospitals with solar plus battery systems were able to operate through periods of wide general blackouts on the island. High levels of autonomy are especially important for disaster prone or underserved areas that require uninterrupted delivery of healthcare in adverse conditions.

Decentralized systems also provide granular control and optimization by means of the IoT enabled smart grids. It allows energy production, consumption, and storage data to be used to increase efficiency, predict demand, and reduce waste in real time. However, these systems can proactively adjust energy consumption based on patient loads, weather, or device status for maximal energy and medical resource use throughout the day (Espinosa, 2021).

Table 1. Overview of renewable energy sources suitable for healthcare applications, highlighting their advantages, typical use cases, and specific healthcare facility applications

Renewable Energy Source	Advantages for Healthcare	Suitable Applications	Example Use Cases
Solar Energy	Low environmental impact, reliable, cost-efficient	Lighting, medical devices, water purification, IoT monitoring	Rooftop solar panels, solar thermal systems for water heating
Wind Energy	Supplementary, resilient in windy regions	Hybrid systems with solar, energy generation in coastal/ elevated areas	Small-scale turbines for healthcare facilities
Bioenergy	Waste management, low-cost power, resilient	Cooking, heating, electricity generation	Biogas plants for rural clinics, waste disposal energy systems

Favorable environmental, economic, operational benefits of integrating renewable energy in healthcare settings have encouraged its momentum. From Table 1, healthcare facilities have the potentials to utilize solar, wind and bioenergy due to unmatched advantages offered in each as shown. For example, solar power is frequently utilized for driving IoT devices and medical products because it is cost effective and eco friendly. Although wind and, to some extent, solar energy are more variable, wind energy complements solar in areas where the wind pattern is stable and bioenergy offers a sustainable energy solution to areas, especially rural areas, which have a lot of organic waste. All of these sources are essential in cutting reliance on non renewable energy while help in keeping up with the rising need for a supply of dependable, decentralized power systems within health care.

3. RENEWABLE ENERGY-POWERED IOT SYSTEMS IN HEALTHCARE

With the advancement to intelligent, interwoven networks of healthcare systems running on Internet of Things (IoT), the crucial requirement for a supply of reliable and perpetual energy also increases. In the healthcare IoT systems real time data delivery, improvement in patient outcome, reduction in operational efficiency and supporting proactive health intervention are designed. Nevertheless, the energy requirements involved in the continuous and distributed nature of these devices become very demanding particularly in the low resource setting or in the emergency situations where grid power may not be reliable or be available at all.

This challenge can be solved by an integration of renewable energy on IoT healthcare systems. Healthcare providers can deliver sustained functionality of critical devices while also harmonizing with environmental goals, by harnessing solar, wind, and bioenergy. It provides an overview of how the renewable energy can be incorporated into healthcare IoT systems, describes real application, addresses the utilization of off grid and hybrid systems, and details the synergy of using IoT and renewable energy for advanced energy management (Mashallahi, Ardalan, & Nejati, 2022).

3.1 Integrating Renewable Energy with IoT Healthcare Systems

For those integrating renewable energy into IoT healthcare system, these do call for a strategic approach that should be in tune with capacity of energy generation as much as the operational requirements of the IoT device. This integration is essentially understood at its core as having three key components: renewable energy generation, energy storage, and energy aware IoT device design.

For instance, solar photovoltaic (PV) panels can be deployed on the rooftop of a hospital, clinic or a mobile medical van to provide electricity during daytime. This energy can be stored in batteries to continue to have power while the night or during low time of sun spots happens. For instance, wind turbines can help with the performance of solar systems particularly in regions where wind patterns follow a more predictable path through the night or at a seasonal change.

Smart controllers and inverters can make the system of renewable energy be directly connected to the IoT network, which in turn creates the link between the supply and quality of power delivered to the devices. So, energy efficient IoT devices including micro controllers and communication modules consume the little energy harvested optimally. These systems embed actual power saving protocols and real-time monitoring so that these systems can automatically regulate their performance according to available energy and hence extend uptime for longer period and also help minimise disruption.

3.2 Real-World Applications of Renewable-Powered IoT Devices

There are many examples of where innovations in the field of renewable powered IoT systems are proving practical for use in actual real world healthcare scenarios. For instance, wearable sensors are also becoming more and more designed with embedded solar panels for solar power or with embedded layers that can harvest body heat and body motion as electricity. The devices monitor vital signs like heart rate and body temperature and blood oxygen levels and send the real time data to healthcare providers through cloud platforms. Continuous monitoring of chronic conditions without needing changes in batteries or recharging have been enabled in remote rural areas with little infrastructure where solar powered wearables are feasible (Bevere & Faccilongo, 2024).

Because there are no current plans for sending astronauts on robotic missions, the technology is also being adapted to function for renewable power, using diagnostic tools such as point of care testing devices and portable ultrasound machines. The concept behind this mini lab mobile unit uses solar power to provide a blood test or pregnancy screening or to perform disease surveillance in an off grid area of sub-Saharan Africa or Southeast Asia. Logistical barriers are removed and an early detection as well as early treatment of medical conditions becomes possible.

Although home healthcare setups use remote monitoring devices like glucose monitors, digital stethoscopes and blood pressure cuffs, they can be made to work on solar charged battery packs or even kinetic energy. Meeting such needs requires these devices to support aging in place models and chronic disease management by enabling patients to transmit health data directly to clinicians without leaving their homes. For patients who do not have grid access - whether because they live in places with prolonged blackouts or the remote area they live in could not get energy - ability to recharge monitoring devices off of renewable energy, is transformational.

3.3 Off-Grid and Hybrid Systems for IoT Device Reliability

In healthcare, reliability is non-negotiable. You need to maintain life critical functions working as they are always in operation; be it monitoring a fetal heart beat or providing an infusion pump which delivers a medication. Off grid and hybrid renewable energy systems have an extremely important role in this.

Off grid system is an electricity system independent of central electricity grid system and is normally found in a remote or disaster prone region. The systems are composed of renewable energy generators as localized power sources (solar panels, wind turbines), energy storage (lithium ion or flow batteries), and power management electronics. If they are designed right, they can be a one stop for an assortment of IoT devices for some period of time when poor power requirements.

On the other hand, hybrid systems incorporate renewable sources and grid power or diesel generators so the transition from power fluctuations will be seamless. In a more common hybrid set up, solar energy can be used during the day but when the night or demand peak hit, the grid or a backup generator pitches in. In such systems, IoT devices can peruse power sources dynamically without any downtime for this device therefore operating continuously. They are especially handy in large hospitals or regional health centers during the needy hours of the day (Nwobodo-Anyadiegwu, Ditend, & Lumbwe, 2022).

What the combined models offer is that both models increase the healthcare resilience so data collection, communication and remote care delivery continues under the challenging environment. Also, they help to minimize operational costs down the line and to scale out with the growth of IoT networks.

3.4 Synergistic Benefits: IoT and Renewable Energy for Energy Management

It is by better integration of IoT and renewable energy systems that new levels of real time energy monitoring and management are created. Energy generation and the battery level, the device power consumption, as the environmental variables offset the sunlight and wind speed are parameterized continuously monitored by IoT sensors. The data is processed though edge or cloud platform to allow for decisions regarding energy allocation, device scheduling and predictive fixed capital.

For instance, an IoT can provide several future weather forecasts and anticipate solar availability soon. With this prediction, it would lower the sampling rate of non critical sensors, delay non urgent device update, or move the charging time for the sake of energy usage. Through such intelligent energy management, the modules prevent overloading and waste of energy while mission critical systems are kept operational.

In addition, dashboard can be accessed to administrators to carry out remote energy audit, carbon footprint tracking and performance benchmarking using this data. The insights also enable healthcare providers to recommend system designs and apply for green certifications, or prove that they comply with environmental regulations (Arora, Rao, & Bhutani, 2020).

Figure 2. Renewable energy-powered IoT healthcare ecosystem

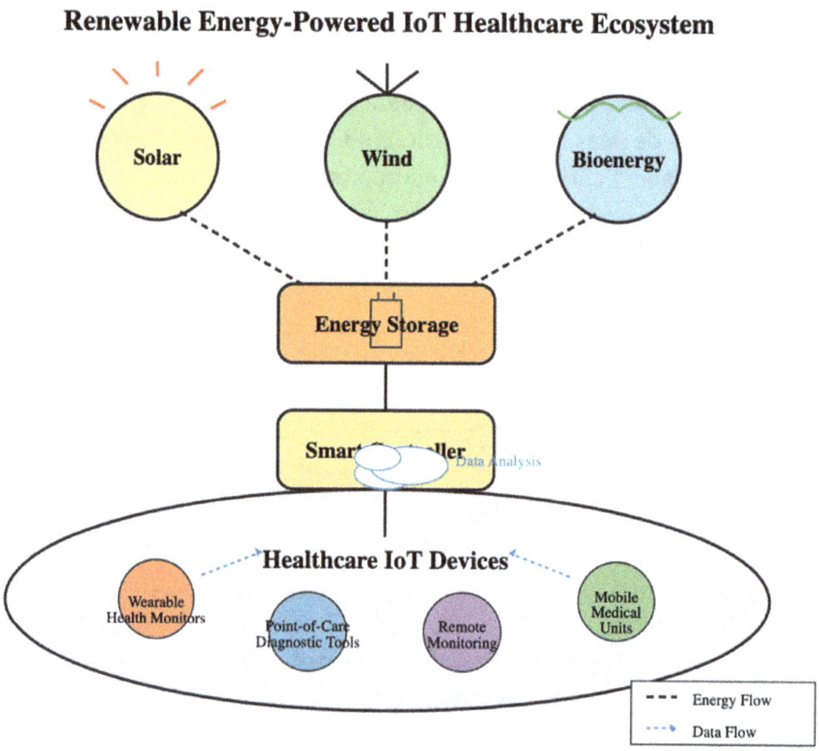

Integration of the health care devices' energy consumption with the solar, wind and bioenergy is achieved by using energy storage and smart controllers to connect them with multiple renewable energy sources (solar, wind, bioenergy) as shown in Figure 2. With bidirectional energy flow, such an ecosystem supports a wide range of applications that include wearable health monitors, point of care diagnostic tools, remote monitoring systems, mobile medical units, etc.

Figure 2 shows just how effective the combination of renewable energy with the IoT healthcare systems will be in giving us a complete ecosystem that will address the issues of sustainability and reliability in one go. This ecosystem feature a mix of diverse renewable energy sources (solar photovoltaic panels, wind turbines and bioenergy systems) and state of the art energy storage solution together with smart controllers to serve a battery of healthcare IoT applications. The figure depicts how energy is fed from renewable sources through storage systems and controllers to support critical healthcare devices while the data is transmitted between the devices and the analytics systems. Conversely, the energy management system is also capable of intelligent functionality — IoT sensors equipped on generation capacity, battery levels, and consumption of power on devices can optimize energy allocation between clinical priorities and resources available. Such integration is especially useful in remote locations or during emergency situations when the grid reliability is affected, thereby ensuring that the essential health care functions continue to function even with the smaller environmental impact (Bamakan, Malekinejad, & Ziaeian, 2022).

4. ENERGY HARVESTING TECHNOLOGIES FOR HEALTHCARE IoT

Given the healthcare digital transformation, the needs to rely on small, wireless, and autonomous Internet of Things (IoT) devices have greatly increased. While, there is no denying that uninterrupted power supply to these devices—particularly in resource constrained or remote settings—is the issue, but. The ambient energy from the environment is an energy source and energy harvesting technologies are promising by converting it into the useful electrical power. Thus, this approach not only mitigates the dependence on the conventional batteries and the power infrastructure but also improves the sustainability and reliability of the healthcare IoT systems. This section then investigates on the various aspects of energy harvesting, its applications to the healthcare IoT, the real world use cases and the innate limitations and promised bright future for these technologies.

4.1 Overview of Energy Harvesting Technologies

Energy harvesting is defined as the ability to capture and store small quantities of energy from surrounding sources like heat, light and mechanical vibrations. There are three main energy harvesting technologies that are both feasible and practical in the realm of healthcare IoT and these are the thermoelectric, the piezoelectric and finally the solar harvesting.

Harvesting of thermoelectricity uses the Seebeck effect that an electric current is generated by temperature gradients between two materials. This method can be used to harness body heat as a power source in a healthcare context. Thermoelectric generators (TEGs) implanted into smartwatches or skin patches generate continuous electricity as long as the device is worn.

In terms of mechanical stress to electricity conversion, piezoelectric harvesting is the most common. Particularly in wearable applications such as the number of steps taken by a patient or flexion of a joint, this technology is well suited. By putting piezoelectric films or crystals on insoles, knee bands, or any other respiratory monitors, power is generated during every day activities (Rosen, Adams, & Geiling, 2021).

However, solar harvesting is still one of the most mature and widely used techniques, where photovoltaic cells 'harvest' (electrical power) from sunlight (and even from ambient indoor light). Current flexible and transparent solar panels have been integrated into wearables, mobile diagnostic tools and facility level systems as a renewable and scalable energy source.

These technologies can be used independently or in combination to meet the energy demands of various healthcare IoT devices, contributing to self-powered, low-maintenance, and environmentally friendly solutions.

4.2 Supporting IoT Devices in Remote and Resource-Constrained Settings

Energy harvesting is one of healthcare IoT's biggest advantages in the sense that no location where the currently provided power source is unreliable or it is expensive or doesn't exist at all. Depending on utility reliability, harboring batteries can be very difficult in many rural and underserved regions since replacement or recharging can be difficult or impossible in many remote regions.

The barrier can be overcome by incorporating energy harvesting mechanisms into IoT devices and delivering them to patients, which thereby, avoids the need for patients to carry a power source incessantly. An example is the biosensors worn on the human body, with thermoelectric or piezoelectric modules integrated into it, which can be acted upon autonomously with little recharging needed. For long term data collection in the case of patients with chronic diseases, continuous health monitoring necessitates this autonomy.

Mobile diagnostic devices employing solar panels can be rapidly deployed to disaster stricken or conflict areas to set up temporary medical unit. Using these systems, an infection can be diagnosed, vital signs can be monitored, or data can be transmitted to remote specialists without the use of the grid. Similar to this is energy harvesting enabled telemedicine kits that can be used for liaison doctor patient consultations and health data exchange in the isolated communities (De Pretis, van Gils, & Forsberg, 2022).

Additionally, energy harvesting enables more deployments of sensor networks within healthcare facilities. With the help of sensors placed in patient beds, wheelchairs, or even hospital floors, this data can be collected operationally and health wise without all the wiring or periodic visits to care for the sensors.

Table 2. Overview of key energy harvesting technologies for healthcare IoT

Energy Harvesting Technology	Principle of Operation	Healthcare IoT Applications	Example Use Cases
Thermoelectric	Converts temperature gradients into electrical power	Wearables, skin patches, biosensors	Thermoelectric generators in smartwatches, body heat-powered ECG sensors
Piezoelectric	Converts mechanical stress/vibrations into electrical energy	Wearables, movement sensors, knee bands	Energy harvesting from patient movements, piezoelectric sensors for neuromuscular stimulators
Solar	Converts sunlight (or ambient light) into electricity via photovoltaic cells	Wearables, mobile diagnostic tools, telemedicine kits	Solar-powered pulse oximeters, solar-powered vaccine storage units

Energy harvesting technologies are now prominent and are effectively engaged in the sustainability of healthcare IoT systems with the provision of via alternative power sources especially in the resource limited environment. In Table 2, three key energy harvesting methods (thermoelectric, piezoelectric, and solar) are summarized with their advantages for different types of healthcare application. Thermoelectric harvesting captures body heat to power wearable biosensors and piezoelectric technology captures mechanical energy from patient movement in the form of powering wearable devices which is another example. Solar harvesting is very mature and scalable, thus used in a lot of the medical grade solar powered diagnostic tools. With energy harvesting systems adding these, they greatly decrease dependence on conventional batteries and grid electricity resulting in the feasibility of deploying such health care solutions in remote areas and also in improvement of efficiency and sustainability of IoT based health care models.

4.3 Case Studies of Energy Harvesting in Healthcare

There has been numerous pilot programs and research initiatives to establish energy harvesting in healthcare settings to be viable and effective.

Among others, one prominent example is the solar powered wearable pulse oximeter developed for remote maternal care in rural India. A flexible solar panel, Bluetooth connectivity and no need to recharge make for a backpack that combines solar technology with a high school graduation video game robot to help midwives monitor oxygen saturation and heart rate during prenatal visits.

In a separate example, in Japan, researchers developed an ECG sensor with a thermoelectric power that can be used to power it from the patient's body heat. As a lightweight skin conforming patch, it continuously monitors heart activity that can be fed to a smartphone app, for urban and remote healthcare systems.

In a European project (M2Neural) piezoelectric sensors were explored for using to harvest energy from patient movements into power neuromuscular stimulators. However, these devices help people who have movement disorders by delivering electrical impulses to the muscles. The system frees the system from battery pack dependance and increases user comfort by using the patient's own movements as the power source.

Solar powered vaccine storage units with advanced IoT temperature sensors are also very popular with the deployment happening across Africa and Southeast Asia. To maintain a stable cold chain and to be alerted in real time to temperatures falling outside of acceptable ranges, these units help secure efficacy of immunization programs. The energy harvesting enabled Healthcare IoT to become both accessible, efficient and sustainable (Gagneja, Lall, & Bhutani, 2024).

4.4 Limitations and Future Potential

While technology for mapping the activity pattern of a mobile device on the body looks promising, energy harvesting technologies do have limitations that need to be addressed in order to be widely adopted in healthcare. Power output is key drawback as of now. The majority of energy harvesting methods generate an amount of energy that is only too small to support power hungry devices, for example image equipment or data intensive wearables. This constraint necessitates the use of ultra-low-power electronics, energy-efficient communication protocols, and energy-aware software design.

Environmental dependency is another challenge. In well-lit environments, harvesting of solar is very good, but if in along backstrasse environment, then harvesting of solar becomes minimum. Like thermoelectric devices, for example they also require a fixed temperature gradient that may not be readily provided, and they are inapplicable to sedentary patients as they depend on user movement, which is not the case with all applications.

Integration of energy harvesting components in small as well as complex medical devices can be technically challenging from the scalability perspective. Such issues as the design of form factor, material compatibility, and patient comfort must be attended to during the design process.

However, great breakthroughs in nanotechnology, materials science and flexible electronic devices have been made that are advancing the possibilities of what energy harvesting systems can tolerate. Ambient energy sources that previously have not been considered in medical applications, such as triboelectric nanogenerators (TENG) and bioenergy harvesting (e.g., glucose powered biofuel cells), are also emerging.

Additionally, hybrid energy harvesting systems that merge several sources (such as solar and piezoelectric) into one single device have a very high possibility of existence in the future. These systems overcome the shortcomings of specific methods to make a system that is both reliable and capable in many applications (Bhutani, Lall, & Agrawal, 2023).

5. DESIGNING THE HEALTHCARE IoT MODELS FOR RENEWABLES INTEGRATION

Integration of renewable energy to the mechanisms used in the healthcare IoT models is a significant step in the direction of sustainable, reliable and energy resilient healthcare systems. This can be attributed to the fact that available digital health technologies such as wearable biosensors, smart hospital infrastructure have intensified the demand for constant power supply sources that are affordable, reliable but eco friendly. Development of scalable health care systems that scale down carbon footprints, reduce energy costs and improve service reliability in remote or underserved areas is enabled by renewable energy in conjunction with energy awareness in IoT design principles.

In this section, we explore both the key design considerations for such energy efficient IoT based models and showcase some real world examples, and lastly highlight some practical principles for hybridizing the renewable energy with the healthcare IoT devices and systems.

5.1 Key Considerations for Designing Energy-Efficient IoT Models

As interesting as it is, simply attaching a solar panel to a sensor is not enough to create energy efficient healthcare IoT systems. This calls for a comprehensive knowledge of device power needs, the amount of energy harvesting possible, energy storage, and the scaling of the system (Psillaki, Apostolopoulos, & Makris, 2023).

Device Power Requirements: Whatever IoT component that we have, whether it is a sensor, a microcontroller, a communication module, each one of them has their power demands. The failure in accurately profiling these needs will lead to an incorrect determination of suitable renewable energies sources to be selected and the design of the corresponding energy management system. Given the huge power consumption, low power design strategies, like duty cycling (periodic activation and sleep modes), energy-efficient data transmission protocols (e.g., LoRaWAN, Zigbee), and local data processing (edge computing) are important to reduce overall power.

Energy Storage: However, since renewable energy sources such as solar or wind are intermittent, efficient energy storage system is required to provide continuous device operation. Harvested energy can be buffered by batteries, supercapacitors, or hybrid storage systems and delivered to power when energy supply is low. But commonly used that are high energy density, rechargeable are lithium ion and lithium polymer batteries, newer is solid state battery and flexible storage for wearables and implantables.

System Scalability: The addition of new devices or functionalities in scalable IoT architectures must be supported without overwhelming of energy infrastructure. Easy expansion of modular design approaches, can be seen while adaptive energy allocation systems provide balanced delivery of power among the interconnected devices. On the other hand, cloud and edge integration is also used to keep system efficiency by offloading cumbersome computational tasks from local devices.

These considerations constitute the core of healthcare IoT architectures that can be and are being deployed to sustainably and resiliently power the various care facilities with renewable energy.

5.2 Examples of Renewable Energy-Powered IoT Models

Across the globe, healthcare institutions are increasingly adopting renewable energy-integrated IoT systems to enhance care delivery while promoting environmental sustainability. Smart Solar-Powered Clinics: Solar powered modular clinics with solar based IOT based health monitoring systems have made rural health care a revolution in sub-SAHARA AFRICA. The equipment that runs a patient's monitors, as well as accept called patient data through telemedicine and power systems to store the vaccine cold chain systems all run off photovoltaic panels. Solar panel output, battery health, and energy consumption are all monitored by IoT enabled sensors to use the most amount of energy while also preserving the system downtime to the absolute minimum (Goyal, Elawadhi, Sharma, Bhutani, & Jain, 2024).

Hospital Infrastructure Monitoring: Solar power enabled IoT systems are being deployed at several hospitals in India and the Middle East to monitor real time air quality, HVAC efficiency and occupancy in hospitals. The rooftop solar arrays cut into grid reliance, and in the case of smart ventilation systems that regulate airflow based on occupancy data, cut into grid consumption as well. But the fact that energy efficiency and patient comfort are both improved by renewable energy highlights how the operation could be enhanced by renewable energy in large healthcare environments.

Home Healthcare and Wearables: Solar charged wearable devices and portable diagnostic tools have started to gain uptake at the patient level. Small scale solar panels or kinetic energy harvesters are now incorporated in smart health watches, ECG patches as well as glucometers. Thus, these devices continuously monitor vitals and send data to cloud-based healthcare platforms, making them the most preferred devices for elderly monitoring and home based care.

Renewable-Powered Ambulance Systems: Solarturbines and solar panels are being installed aboard some smart ambulances as a way of powering onboard diagnostic devices and IoT communications systems. Assuredly, this renewable setup sees to it that ambulances are operational for long hours on the road or in places with no electricity to play down on the care delivery during emergences.

These examples demonstrate how renewable energy can be utilized in IoT eco systems in healthcare to enable efficiency, autonomy and environmental friendly delivery of healthcare services.

5.3 Integrating Renewable Energy and Energy Management Systems

Healthcare IoT models must contain strong energy management systems (EMS) composed of renewable sources, storage and load balancing that work seamlessly and reliably.

Real-Time Energy Monitoring: Renewable energy input, battery levels, and energy consumption across all connected IoT devices should also be tracked in a continuous way in an EMS. It facilitates proactive energy distribution and performance optimization, as well as fault detection.

Load Prioritization and Balancing: Energy delivery will have to be prioritized for critical healthcare functions, e.g., (life supporting systems, real time monitoring). EMS in turn can classify devices by priority and power outlet automatically, without oversight, assigns power that cannot interrupt mission critical services.

Predictive Analytics for Energy Availability: EMS can predict future energy generation and consumption by using the weather forecast and some usage trends. The system is also able to predict when charging cycles can occur and when non-urgent tasks can be scheduled during peak solar hours and can prevent energy shortages.

Integration with Smart Grids: Renewable powered IoT systems can converse with smart grids so that they can have energy exported or used when there is a shortage. This bidirectional capability provides resilience and adds up to nicer energy efficiency.

The embedded intelligent energy management systems in healthcare facilities allow healthcare facilities to confidently adopt and adopt renewable-integrated IoT models while continuing to provide consistent service delivery and contribute to a broader movement towards sustainability.

6. ENERGY MANAGEMENT TECHNIQUES FOR HEALTHCARE IoT

The use of renewable energy and Internet of Things (IoT) technologies is becoming a common mode in healthcare systems and managing energy consumption is an important aspect in designing and operating such systems. Healthcare IoT devices include wearable biosensors and often monitoring systems for entire hospitals which have to work reliably and efficiently while often operating with restricted or varying power supply. Thus, energy management is not just about conservation but also about the assurance of uninterrupted supply of the basic essential health services (Goyal, Sharma, Jain, Elawadhi, & Bhutani, 2023).

Smart energy management techniques can include all the use of renewable resources, minimizing operational costs and increasing energy resilience that health care facilities can incorporate. This section looks into the energy consumption management techniques applied to healthcare IoT systems, the role of IoT in energy management strategies at large, and case studies of the latter.

6.1 Techniques for Optimizing Energy Consumption in Renewable Energy-Powered IoT Systems

Right at the device and all the way up to system infrastructures, effective energy management practice in healthcare IoT begins. A number of known ways to minimize power usage can be used without compromising performance for critical health applications.

Power Management Protocols: For example, BLE, Zigbee, and LoRaWAN are wildly used protocols in healthcare IoT due to their energy efficiency. With these protocols, devices can go on and off within each other's range intermittently, and data is only sent in bursts, so power drain is reduced. Sensors and microcontrollers also stay in low power mode using custom designed sleep wake cycles until triggered for activity.

Energy-Efficient Algorithms: How and when devices are being managed, has a lot to do with software algorithms. For example, energy aware routing guarantees a data is taken by the path that would use the least amount of energy in a sensor network. In the same way, edge computing minimizes the energy consumption caused by the need to transmit large amounts of data to the cloud. Sampling rates, data transmission intervals, and computational tasks are all algorithmically adaptively changed to extend device life and system sustainability.

Dynamic Voltage and Frequency Scaling (DVFS):A processor that supports DVFS techniques allows it to adjust its voltage and frequency in real time according to the workload. For example, in the healthcare IoT this is very useful to manage energy consuming intensive computations for diagnostic or real time monitoring, keeping the energy saved to other critical period (Adibi, Rajabifard, Shojaei, & Wickramasinghe, 2024).

Energy Harvesting-Aware Scheduling: Given that IoT systems powered by solar –or any other renewable source–must operate in accordance with available energy, it becomes indispensable to minimize the use of these decentralized sources and instead, maximize the use of centralized electrical grids to power the IoT devices. Multiplexing data and allowing software updates to be done during times of peak energy generation can also be handled by intelligent scheduling algorithms. So it keeps the stored energy ready for critical operations during times of low generation.

Low-Power Hardware Design: The longer device operation on limited power is enabled by the hardware innovation of ultra low power processors and energy efficient memory. For instance, adding non volatile memory, such as data storage, means that information can be saved even if there is no energy input.

These techniques, combined together, will guarantee that the healthcare IoT are reliable and use renewable energy.

6.2 The Role of IoT in Energy Management: Smart Grids, Real-Time Monitoring, and Predictive Analytics

IoT technologies are also critical for controlling energy use at a system and facility level beyond device level management. As smart grids and intelligent building systems are being deployed in healthcare institutions, more and more data is available regarding energy use that healthcare institutions are able to track, predict and optimize in real time.

Smart Grids and Bi-Directional Energy Flow: Smart grids enabled by an IoT can dynamically control energy interactions between health care facilities and the main power grid. Solar panels and other renewable sources can supply excess power to the grid and receive credits, or can draw power when it is in short supply. With smart meters, grid connected sensors, hospitals have the flexibility to respond to real time energy price and grid demands; and even seamless bi directional energy exchange.

Real-Time Energy Monitoring: Hospitals can monitor it at granular levels using interconnected IoT sensors, down to the granular levels of devices, departments. The data available in real time let administrators know what energy intensive operations are going on, how the load distribution is being optimized and even when there is a problem, such as a power leak or equipment malfunction.

Predictive Analytics for Energy Demand Forecasting: Using them, machine learning & AI enabled platform can look into historical patterns of energy consumption with respect to the variables such as weather, patient load, and equipment usage to predict the future need of energy. According to this, healthcare providers can make pro-active changes in system operations, plan for peak energy demand periods, schedule maintenance for high energy devices to continue uninterrupted service (Vergara, Théate, Boor, & Gordon, 2025).

Automated Load Management: Rest & presented control of energy consuming systems like HVAC, lighting and sterilization units are automated using IoT platforms. These systems adjust their operation based on occupancy data and energy availability to use minimum amount of energy for optimal conditions.

The use of these capabilities allows healthcare agencies to not only follow on with energy supply and business continuity; they also provide insight into decisions that support the sustainability goals.

7. CHALLENGES IN IMPLEMENTING RENEWABLE ENERGY-POWERED IoT MODELS

Despite huge potential of integrating renewable energy with IoT technologies in healthcare, such integration comes with many challenges. Real world deployment of a range of obstacles, technical, regulatory, economic, and sociocultural, though, occurs despite the clear environmental and operational benefits. However, that can make adoption difficult, delay scalability, and negatively affect the sustainability of such systems in the long term. This helps in identifying strategies for a successful implementation and optimization in a highly varied environments of healthcare. In this section, various problems associated in deploying renewable energy powered IoT models in healthcare environment are discussed and solved through policy intervention, technological improvement and stakeholder engagement.

7.1 Technical Challenges

A collection of technical limitations at the heart of IoT based on renewable energy must be accounted for, and largely affect its feasibility and reliability in the context of healthcare applications.

Energy Storage Limitations: The limited capacity and the short lifespan of the energy storage systems like batteries and capacitors are the main technical challenge. Many renewable sources—especially solar and wind—are intermittent and unpredictable. As it pertains to critical healthcare IoT devices, variability is a must for an efficient storage solution to guarantee the continuous delivery of power. Unfortunately, most battery technologies are not as durable, do not provide high energy density, or are environmentally unsustainable, which forces users to replace batteries frequently, making the whole operation significantly more complicated.

Power Fluctuations and Reliability: Given the need for high reliability for a lot of the devices in healthcare environments, such as life supporting and diagnostic devices, we describe DCAC, a DC redux architecture, below. Renewable energy sources are naturally variable due to the fact that solar panels generate less energy on cloudy days or at night and wind turbines are also impacted by changing winds depending on geography and season. But these fluctuations can put IoT devices' stability at risk unless huge system or grid backup is used, and doing so increases the complexity and cost of the infrastructure (Wicklum, Nuique, Kelly, & Nesbitt, 2023).

Integration and Interoperability Issues: This also makes the integration of renewable energy system with IoT infrastructure even more of a challenge. Technical efforts are needed to implement seamless communication, power delivery, and data delivery across heterogeneous devices that each have different energy requirements, communication protocols, and data formats. Additionally, despite utilization of renewable components, there is also a lack of well defined interfaces and protocols implemented to link renewable components with Healthcare IoT platforms making the integration ill performing, yet in most cases, resulting in incompatibility, reduced efficiency and fragmented ecosystems.

7.2 Regulatory and Policy Challenges

However, besides technical barriers, successful deployment of renewable energy powered IoT in healthcare depends upon supportive regulatory, legal and policy frameworks that are enabling innovation without compromising safety and compliance.

Lack of Incentives and Funding Mechanisms: However, there is a serious 'lack of structured incentives like tax rebates, subsidy or grant' in many regions, especially in developing economies, for healthcare facilities seeking to adopt the Renewable Energy Systems. From a lack of financial or regulatory support, it may be challenging for institutions to find incentive to spend on potentially costly upgrades. In jurisdictions that have renewable energy incentives, healthcare applications are often excluded or, at the best, not prominent.

Regulatory Compliance and Standards: The healthcare is a regulated sector, and any new technology should pass the very high standards about patient safety, data security and operational reliability. With the integration of IoT devices and the renewable energy systems we have a new set of regulatory complexities, once again including but not limited to HIPAA compliance in the US, Electromagnetic interference relating to a more traditional device usually situated in office buildings and have additional BI-compatibility implications for wearables. Unification in the global standard of RN renewable medical IoT devices also makes cross border deployments difficult.

Policy Gaps and Bureaucratic Delays: There are even some regions where the use of outdated or archaic policies hindrances the proper use of the modern technologies. For example, though bureaucratic regulatory might mean the time necessary to install solar panels on a hospital roof or might foreclose the importation of a more advanced energy storage system. If governance of health ministry, energy ministry, and technology ministry are separated, the goals could be divergent and the initiatives more loosely coordinated.

Table 3. Key challenges in implementing renewable energy-powered IoT systems in healthcare

Challenge Type	Description	Potential Solutions
Technical Challenges	Issues with energy storage, power fluctuations, and system integration	Use of hybrid systems, improved energy storage technologies, standardized interfaces
Regulatory and Policy Challenges	Lack of incentives, compliance issues, and bureaucratic delays	Financial incentives, clear regulatory frameworks, and cross-border standards
Economic Challenges	High initial investment, ongoing maintenance costs, and uncertain ROI	Government funding, cost-benefit models, and long-term investment strategies
Social and Cultural Challenges	Resistance to new technology, digital literacy gaps, and cultural resistance	Training programs, community engagement, and transparent practices

As shown in Table 3, many key challenges exist in the deployment of renewable energy based healthcare IoT models. The solution to sensory substitution for the blind and visually impaired lies in the overcoming of several barriers, both technical, regulatory, economic and sociocultural, barriers which all must be overcome for the success of any sensory substitution program. From the perspective of technical issues like energy storage limitation, power fluctuation and coupling (system integration) hybrid systems or new in advance energy storage technologies must be addressed. There's also regulatory hurdles such as that without clear framework and financial support there's no incentive or fragmented policies that need to be in place to foster innovation. By establishing targeted funding mechanisms and long term investment strategy, economic barriers of high initial costs and uncertainty of return on investments can be overcome. Finally, the social and cultural challenges, such as resistance to change in the technological field and cultural perceptions about digital health monitoring demand a comprehensive training and engagement with stakeholders. Renewable energy enabled Internet of Things can further improve healthcare delivery in underserviced condition and with resource constrained by addressing the various challenges faced above.

8. CONCLUSION

The healthcare domain in the fusion of renewable energy and IoT technologies is a transformational approach to building such medical systems which are sustainable, resilient and environmentally oriented. IoT models propelled by renewable energy are deemed promising in terms of overcoming these challenges for healthcare facilities which will increasingly struggle with rising energy requirements, significantly enhanced operational costs and the urgent need to decrease carbon footprints. By linking together clean energy sources (such as solar, wind and bioenergy), and with smart, connected devices, clinicians can ensure continuous market, improved diagnostics and real time health management, while greatly decreasing the reliance on traditional power grids. It has been described how comprehensive the integration of this type of renewable energy sources can be and how it can be achieved through the types of renewable energy sources suitable for healthcare applications, to the design and energy management of IoT systems. For instance, it has come up with practical application like solar powered wearable sensors, remote monitoring tool and decentralized energy solutions for rural or off grid healthcare center. It also has discussed important enablers such as energy harvesting and smart grid integration for continuous operation and energy efficient operation.

However, there are technical constraints to energy storage, regulatory barriers, economic barriers as well as sociocultural barriers. These barriers must be overcome with the help of technologists, healthcare professionals, policymakers and the rest of the community. In view of this, we can say that policy reforms, training initiatives and system design that is inclusive require priorities so that widespread adoption is achieved by stakeholders. In the end, renewable energy based IoT models have the potential to transform the way the world approaches global healthcare in creating a smarter, more efficient and more equitable, sustainable one. Ranging from how you obtain your healthcare (doctors, hospitals, stylists, nurses, clinics, and others), to how you share this data (plotting it on Google maps or maps within EHRs of your doctors or hospitals), these models will become an increasingly important piece of the puzzle of what the healthcare sector of the future will look like in both urban and rural contexts.

9. FUTURE DIRECTIONS AND INNOVATIONS

The future holds great future for developing new ideas and implementation in terms of renewable energy integrated IoT in healthcare. The most promising track is the utilization of algorithms of the emerging AI energy management systems that make predictive analytics and machine learning predictions to determine the dynamic energy supply and assignment among the IoT devices. By allowing healthcare providers to operate more efficiently, these intelligent systems will minimize energy waste and helps to prepare to react to energy shortages or demand surges. Solid state batteries and hydrogen fuel cells will also have an important role in development of next generation energy storage solutions. Newer energy denser, faster charged, longer cycle life and safer performing technologies are quite suitable for critical healthcare applications. Integration into IoT enabled systems will allow their uninterrupted operation especially in emergency or off grid settings.

Minaturizing and enhancing energy harvesting technologies is another key area. New materials and design enable wearables and implantables sensors able to harvest energy from body heat, movement, or ambient light. This could facilitate uninterrupted health monitoring without the requirement of recharging or maintenance and would be particularly valuable is remote patient care and management of chronic disease. On a more extensive scale, blockchain technology can be used to safeguard data and transaction of energy in decentralized healthcare network. This would promote transparency, guarantee data integrity, and aid in the smooth exchange of energy credits in microgrid systems where hospitals or clinics may exchange or trade renewable energy resources. The last thing is that policy and infrastructure innovation is required to spur adoption. It involves setting of global standards for the use of renewable energy in the medical IoT devices, cross sectors partnerships for funding of infrastructure and including green energy goals in in the national healthcare strategies.

REFERENCES

Adibi, S., Rajabifard, A., Shojaei, D., & Wickramasinghe, N. (2024). Enhancing healthcare through sensor-enabled digital twins in smart environments: A comprehensive analysis. *Sensors (Basel), 24*(9), 2793. DOI: 10.3390/s24092793 PMID: 38732899

Arora, A., Rao, A., & Bhutani, M. (2020). A MATLAB simulation model for MAC layer of visible light communication. In *2020 7th International Conference on Signal Processing and Integrated Networks (SPIN)* (pp. 941-945). IEEE. DOI: 10.1109/SPIN48934.2020.9071254

Bamakan, S. M. H., Malekinejad, P., & Ziaeian, M. (2022). Towards blockchain-based hospital waste management systems: Applications and future trends. *Journal of Cleaner Production, 349*, 131440. DOI: 10.1016/j.jclepro.2022.131440

Bevere, D., & Faccilongo, N. (2024). Shaping the future of healthcare: Integrating ecology and digital innovation. *Sustainability (Basel), 16*(9), 3835. DOI: 10.3390/su16093835

Bhutani, M., Lall, B., & Agrawal, M. (2023). An efficient and adaptive superframe structure for IEEE 802.15.7-based real-time sensor networks.

De Pretis, F., van Gils, M., & Forsberg, M. M. (2022). A smart hospital-driven approach to precision pharmacovigilance. *Trends in Pharmacological Sciences, 43*(6), 473–481. DOI: 10.1016/j.tips.2022.03.009 PMID: 35490032

Elendu, C., Elendu, T. C., & Elendu, I. D. (2024). 5G-enabled smart hospitals: Innovations in patient care and facility management. *Medicine, 103*(20), e38239. DOI: 10.1097/MD.0000000000038239 PMID: 38758872

Gagneja, A., Lall, B., & Bhutani, M. (2024). Unveiling CM-Det: Leveraging ConvMixer architecture for advanced object detection. *International Journal of Information Technology : an Official Journal of Bharati Vidyapeeth's Institute of Computer Applications and Management, 16*(7), 4273–4278. DOI: 10.1007/s41870-024-02069-5

Goyal, R., Elawadhi, O., Sharma, A., Bhutani, M., & Jain, A. (2024). Cloud-connected central unit for traffic control: Interfacing sensing units and centralized control for efficient traffic management. *International Journal of Information Technology : an Official Journal of Bharati Vidyapeeth's Institute of Computer Applications and Management, 16*(2), 841–851. DOI: 10.1007/s41870-023-01527-w

Goyal, R., Sharma, A., Jain, A., Elawadhi, O., & Bhutani, M. (2023). Adaptive density-based traffic management system utilizing ultrasonic units and RODEM algorithm. In *2023 10th International Conference on Signal Processing and Integrated Networks (SPIN)* (pp. 571-577). IEEE. DOI: 10.1109/SPIN57001.2023.10116660

Karrouk, Y., Debasa, F., & Sanchez, L. M. F. (2025). The digital transformation of smart hospitals: Challenges and opportunities. *Utilizing AI of Medical Things for Healthcare Security and Sustainability*, 1-54.

Mashallahi, A., Ardalan, A., Nejati, A., & Ostadtaghizadeh, A. (2022). Climate adaptive hospital: A systematic review of determinants and actions. *Journal of Environmental Health Science & Engineering*, *20*(2), 983–1013. DOI: 10.1007/s40201-022-00810-5 PMID: 36406601

Nwobodo-Anyadiegwu, E. N., Ditend, M. N., & Lumbwe, A. K. (2022). The benefits and challenges of implementing smart hospital projects: A systematic review. In *2022 IEEE 28th International Conference on Engineering, Technology and Innovation (ICE/ITMC) & 31st International Association for Management of Technology (IAMOT) Joint Conference* (pp. 1-7). IEEE. DOI: 10.1109/ICE/ITMC-IAMOT55089.2022.10033238

Pradere, B., Mallet, R., de La Taille, A., Bladou, F., Prunet, D., Beurrier, S., Bardet, F., Game, X., Fournier, G., Lechevallier, E., Meria, P., Matillon, X., Polguer, T., Abid, N., De Graeve, B., Kassab, D., Mejean, A., Misrai, V., & Pinar, U. (2023). Climate-smart actions in the operating theatre for improving sustainability practices: A systematic review. *European Urology*, *83*(4), 331–342. DOI: 10.1016/j.eururo.2022.01.027 PMID: 35151515

Psillaki, M., Apostolopoulos, N., Makris, I., Liargovas, P., Apostolopoulos, S., Dimitrakopoulos, P., & Sklias, G. (2023). Hospitals' energy efficiency in the perspective of saving resources and providing quality services through technological options: A systematic literature review. *Energies*, *16*(2), 755. DOI: 10.3390/en16020755

Rosen, J. M., Adams, L. V., Geiling, J., Curtis, K. M., Mosher, R. E., Ball, P. A., Grigg, E. B., & Grigg, E. B. (2021). Telehealth's new horizon: Providing smart hospital-level care in the home. *Telemedicine Journal and e-Health*, *27*(11), 1215–1224. DOI: 10.1089/tmj.2020.0448 PMID: 33656918

Singh, Y., Kumar, R., Kabdal, S., & Upadhyay, P. (2023). Youtube video summarizer using nlp: A review. *International Journal of Performability Engineering*, *19*(12), 817.

Upadhyay, P., Agarwal, R., Dhiman, S., Sarkar, A., & Chaturvedi, S. (2024). A comprehensive survey on answer generation methods using NLP. *Natural Language Processing Journal*, *8*, 100088. DOI: 10.1016/j.nlp.2024.100088

Upadhyay, P., & Tomar, P. (2022, December). Alzheimer's disease Classification using various machine learning approaches: A Review. In 2022 4th International Conference on Advances in Computing, Communication Control and Networking (ICAC3N) (pp. 232-237). IEEE. DOI: 10.1109/ICAC3N56670.2022.10074339

Upadhyay, P., & Tomar, P. (2022, August). Alzheimer's Disease Classification Using Ensemble Methods. In International Conference on Signal Processing and Integrated Networks (pp. 3-15). Singapore: Springer Nature Singapore.

Upadhyay, P., Tomar, P., & Yadav, S. P. (2024). Comprehensive Systematic Computation on Alzheimer's Disease Classification. *Archives of Computational Methods in Engineering*, *31*(8), 4773–4804. DOI: 10.1007/s11831-024-10120-8

Upadhyay, P., Tomar, P., & Yadav, S. P. (2024). Advancements in Alzheimer's disease classification using deep learning frameworks for multimodal neuroimaging: A comprehensive review. *Computers & Electrical Engineering*, *120*, 109796. DOI: 10.1016/j.compeleceng.2024.109796

Upadhyay, P., Tomar, P., & Yadav, S. P. (2025). ADNet: Deep learning based model for Alzheimer's disease classification. *Iran Journal of Computer Science*, *8*(2), 1–13. DOI: 10.1007/s42044-025-00243-x

Velmurugan, J. S. A. (2024). A review on green human resource management in hospitals: Paving the way for sustainable development goals.

Verdejo Espinosa, Á., Lopez, J. L., Mata Mata, F., & Estevez, M. E. (2021). Application of IoT in healthcare: Keys to implementation of the sustainable development goals. *Sensors (Basel)*, *21*(7), 2330. DOI: 10.3390/s21072330 PMID: 33810606

Vergara, R., Théate, I., Boor, P., Gordon, I. O., West, J., Abdelmoula, S., Tilmant, C., Wiseman Pinto, R. G., Gaillot-Durand, L., Scott, S., Trecourt, A., & Rullier, A. (2025). Surgical pathology and sustainable development: International landscape and prospects. *Journal of Clinical Pathology*, *78*(4), 233–239. DOI: 10.1136/jcp-2024-209555 PMID: 39638426

Wicklum, S. C., Nuique, K., Kelly, M. A., Nesbitt, C. C., Zhang, J. J., & Svrcek, C. P. (2023). Greening family medicine clinic operations and clinical care: Where do we start? A scoping review of toolkits and aids. *Family Practice*, *40*(3), 473–485. DOI: 10.1093/fampra/cmad006 PMID: 36730055

Woll, A., & Tørresen, J. (2022). What is a smart hospital? A review of the literature. *Human-Automation Interaction: Manufacturing, Services and User Experience*, 145-165.

Chapter 9

Edge-AI for Energy-Efficient Real-Time Health Monitoring in IoT Medical Devices

Koppula China Busi
https://orcid.org/0009-0003-6217-484X
Tirumala Engineering College, India

Rama Krishna T. Siva
https://orcid.org/0000-0003-3351-4393
University College of Engineering Kakinada, JNTUK, India

Bhanu Prakash Battula
https://orcid.org/0000-0002-1944-9727
KKR & KSR Institute Technology and Sciences, India

P. Venkata Siva Naga Jyothi
https://orcid.org/0009-0000-4173-3348
Tirumala Engineering College, India

Kuppani Sathish
https://orcid.org/0000-0002-4651-8310
Tirumala Engineering College, India

Y. V. Narayana
Tirumala Engineering College, India

ABSTRACT

The combination of Edge Artificial Intelligence (Edge-AI) with Internet of Things (IoT) medical devices is revolutionizing real-time health monitoring as it is made possible with low-latency energy-efficient, and privacy-preserving real-time data processing at the edge of the network. In this chapter we will outline the architecture and implementation of Edge-AI systems that are tasked with the work of continuous intelligent health monitoring. It dissects the key challenges (energy bottlenecks, model optimization, and secure data management), and offers state-of-the-art hardware and software tools. With use cases such as wearable ECG monitoring and fall detection,

DOI: 10.4018/979-8-3373-4134-7.ch009

the chapter reveals how Edge-AI improves healthcare delivery and outcomes. It also calls out the future trends – like federated learning and explainable AI – that would bring a new era of proactive and personalized medicine.

1. INTRODUCTION

The Internet of Things (IoT) which emerged has dramatically changed the face of healthcare especially in the aspect of real time health monitoring. The combination of interconnected sensors, actuators and communication networks have enabled the IoT enabled medical devices to empower the continuous monitoring of physiological metrics including heart rate, glucose levels, oxygen saturation, and body temperature. These systems promise a lot in improving the outcomes for the patients through timely diagnosis, preventing, and individualized treatment approaches. However, conventional IoT health monitoring frameworks tend to use cloud-centred architectures extensively, which are inherently laden with some issues on latency, energy consumption and usage, data privacy, and dependence on the network. Edge Artificial Intelligence (Edge-AI) appears as a revolutionary introduction to these restraints by bringing data transmission, and intelligent decision making closer to the origin of the data—the "edge" of the network. In the case of medical-IoT (MIoT), the term Edge-AI defines the execution of AI algorithms at the location of medical sensors, wearable, or near edge-gateway. This paradigm shift fosters faster inference, low energy consumption and minimal dependence on a continuous need to connect to centralized servers (Sathiya & Angel, 2025, January).

Edge-AI combines the strengths of embedded systems, machine learning (ML), and edge computing to construct system to help monitor health, which are smart, adaptive, energy-aware, and privacy conscious. For instance, a wearable ECG patch with a light-weight CNN would detect arrhythmias locally and give alarms without transferring on a continuous stream of data to the cloud. This saves not only bandwidth but also battery life, which is an important requisite for wearable and implantable medical devices. Out of the cloud, healthcare systems, with all their computational promiscuity and storage space, are not a perfect blend for all situations, but particularly not for the cases of real-time analysis of high frequency physiological data. It may be harmful to the performance of critical health events in terms of the round-trip delay that data to the cloud takes to be transmitted, processed, and for actionable insights to be returned. Consider for instance the example of an epileptic patient: a few milliseconds can mean a difference between responding/preflattenor alerting.

Further, unending transfer of sensitive health data to the cloud creates a lot of privacy and security issues. Even though encryption and access control methods are employed by the cloud providers, data breaches; and unauthorized access are still possible. By contrast, edge-based processing keeps the patient's data locally thus reducing exposure to external threats, and better compliance with data protection requirements like HIPAA (Health Insurance Portability and Accountability Act) and GDPR (General Data Protection Regulation). Another concern is energy consumption. Wearable and implantable medical devices are usually powered from limited power supply including small batteries or harvesting units. The requirement to transmit enormous numbers of data continuously to remote servers draws a lot of these power reserves, and impacting on the longevity of the device and the patient's comfort. Built with energy efficient algorithms and low power hardware, the edge AI systems can intelligently share the resources by carrying out the computation at local level and transmitting only required data like abnormal reading or periodic summarizations (Izhar & Naqvi, 2023).

Timeliness and energy efficiency are two cornerstones of proper health monitoring. Real time monitoring helps to prevent and intervene as early as possible for chronic conditions such as cardiovascular diseases, diabetes and etc. At the same time, energy efficiency allows devices to operate for longer periods, especially in situations where recharging or maintenance is impractical, including remote locations or follow-up outpatient care for the elderly or in a hospital's outpatient rooms. To provide real-time performance on device constrained systems, Edge-AI uses optimization methods like model quantization, pruning, and knowledge distillation, minimizing computational load without penalizing accuracy. Higher-level signal processing and feature extraction methods allow further condensing of raw physiological signals to meaningful forms that can be processed by AI models more easily.

Figure 1. System architecture of Edge-AI for real-time health monitoring showing the flow of information from patient wearables through edge devices to cloud platforms and healthcare providers

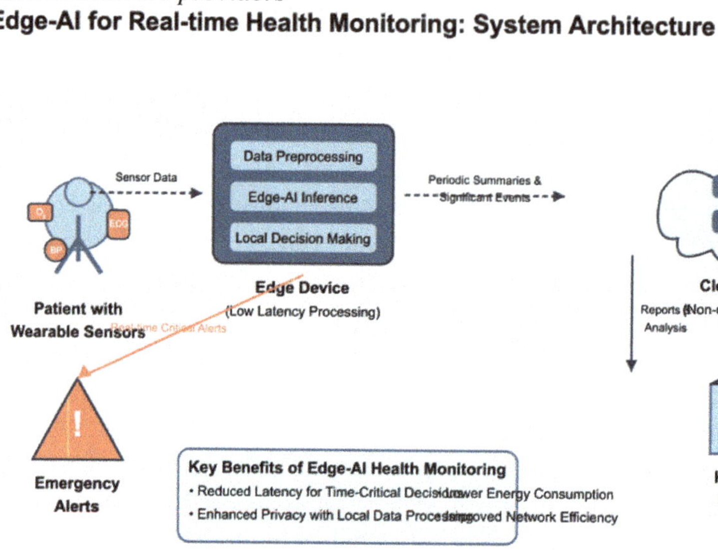

The figure below in Figure 1 depicts the multi-tiered approach of data processing and decision making framework through the Edge-AI architecture that is used for real-time health monitoring. The system starts with the acquisition of physiological parameters of patients via the wearable sensors, and these are processed locally, on the edge device that also supports AI inference. This local process also makes it possible to instantly detect key events such as arrhythmias or hypoglycemia episodes which would alarm real time without cloud dependency. Meanwhile, only summarized data and big events are sent to cloud platforms for long-term storage, in-depth analytics and integration with the systems of healthcare providers. This architecture strikes an effective balance between the opposing demands for intervention speed, energy effectiveness, privacy of data, and maintenance of a complete health record. Edge-AI makes it possible, even in situations of poor connectivity or power restrictions, to provide responsive healthcare deployment through spreading the intelligence across the system (Gagneja & Lall, 2024).

The figure visually captures the end-to-end workflow starting from data collection and ending with healthcare delivery, and in doing so, it emphasizes Edge-AI as the critical first layer of intelligence in modern health monitoring systems. It highlights the separate roles of edge and cloud components, where the time-critical processing is done locally while the non-urgent analyzed data aspects are relegated to cloud infrastructure – this will result in the perfect performance and power consumption/ privacy balance.

From an overall system perspective, real-time and energy effective health monitoring needs integrated fusion at three levels: from sensing hardware to embedded processors to communication modules to AI inference engines. Not only physically but also clinically effective and intuitive interfaces should be created through collaboration between manufacturers of devices, AI researchers, and healthcare practitioners in the development of holistic solutions. Edge-AI is not an endemic solution, but rather a critical enabler in the context of digital healthcare ecosystem. It complements cloud-based analytics, Electronic Health Records (EHR) and Telemedicine platforms as the first line of intelligence and decision making. For instance, an edge-based wearable is capable of analyzing vitals of a patient in real time and report only significant events to a cloud platform for further analytics and integrating to a medical records held by that patient (Choudhary & Vijitha, 2025).

Edge-AI also abets decentralized healthcare implementations i.e. remote patient monitoring (RPM) and home-based care that have become of much significance in the post pandemic era. Through minimization of the necessity of regular visits to the hospitals and offerings of healthcare service at the point of need, Edge-AI provides sustainable models of healthcare which are both affordable and scalable. Even in many low resource settings access to advance healthcare infrastructure is constrained and continuity of Internet connection can not be assured. Edge-AI offers a valid alternative through supporting intelligent health monitoring systems that are autonomous with a small dependence on external infrastructures. This makes quality healthcare accessible to all including in rural or underserved areas.

For instance, community health workers armed with diagnostic tools equipped with Edge-AI can perform initial screenings for, say, pneumonia or malnutrition that can be performed with portable, hand-held devices that can provide on-device analysis and decision support. The purpose of this chapter is to discuss the convergence of Edge-AI and Internet-of-Things (IoT) medical devices by concentrating on energy-efficient, real-time health monitoring. Related aspects to be discussed in the following sections include the technical aspects and goals in this domain, state-of-the art of hardware/ software architectures used, as well as lessons on AI that are applicable to the resource-constrained environment. Case studies and real-world illustration will show how Edge-AI is changing healthcare delivery (Yallamelli, 2023).

We will also consider energy efficiency enhancement strategies including system level optimization and power management approaches. Issues of security and privacy will be addressed in order to ensure the compliance with the regulations related to healthcare. Finally, the chapter will end with some insight into future directions related to the use of federated learning, explainable AI, and the next generation of communication technologies, 6G. By the end of this chapter, readers will learn in detail how Edge-AI can transform IoT based health monitoring providing solutions which are not only smart and efficient, but also sustainable and human-centric.

2. REQUIREMENTS AND CHALLENGES IN REAL-TIME HEALTH MONITORING

The successful roll out of the real-time health monitoring systems based on Edge-AI in the IoT medical devices depend upon two vital technical and operational challenges. With more and more people demanding continuous, real-time physiological monitoring, in remote, resource-poor areas, the performance characteristics of these systems must be carefully tailored to address responsiveness, reliability and resource efficiency tradeoffs. Health monitoring on-the-fly, especially for life-critical applications like the arrhythmia detection or prediction of seizures, requires as little time as possible passed between processing and reacting to incoming data. However, the limitations of wearable and implantable devices (limited battery life, low computational ability, and transmission limits) ensure that this task is not trivial. In order to manage these complexities, the following are considered as four major pillars of consideration under which the following sub-sections proceed: latency, energy, privacy, and accuracy.

Table 1. Key requirements and challenges in real-time health monitoring using Edge-AI

Requirement	Description	Challenge
Low Latency	Instantaneous response for critical health events	Processing complex data locally without delay
Energy Efficiency	Long battery life in wearable and implantable devices	Balancing computational load and power consumption
Data Privacy and Security	Protection of sensitive medical information	Secure storage, transmission, and compliance with regulations
Accuracy and Reliability	Consistent and correct health insights from edge models	Maintaining model performance on constrained hardware
Real-Time Processing	Continuous monitoring and timely decision-making	Managing data streams without overloading system resources
Model Adaptability	Updating models based on individual health patterns	On-device learning without compromising data privacy

Table 1 describes the fundamental requirements and associated challenges necessary for the implementation of edge-AI practices in real-time health monitoring. The need for ultra-low latency, energy efficiency, and strong security of client data is fundamental for health care where, as delayed or erroneous responses can lead to adverse outcomes, the demand for ultra-low latency, energy efficiency, and strong security of client data is crucial. In addition, edge devices need to sustain high model accuracy and reliability levels while adhering to stringent limitations on power and resources. Such challenges require innovation from hardware to algorithm to make this real time monitor system provide intelligent, reliable and safe healthcare solutions on the point of care.

2.1 Low-Latency Data Processing

The ability to process data with minimum delay is one of the basic needs of real time health monitoring systems. The latency in this context is defined as the time taken for data collection to culminate into decision making or action taking. In healthcare situations where time is a determining factor of clinical results, latency directly affects system efficiency. For instance, identifying an incoming cardiac arrest or a fall in a person of old age should lead to immediate notification or emergency response. The legacy cloud-based architectures that involve transmitting data to long distant servers for analysis introduce delays, through network congestion, transmission delays, and server queues in processing. Comparatively, Edge-AI relocates the computational load to local devices, and the analysis of the data is only done as close to the source as possible. In this way not only latency, but also system responsiveness are reduced as devices can take decisions in real time in this edge-based processing architecture. Nevertheless, practical low-latency AI inference on resource-starved edge devices requires streamlined algorithms, lower model complexities, and smart memory management, yet it should not compromise the accuracy of clinical judgments (Putra & Arrayyan, 2024).

2.2 Energy Constraints in Wearable and Implantable Devices

Although real-time operation is crucial, this should not mean that the energy requirements will be too high. The majority of wearable and implantable medical devices rely on small battery supplies which are hard or even impossible to recharge on a regular basis. This calls for a rigorous demand for energy efficiency in energy-related hardware and software elements. Extended device uptime is critical to enable unbroken monitoring, especially for chronic disease patients or the aged based on whom these devices can monitor continuously. Edge-AI is helpful here in decreasing the need for constant wireless data transmission which is the most energy intensive operation in the traditional IoT systems. Nevertheless, doing inference on the edge devices itself also uses energy especially if the AI models are complex, and sensor data acquisition occurs ceaselessly. Hence, edge systems need to adopt such techniques such as dynamic voltage and frequency scaling (DVFS), dynamic sampling and event-based wake-up to save power. In addition, AI models themselves have to be optimized by methods such as quantization and pruning to reduce their computational footprint. A careful balancing act between real time performance and energy efficiency is called for, often requiring radical co-design of the hardware and software from scratch(Arthi & Krishnaveni, 2024).

Figure 2. Key challenges and considerations in implementing Edge-AI for health monitoring, balancing data privacy protection with AI model reliability in resource-constrained environments

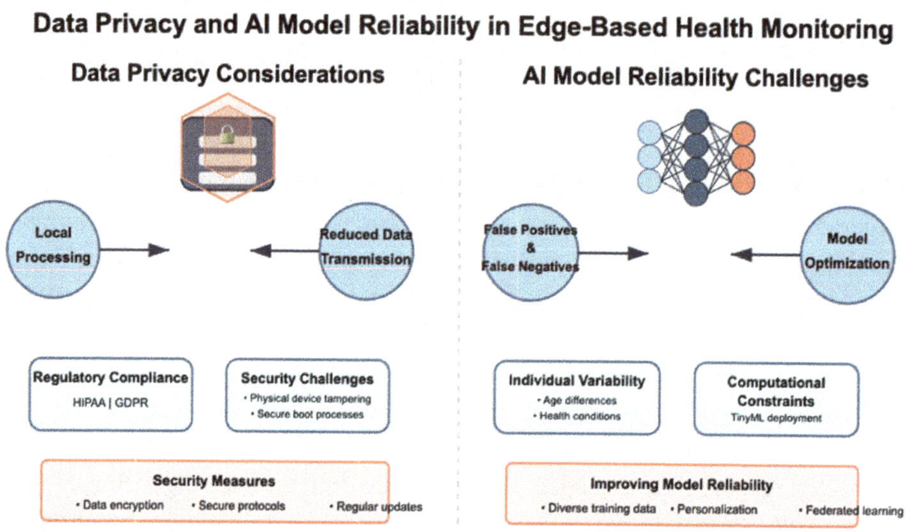

As shown in Figure 2, the implementation of Edge-AI for health monitoring systems calls for proper balancing between solid data privacy protection and trustworthy AI model performance. The left side is a picture of a concept that fundamentally improves patient privacy through edge processing, reducing data that is transferred, and complying with medical regulatory structures such as HIPAA and GDPR, while still necessitating full-scale security for physical tampering and unauthorised access. The right side has the intrinsic tensions of AI model deployment, where accuracy had to be compromised in the face of computational limits, the differences among individuals and the very important fact that there both the false positives and false negatives ought to be minimized in health monitoring context. This dual challenge paradigm requires creative methods ranging from TinyML optimization practice to multiple training datasets, personalization ability, and federated learning structure to guarantee privacy while on the other hand preserving clinical validity. The pattern emphasizes that whereas the successful implementation of edge-based health monitoring systems must concurrently consider security shortcomings found at the device level while providing an assurance of reliable performance in the real world even under constrained resources (Sharma & Solanki, 2024).

2.3 Data Privacy and Security

Given the sensitivity level of health data among personal data groups, the private and security concerns are of upmost importance in real time monitoring systems. Patients should feel assured that their physiological data – heart beat patterns, sleep cycles, even indicators related to mental health – is subject to the utmost non-disclosure. In the case of cloud based systems, data frequently passes through several intermediaries with the corresponding number of possible vulnerabilities. Edge-AI reduces this by localizing the majority of the data processing at the device end, thereby decreasing the level and need of data transmit. But the edge paradigm brings its own security problems. Edge devices are easier to physically access or tamper with compared to cloud servers therefore they are vulnerable to hardware attacks, malware or malicious firmware change. Data security when rested and in transit, secure boot processes, and the regular updating of device firmware are vital in ensuring data integrity. Moreover, regulatory compliance (in the form of HIPAA in the US or GDPR in Europe) requires open and responsible data contraction approaches. Therefore, the decentralization provided by Edge-AI must be enhanced by strong security protocols to secure user trust and safe medical information.

2.4 Accuracy and Reliability of AI Models in Health Monitoring

In real time monitoring of health, decisions by AI models could have direct impacts on patients' health, and therefore accuracy and reliability are pivotal measures for system performance. < False positives may create undue panic and medical interventions while false negatives may cause unperceived emergencies. This is especially difficult considering the variability between physiological signals of individuals based on age, health status, life style and environmental context. Edge-AI systems will need to be trained over a wide variety of representative datasets that generalize well across populations, but that also must be adaptive enough to personalize over time. In addition, the decreased computational capability of the edge devices requires the use of lightweight AI models that still need to maintain high level of accuracy on unoptimized architectures. Ideas such as TinyML allow developers to deploy optimized models on microcontrollers, but developers must be keen about negotiating between model complexity and interpretability. In addition, the issue of reliability over time is present – models should carry through performance despite sensor drift, noise or modifications of user behavior. Frequent recalibration or federated learning-based methods could be used to maintain a model up to date without harming the privacy of the user. Ultimately, the clinical validity of health monitoring driven by AI depends on attainment of steady, verifiable model performance in the real world.

3. HARDWARE AND SOFTWARE ARCHITECTURES FOR EDGE-AI

For practical incorporation of Edge-AI to enable real-time health monitoring in IoT medical tools requires a combination of purpose executed hardware and optimised software frameworks. With the harsh power, latency and size requirements for wearable and implantable devices, the architecture at the edge must facilitate efficient, dependable, and scalable AI processing(Pandey, 2025). This section provides a detailed overview of the top-edge hardware platforms, energy-efficient design strategies, associated software ecosystems, and AI model optimization strategies that collectively support robust and intelligent health monitoring systems.

Table 2. Key Edge-AI hardware platforms and their characteristics for IoT medical devices

Device	Processor Type	Key Features	Targeted Applications
ARM Cortex-M Series	Microcontroller (MCU)	Ultra-low power, small footprint, basic AI capabilities	Wearable health monitors, simple health anomaly detection
Google Edge TPU	AI Accelerator	High throughput for CNNs, low power consumption	Portable diagnostic tools, image-based health monitoring
NVIDIA Jetson Nano	AI Accelerator	GPU-based, supports complex AI models	Real-time health data processing, complex anomaly detection
Intel Movidius Myriad X	AI Accelerator	Dedicated hardware for vision and AI tasks, energy-efficient	Activity recognition, real-time image analysis for health
Arduino Portenta H7	MCU + AI Accelerator	Dual-core MCU, integrated AI capabilities	Rapid prototyping of wearable medical devices

Choosing the right hardware platforms for Edge-AI in IoT medical devices fundamentally requires a strike between computational performance, energy efficiency, and the size constraints. As it can be observed in the table 2 above microcontrollers such as ARM Cortex-M series are highly used for basic AI functions in low power applications such as the wearables for health monitoring. On the other hand, AI accelerators like the Google Edge TPU and NVIDIA Jetson Nano offer dramatically increased computation power, thus allowing for such things as real-time image analysis or sophisticated anomaly detection, with minimal energy expenditure. Integration of advanced hardware such as the Intel Myriad X boosts real-time processing capability, with particular emphasis on applications demanding high throughput on computer vision related activities. Combined, these platforms provide the framework for contemporary energy smart and intelligent health monitoring devices.

3.1 Edge Devices: Microcontrollers and AI Accelerators

At the heart of any Edge-AI-enabled health monitoring system lies the edge device, which must be capable of real-time sensing, processing, and decision-making. These devices typically fall into two categories: ultra-low-power microcontrollers (MCUs) and specialized AI accelerators(Santoso & Surya, 2024).

Microcontrollers (MCUs) such as the ARM Cortex-M series are commonly used in medical wearables due to their compact size, low energy consumption, and sufficient computational capabilities for basic AI tasks. Paired with analog front ends and biomedical sensors (ECG, PPG, accelerometers), MCUs can execute lightweight AI models like decision trees or quantized neural networks.

AI accelerators, on the other hand, provide enhanced computational throughput for more complex AI inference. Devices like the Google Edge TPU, NVIDIA Jetson Nano, and Intel Movidius Myriad X are increasingly being integrated into portable diagnostic tools and home-care monitoring devices. These platforms are particularly suited for convolutional neural networks (CNNs) and recurrent neural networks (RNNs) used in tasks like cardiac anomaly detection or activity recognition.

TinyML platforms such as the Arduino Portenta H7, SparkFun Edge, and Seeed XIAO offer integrated development environments for deploying AI on ultra-low-power hardware. These platforms bridge the gap between limited-resource MCUs and the AI performance demands of healthcare applications, making Edge-AI accessible even to small-scale developers.

3.2 Energy-Efficient Hardware Design Principles

Designing for energy efficiency is a paramount consideration for edge medical devices, especially those intended for continuous monitoring. Several hardware design principles are employed to reduce power consumption without sacrificing real-time performance:

- **Low-Power Modes and Sleep States**: Edge devices implement aggressive power management schemes, including deep sleep and wake-on-interrupt states that allow them to remain dormant until triggered by relevant events (e.g., abnormal heart rate).

- **Event-Driven Architectures**: These architectures ensure that sensors and processing units are only activated when specific thresholds or anomalies are detected, minimizing unnecessary data acquisition and processing.
- **Dynamic Voltage and Frequency Scaling (DVFS)**: This technique dynamically adjusts the processing speed and voltage levels based on the workload, conserving energy during less intensive operations.
- **Hardware Accelerators for Specific Tasks**: Instead of general-purpose CPUs, edge devices increasingly incorporate hardware accelerators tailored for matrix multiplication or FFTs, which are common in biomedical signal processing. This specialization reduces the energy-per-operation cost.

In addition to component-level optimizations, **system-level co-design**—where hardware and software are developed in tandem—ensures that power consumption remains minimal while supporting AI workloads efficiently.

3.3 Software Frameworks: Enabling AI at the Edge

Software plays a critical role in deploying, managing, and updating AI workloads on edge devices. Several software frameworks and development environments have been specifically designed to support Edge-AI, with a focus on lightweight execution, ease of model conversion, and support for real-time processing(Bhutani & Lall, 2022).

- **TensorFlow Lite for Microcontrollers (TFLM)**: A subset of TensorFlow Lite optimized for MCUs, TFLM enables developers to run quantized neural networks on devices with less than 100KB of memory. It supports various operators common in health-related AI models and is compatible with numerous embedded toolchains.
- **Edge Impulse**: A cloud-based platform that streamlines the end-to-end development of AI on edge devices, Edge Impulse provides tools for data acquisition, model training, and deployment. It supports a wide range of sensors and MCUs, making it ideal for rapid prototyping and deployment in medical applications.
- **CMSIS-NN and uTensor**: These ARM-based libraries provide optimized kernels for neural networks on Cortex-M devices. They enable faster inference while maintaining a small memory footprint, which is crucial for continuous health monitoring applications.

- **TinyML Ecosystem**: Encompassing both hardware and software, the TinyML community supports a range of libraries and tools aimed at running machine learning models on kilobyte-scale memory devices. Platforms like TF Lite, TVM, and microTVM offer model compression and deployment utilities tailored to health AI workloads.

3.4 Optimization Techniques for Edge AI Models

AI models originally trained in cloud environments are often too large and resource-intensive for direct deployment on edge devices. To overcome this, various optimization techniques are employed to reduce model size, increase inference speed, and lower power consumption, without compromising accuracy.

- **Quantization**: This technique converts floating-point numbers to lower-precision formats (e.g., 8-bit integers), significantly reducing model size and memory requirements. Post-training quantization and quantization-aware training are both widely used in health monitoring applications.
- **Pruning**: Pruning removes redundant or non-contributory weights from neural networks. Structured pruning (removal of entire filters or neurons) is often preferred in edge environments due to its compatibility with hardware acceleration.
- **Model Distillation**: In this technique, a smaller "student" model is trained to replicate the outputs of a larger "teacher" model. This allows for efficient inference while retaining the performance benefits of larger, more complex models.
- **Graph Optimization and Operator Fusion**: Tools like TensorFlow Lite's flatbuffer converter and Apache TVM optimize computational graphs by merging consecutive operations and reducing memory allocation overhead.
- **Adaptive Inference**: Some models include early-exit branches that allow decisions to be made earlier in the inference pipeline if confidence thresholds are met, reducing unnecessary computation.

These optimization strategies collectively enable real-time AI processing within the tight memory, compute, and power budgets of IoT medical devices. For instance, using a quantized CNN for ECG classification on a Cortex-M4 MCU can reduce inference time to under 100 ms with under 50 KB RAM usage, making it viable for deployment in wearable cardiac monitors.

4. AI MODELS AND TECHNIQUES FOR HEALTH MONITORING

Artificial Intelligence (AI) is now a disruptive force in the field of health monitoring systems, which can analyze patterns, diagnose and predict critical events from physiological data. When deployed at the edge, these AI models need to run under very strict limitations of memory, processing capability and availability of energy, but produce accurate and timely output. The creation of lightweight efficient AI models able to conduct some form of meaning analysis on raw sensor data is therefore fundamental to the implementation of intelligent IoT medical devices (Goyal & Elawadhi, 2024).

Corpus of physiological data, including electrocardiograms (ECG) and photoplethysmograms (PPG), body temperature, blood oxygen saturation (SpO_2), and accelerometry signals, contains plenty of information for a person's physical state. However, the data is often noisy, and high frequency, and needs a lot of preprocessing before it can be usable for inference. In usual cases, filtering, Fourier Transforms, Wavelet Analysis are used in the introductory steps to clean data and extract salient features. Such features --like heart rate variability, peak intervals, or frequency-domain coefficients -- act as input to the models of machine learning which perform classification, regression, and anomaly detection.

In order to fit the scarce resources of edge devices, AI models should be designed with simplicity and efficiency in their minds. Classical machine learning models such as decision trees, support-vector machines (SVM), and k nearest neighbors (KNN) are still relevant because of their compact size of memory and interpretability. An example, such as a decision tree, can efficiently classify types of physical activity or detect abnormal heartbeat rates with almost zero computational workload. Logistic regression and linear classifiers are also very common in the early detection tasks because relationships between the features and the labels are either linear or easily separable.

Edge application repurposing of deep learning models, facilitated by efficiency and compression mechanisms have gained increasing traction in the past few years. Convolutional Neural Networks (CNNs) are especially useful for the study of structured signals such as ECGs were temporal and mor-phological patterns needed to be identified. Aware of the approach of edge/ wearable computing, lightweight CNN architectures such as MobileNet, SqueezeNet, or even custom shallow networks have tailored for wearable deployment, high accuracy with minimal computational cost. Recurrent Neural Networks (RNN) and its gated forms (like the now most-regarded GRU and LSTM) are particularly well suited for the modeling of temporal dependences in time series data such as PPG or respiration if more complex dependences in time are preserved. Such models also can detect long-term irregularities or transition from one physiological state, for example, from sleep stages to pre-seizure states, etc.

Table 3. AI models and techniques for edge health monitoring in IoT devices

Model/Technique	Application	Strengths	Constraints for Edge Devices
Decision Trees	Classification of activity types, anomaly detection in heart rhythms	Simple, interpretable, low computational requirements	Limited to simple tasks, may not handle complex patterns effectively
Support Vector Machines (SVM)	Classifying physiological conditions (e.g., arrhythmia detection)	Good for binary classification, relatively small memory footprint	Requires careful tuning, may be slower with larger datasets
K-Nearest Neighbors (KNN)	Health event prediction, activity recognition	Easy to implement, no training phase required	Computationally expensive as dataset grows, needs sufficient memory
Logistic Regression	Early detection tasks, linear relationships between features and labels	Fast, lightweight, suitable for linear classification problems	Cannot handle non-linear patterns effectively
Convolutional Neural Networks (CNN)	ECG analysis, PPG analysis, image-based health monitoring	Excellent for feature extraction from structured signals, high accuracy	Requires significant memory and computational power, even in lightweight variants like MobileNet
Recurrent Neural Networks (RNN), LSTM, GRU	Modeling temporal dependencies in time-series data (e.g., respiration, PPG)	Effective for sequential data, detects long-term dependencies	High computational cost, needs optimization for edge deployment
Hybrid Signal Processing + AI	Feature extraction + classification (e.g., hybrid ECG anomaly detection)	Efficient, balances traditional signal processing with AI advantages	May still require careful tuning to balance processing time and accuracy
Transfer Learning	Personalizing models based on individual data	Efficient use of pre-trained models, small on-device datasets	Requires a base pre-trained model, may struggle with very diverse data
Federated Learning	Collaborative model updates across devices, data privacy preservation	Enhances privacy, reduces data transmission overhead	Complex synchronization, requires secure model-sharing protocols
Explainable AI (XAI)	Providing interpretable results for clinicians and users	Increases trust and usability in clinical settings	Can add complexity to model, may impact performance and efficiency
Adversarial Training	Ensuring model robustness, especially in diverse environments	Enhances reliability under varied conditions	Increases training time, requires large and diverse datasets

AI model and technique for edge devices in health monitoring must achieve a trade off of efficiency and effectiveness in computation versus clinic. It is clear from the above table 3, that simplistic classical machine learning models such as decision trees and support vector machines (SVM) are still valid for cheaper tasks because of their low computational pressures. However, to support complex analyses, complex models such as deep learning models such as convolutional neural networks (CNNs) and recurrent neural networks (RNNs) have been optimized to run on resource-constrained environments and allow for advanced capabilities like ECG analysis and activity recognition. Hybrid strategies that merge classic signal processing with lightweight AI environments have demonstrated ability to reduce computational load without loss of accuracy. Other innovations, such as transfer and federated learning can advance adaptability and privacy of these models to make them ideal for personalized health monitoring systems. Lastly, the combination of explainable AI (XAI) and adversarial training mean that these models remain robust and interpretable and are critical for acceptability and reliability by clinicians (Shumba & Montanaro, 2022).

Hybrid strategies joining signal processing with shallow neural networks are often used to diminish computational costs. In these systems, it is the traditional algorithms to extract the feature while, the smaller scale AI model handles the classification hence balancing between accuracy and efficiency. Transfer learning has also gained momentum in the domain of health monitoring where large dataset trained model is tuned with a few pieces of personal data. With this technique high performing models can be tailored to individual users without the need for extensive on device training.

Federated learning is one of the most promising innovations in the field that makes it possible for devices to collaboratively learn a shared model without sharing any data. In the healthcare, this is especially valuable for privacy preservation and regulatory compliance. Devices will update a central model at intervals by transmitting only the learned parameters or gradients, and not the raw health data themselves. Apart from ensuring protection of access key and other sensitive information, this decentralized approach also minimizes the communication overhead that is customary of cloud based AI.

Practical use cases are already delivering significant impact from edge-AI applications in health monitoring. Wearable ECG monitors use AI to identify arrhythmias and early indicators of cardiovascular disorder that allows early interventions. Based on PPG, devices monitor pulse and measurements of oxygen in streams continuously and are being improved with AI for a prediction of hypertension or dehydration. Accelerometer-based fall detection systems use AI models to discern between normal movements and potentially lethal incidents, which instantly triggers alarms in cases of elderly care. AI-powered thermometers and respiration trackers are assisting in early infection detecting of infections such as COVID-19 or asthma.

Crucially, as these AI models come closer to patients on wearable or implantable devices, the need for interpretability and robustness becomes eminent. XAI techniques are being used earlier in the health monitoring models to have insight on the predictions so that the clinicians and the users of the software understand the basis of the alert or diagnosis. Robustness to the model that is crafted out through data augmentation, adversarial training, and uncertainty quantification, makes this system reliable regardless of the type of users or operating environment.

In a nutshell, the survival and propagation of AI models for health tracking on edge devices calls for a comprehensive strategy, which incorporates device restrictions, data privacy needs as well as provides clinical relevance. As innovation persists with regard to model architecture, training strategies, and optimization methods, Edge-AI is positioned to transform the way health data is analyzed in real time, enabling both patients and healthcare professionals to benefit from proactive and personalized interpretations of the data.

5. ENERGY EFFICIENCY STRATEGIES

The Ziffre of Edge-AI in IoT medical devices also depends on ability to operate under tight energy constraints, besides having accurate inference and responsiveness particularly in battery-powered wearables and implantables. The effect of energy efficiency is immediate on device longevity, usability and feasibility in running continuous health monitoring. Thus, several of hardware-software co-design strategies are used in order to reduce the power consumption without degrading performance. In this section we recommend four techniques to optimize the energy consumption of Edge-AI enabled health monitoring systems.

5.1 Duty Cycling and Event-Driven Processing

Duty cycling is among the most commonly adopted methodologies in low-power embedded systems. It entails the disconnection, or scaling down, of system components at intervals when they are not actively processing data. In health monitoring, for example, sensors and even processing units may be in a suspended state for most of the time, but may need to be activated if certain conditions are satisfied, such as a deviation of heart rate or body movement. Event-driven architectures extend this idea by triggering the activation of processing pipelines only to predetermined physiological triggers such as arrhythmia or unexpected fall of oxygen saturation. This selective activation reduces unnecessary computation and data transfer thus it conserves energy. By integrating duty cycling with the intelligent event detection, systems can lie dormant for long periods of time, and react only to clinically relevant activity, thereby maximizing the efficiency of resource utilization.

5.2 Dynamic Voltage and Frequency Scaling (DVFS)

DVFS stands for dynamic voltage and frequency scale which is a hardware mechanism; it adapts processor voltage and clock frequency dynamically to meet the needs of current calculations. In the health monitoring setting in Edge-AI environments, use of DVFS in the system allows it to decrease the operating characteristics and keep the activity levels low when handling stable physiological data and only when needed should the high performance modes be activated for the AI-Intensive tasks. For instance, when the edge device monitors even and normal heart rhythms, it can decrease processing rate in order to minimize energy consumption. If one could find an irregular signal, the system is able to rapidly increase frequency and voltage to do intensive analysis via AI models. This is specially the case where dynamic resource modulation is incorporated in order to attain the best performance with power conservation which is vital for a continuous long run mobile device operation.

5.3 Energy-Aware AI Model Scheduling

Optimal organization of AI workloads contributes greatly to reduced energy expenditure. Through a focus on energy-aware AI model scheduling, inference tasks are choreographed with the goal of achieving efficient power utilization without sacrificing the need to maintain critical real time responsiveness. Important aspects are timing the computation of demanding components efficiently, prioritizing AI models based on their criticality and moving tasks to the hardware with best energy saving yet keeping performance. For instance easy anomaly detection can offload less demanding activity onto low power Homes, whereas sophisticated pattern recognition applications are only to be run on more powerful hardware as needed. Some of the more advanced systems have attached runtime intelligence where inference scheduling is altered in real-time to match changes in battery or usage condition. This stratified scheduling approach promotes longer functionality of the devices, specifically supporting applications that need constant monitoring, including continuous ECG tracking and unremitting sleep pattern monitoring (Nijhawan & Bhatia, 2023).

5.4 Adaptive Sampling and Data Compression

Another prime strategy towards energy savings is also the adaptive sampling, which allows varying the sampling frequency for sensors based on the specific physiological case or changes in the measured variable. By changing the sampling frequency depending on physiological stability edge devices can save energy and unnecessary data is collected by frequent high rate sampling. By reducing the size of the dataset, energy consumption and memory access cost savings are achieved due to reduced computational and memory usage requirements. In addition, data compression is also used to reduce the weight of data payload from sensors before transmission or more digital processing. Compression techniques tailored towards biomedical signals preserve the most important clinical information while conveying very little of the data's storage burden. The combination of adaptive-sampling and compression approaches is very effective towards improving the performance of a battery and suitable for long-term health monitoring devices that necessitate continuous data collection.

6. SECURITY AND PRIVACY CONSIDERATIONS

With Edge-AI being at the center of real-time health monitoring on IoT medical devices, today sensitivity in securing and protecting personal data is more critical than ever. These devices control vital personal health information on a persistent and immediate basis, often while operating over distributed networks that are resource starved. Theoretical threats to data confidentiality, integrity, and availability may not only undermine privacy, but also create serious threats to health and safety. That is, the imperative of security and privacy must be strong but more than that a technical imperative it has an ethical and legal necessity in the area of intellectual health care development and implementation (Manjula Devi & Sivakarthi, 2025).

The advent of Edge-AI solutions makes data management on health securely more complex but viable. Processing data locally within the edge reduces the volume of network traffic thus reducing the vulnerability of an issue to cyberattacks. However, due to the large numbers of endpoints possible within the edge systems, each enabled with reduced computational capabilities and varying network connectivity, such systems create a wide and varied set of security exposures that need to be effectively handled. The main themes in this section outline core measures to ensure privacy and security, which include encryption, secure data storage, model security, and HIPAA and GDPR-regulated compliance, as well as privacy-friendly practices of machine learning.

6.1 Data Encryption and Secure Storage on Edge Devices

Encryption, in the form of both data at rest as well as in transit, is an absolute necessity for the secure operations of edge-based health monitoring. Health information from wearable or implantable sensors must be encoded as soon as it is taken from a device in case it gets damaged or signals are hacked wirelessly. The use of Advanced Encryption Standard with 128 or 256-bit keys for symmetric encryption in edge devices, is wide spread due to strong security and reasonable performance.

Real-time encryption is very feasible in such edge devices with dedicated hardware cryptographic functions minimizing both their power consumption and delay.

Secure storage of data that is kept locally – either for current processing or long-term preservation – necessitates more than encryption and requires credible access controls. This entails validating the identities of data subjects, tracking access attempts through an audit trail and creating segmented boundaries around sensitive data/applications. Trade-off management gets progressively complex in devices limited in memory or computational ability. Protocols such as ECC (Elliptic Curve Cryptography) are applicable in cases with reduced memory and processing power without compromising the security of the systems. The latter saves resources taken by these systems.

Another form of protection that comes into play with the use of secure boot entails verifying the firmware to be exclusively loaded from verified, digitally signed firmware from which edge devices boot, which use cryptographic hashes to ensure these firmware are intact at the time of booting. Such initiatives prevent efforts by intruders to implant malevolent code that can control data or reveal sensitive data. The ultimate value of combining these encryption and secure storage techniques is that organizations can provide effective protection at the edge to patient data.

6.2 Secure Model Updates and Code Integrity

Providentia of safe delivery and periodic updating of AI models is another critical aspect often under discussion in Edge-AI security. In real world health monitoring systems, AI models require updating in order to improve accuracy, apply to trends in data or counter new risks. Unsecured updates become an avenue of vulnerability to attackers. Threat actors could inject malicious models or overwrite existing models with compromised versions or leak sensitive intellectual property.

In order to safeguard at such potential compromises, there is need for edge devices to provide secure OTA update solutions. Secure OTA updates involve use of encrypted connections, such as TLS or DTLS and encryption employing digital signatures or certificates from trusted authorities for authentication. Update packages must contain integrity verification techniques such as comparing hashes, to protect against changes to the data in transmission. A number of systems use version control and rollback protection mechanisms to prevent downgrades attacks thus protecting devices from reverting to legacy, vulnerable software (Muhoza & Bergeret, 2023).

Through code integrity and runtime attestation, the model integrity is preserved during the runtime. The use of such TEEs as ARM Trust Zone bolsters the protection of sensitive code and data within a secure boundary thus separating it from the areas of a system that can be compromised. In safety critical applications, redundancy, model diversity strategies (i.e. running multiple models in parallel and validating outputs against each other) can be used to find and mitigate unexpected results from adversarial activity.

6.3 Compliance with Healthcare Regulations: HIPAA, GDPR, and Beyond

Regulatory compliance is important not only to run health monitoring system legitimately but also to build trust of users in it. In the USA, the a nation-wide guideline is set under the Health Insurance Portability and Accountability Act (HIPAA) which is intended to ensure the protection of the protected health information. Within the European Union, the General Data Protection Regulation (GDPR) governs the protection of personal data of which rules are governing that prevent data minimization, consent, and user rights. Privacy by design is how Edge-AI systems should be constructed from their inception forward taken into account privacy legislation. This principle requires that identifiable information be limited if possible, with the data anonymized or pseudonymized and that the users are given the ability to control their personal information. Consent management is particularly critical; It is vital that the people know what personal data will be gathered and how it will be used and preserved. People need to consent voluntarily, in several reasons, and can revoke their consent at any time.

It takes cooperation from the legal, technical, and procedural worlds to meet regulations relating to IoT medical devices to fully comply. Possible solutions include protecting keys with firm measures, defining exact access limitations, and constantly tracking the use of data. Important procedural measures include training of workers, development of standard incident procedures as well as routine security evaluation. Furthermore, specific regions have distinctive regulation mechanisms, including India's DISHA, or Canada's PIPEDA, which must be adapted for implementing at a global level. Compliance advantages inherent to the Edge-AI include reducing the reliance on central cloud storage, along with enabling on-site data processing. However, it introduces challenges relating to managing audit processes, data processing consent, and safeguarding of users' rights when there is operation at the edge. In order to obtain this, a transparent and modular system design needs to be provided to allow for secure data transfer between edge devices and central authorities without compromising performance and usability.

6.4 Privacy-Preserving Learning and Federated AI

There is now a stronger need for privacy-preserving learning approaches as health data increases, and personalized models grow in popularity. Traditional centralized machine learning jets require pushing raw data to the far-off cloud servers, raising very serious privacy issues. By supporting decentralized training, federated learning allows for data analysis across spread devices without the risk of exposing raw data to one single node. Individual devices operate on the model with their own data, uploading only model updates to a central server which merges these updates with the rest into a global model. Healthcare uses of federated learning avoid sensitive data leaving the device, which greatly reduces privacy risk exposure. Additional privacy can be gained by using differential privacy in updating the models, by adding the controlled noise that discourages reconstruction of original data without affecting the performance of the models. Moreover, homomorphic encryption and secure multi-party computation are being researched to make sure that medical data is kept encrypted when it undergoes distributed computations.

Although federated learning has advantages, employing just federated learning onto edge devices brings in new complexities. Problems include addressing non-IID datasets, synchronizing across numerous devices and counteracting the effects of stragglers, i.e., devices that have limited access to the network or are not always connected to it. Moreover, edge devices need sufficient processing power and energy to carry out the more demanding job of local training instead of inference. To overcome these difficulties, some lightweight optimization algorithms, such as FedAvg and personalized federated methods are being developed. In addition to the use of federated learning, split learning is a complementary approach that increases data privacy because a model divided into a device and a centralized server. Only partial model states or feature presentations are exchanged through this method between the device and the server, rather than exchanged data, or completed models. In so doing, this solution ensures efficient computation is maintained, privacy is preserved and satisfactory model outcomes are delivered.

6.5 Threat Landscape and Risk Mitigation

Although many measures have been put in place, edge health monitoring systems continue to be vulnerable to numerous cyber vulnerabilities. Cybersecurity threats covered in this include data poisoning and model inversion attacks, adversarial examples, side-channel attacks, and physical compromising of a device. Opponents may corrupt sensor data in order to trigger unnecessary alerts, or hide underlying health problems for patient peril. Adversarial methods can obtain sensitive information from AI model responses completing the decision and playing with input signals to deceive AI systems into providing false diagnoses.

As this risk may have multiple triggers it needs strict threat modeling and risk assessment. To provide robust security developers should examine the full range of vulnerabilities from sensor input to AI output and communications which is best done using a cascading defense model. Such systems could raise flags on anomalies in sensor read-outs, or behavior of the model that may deserve to be suspected as result of unauthorized tampering. Edge device architecture requires the adoption of hardware security modules (HSMs), key encryption policies, and intrusion warning mechanisms.

Keeping up to date with standard software updating and security patching is the key to fighting the appearance of vulnerabilities. For this reason, mechanisms for updating should be built keeping security, dependability, and little-to-no, interference with device performance in perspective. Is the continuous security certifications as ISO/IEC 27001, penetration test, participation in the bug bounty programs, which also helps in maintaining the integrity of edge-AI through its operational lifetime.

6.6 Ethical and Social Considerations

The security and privacy of health monitoring need to be secured and this is as much an ethical problem, as it is technical. Patients disclose very personal health information unsupervised to such systems, often not understanding the consequences on a broad scale. Therefore, it is important that system design should focus on transparency, accountability, and equity. Patients should be able to comprehend what data is collected and how and why it is processed, and which aspects of the analyses it affects. There is a need for clear transparency of decisions based on AI, and capability of users to challenge decisions that affect their health directly.

Systemic inequities or inaccurate interpretations of patient health exist when AI systems have been trained with data that has a lack of diversity. Most of all, bringing the results on an equal plane requires extensive data collection, extensive validation, and continuous observation of the model's behavior in the actual healthcare setting. In AI-based systems of health monitoring, developers need to not only concern themselves with data protection, but also defend the dignity, autonomy and right to informed consent of users.

Table 4. Key security and privacy considerations for Edge-AI health monitoring

Security/Privacy Aspect	Key Strategies	Challenges
Data Encryption and Secure Storage	AES encryption for data in transit and at rest, ECC for lightweight encryption	Balancing encryption with power/ latency constraints, secure access control for limited memory devices
Secure Model Updates and Code Integrity	Encrypted OTA updates, integrity checks, version control, and rollback protection	Ensuring integrity in remote updates, managing trust in OTA channels
Compliance with Regulations (e.g., HIPAA, GDPR)	Privacy by design, consent management, anonymization, local data processing	Adapting regulations to edge contexts, auditing across decentralized systems

continued on following page

Table 4. Continued

Security/Privacy Aspect	Key Strategies	Challenges
Privacy-Preserving Learning (Federated AI)	Federated learning, differential privacy, secure multi-party computation	Addressing non-IID data, computational and energy demands on edge devices
Threat Landscape and Risk Mitigation	Threat modeling, anomaly detection, intrusion detection, hardware security modules (HSMs)	Protection against data poisoning, adversarial attacks, and physical tampering
Ethical and Social Considerations	Transparency, fairness, explainable AI, bias mitigation, user control over data	Ensuring equitable treatment and privacy in diverse populations

Table 4 shows that Edge-AI has an important role to play on real-time monitoring of health, security and privacy are of the upmost importance. Human data protection pillars from the heart of the matter include; applied encryption (AES, ECC), secured model (guaranteed) and consistent compliance to health care data regulations (HIPAA, GDPR). However, the combination of such methods as federated learning and differential privacy enables models to learn from local data without the concern for external threat. The capacity of Edge-AI to reduce data centralization and transmission is an advantage, but the decentralized nature of edge devices raises new security challenges. There are threats including data poisoning, model inversion and adversarial attacks. Therefore, continuity of data integrity and patient safety demands continuous risk mitigation in the areas of secure boot process, anomaly surveillance system, and hardware security modules. In addition to technical challenges, the ethical aspects highlight the need for transparency, fairness, and the involvement of users guaranteeing that AI technologies strengthen the patient autonomy and provide trustworthy and fair health provision.

7. CONCLUSION

The intersection of Edge-AI and IoT medical devices is a paradigm shift in the partnership between technology and healthcare, particularly in health monitoring in real time. By allowing intelligent data processing at the point of care, Edge-AI decreases reliance on cloud infrastructure, drastically decreasing latency, bandwidth consumption and energy consumption while also improving patient privacy. These improvements are especially disruptive in situations where prompt intervention is required – e.g., cardiac abnormalities, falls, or a rapid decrease in oxygen saturation. In this chapter, the complex architecture of Edge-AI systems for health monitoring has been discussed, reviewing the hardware and software components necessary for developing efficient, secure, and responsive medical devices. And from energy-aware microcontrollers to purpose-built AI chips, lightweight frameworks and model optimizations – every call makes a part of satisfying the severe demands of the contemporary healthcare technologies.

Critical challenges like energy efficiency, data privacy, model robustness, and regulatory compliance must be addressed as a whole. Dynamic power management, federated learning, secure model updates and privacy-preserving AI have developed as possible solutions. Concomitantly, compliance with global rules such as HIPAA and GDPR doubles down on ethical and legal accountability and the trust in these technologies. Edge-AI is not just a technical invention but a push towards personalized and pro-active health care. It gives the strength for continual monitoring without reducing independence or data sovereignty, it is therefore ideal for chronic disease management, elder care and remote health services. By getting intelligence closer to the patient it provides timely insights and interventions – ultimately saving lives and cutting healthcare cost. In short, Edge-AI is full of potentials that could be harnessed to translate evidence-based real-time, intelligent, and energy efficient health monitoring into a practical reality to shape healthcare in the future. However, the implementation of such systems entails a synergistic approach that combines engineering eminence with man-centred design, regulatory prophetic approach, and moral accountability.

8. FUTURE DIRECTIONS

As the Edge-AI matures, so too will its ability to change healthcare. Future changes have the potential to make it even more effective in terms of health monitoring and disease management. Federated learning and on-device training, among them, are outmost directions. These strategies make decentralized model enhancement possible, and protect the data privacy, thus making it possible for the global AI model to learn from the patterns of the individual user without taking away the health-sensitive data from device. Another hope on the horizon is explainable AI (XAI), aiming to render the decision-making process of AI models open and understandable. In healthcare, this power is extremely important in terms of the development of clinician trust and decision-making. XAI tools embedded in edge devices might justify alerts or predictions, enabling the clinicians to understand the trends underlying health issues while enabling patients to make better engagements on their care.

Edge-cloud collaboration is thus expected to be more dynamic in forms of hybrid architectures. The edge will take care of the real-time, energy-sensitive tasks while the cloud can be used for complex analytics, long term data storage and population level insights. Elastic orchestration between edge and cloud will maximize performance and cost compared to maintaining data locality as needed. Other achievements in neuromorphic computing and bio-inspired architectures are likely to bring up a revolution in terms of energy efficiency of AI on the edge devices as well. These technologies mimic the brain's capability of handling low powers and are able to execute real-time inference with the least of hardware resources. The same solutions coupled with innovations in battery technology and energy harvesting may result in autonomous, maintenance free medical devices. Moreover, fusing multimodal sensing, like, ECG, PPG, accelerometery, and thermal, will allow for more comprehensive and effective health tracking. Future systems will detect anomalies and contextualize them as concerns about a person's behavior, environment and medical history. Regulatory frameworks and ethical AI governance will also be at the forefront. With the growing complexity and autonomy of Edge-AI systems, serving the interest of fairness, accountability, and transparency will therefore become critical. Constructive interactions among technologists, clinicians, and policymakers will be the cornerstone for ensuring that these innovations are safe, fair, and useful to everyone.

REFERENCES

Arthi, R., & Krishnaveni, S. (2024). Optimized Tiny Machine Learning and Explainable AI for Trustable and Energy-Efficient Fog-Enabled Healthcare Decision Support System. *International Journal of Computational Intelligence Systems*, *17*(1), 229. DOI: 10.1007/s44196-024-00631-4

Bhutani, M., Lall, B., & Agrawal, M. (2022). Optical wireless communications: Research challenges for MAC layer. *IEEE Access : Practical Innovations, Open Solutions*, *10*, 126969–126989. DOI: 10.1109/ACCESS.2022.3225913

Choudhary, S., Vijitha, S., Bhavani, D. D., Bhuvaneswari, N., Tiwari, M., & Subburam, S. (2025). Edge AI Deploying Artificial Intelligence Models on Edge Devices for Real-Time Analytics. In *ITM Web of Conferences* (Vol. 76, p. 01009). EDP Sciences. DOI: 10.1051/itmconf/20257601009

Gagneja, A., Lall, B., & Bhutani, M. (2024). Unveiling CM-Det: Leveraging ConvMixer architecture for advanced object detection. *International Journal of Information Technology : an Official Journal of Bharati Vidyapeeth's Institute of Computer Applications and Management*, *16*(7), 4273–4278. DOI: 10.1007/s41870-024-02069-5

Goyal, R., Elawadhi, O., Sharma, A., Bhutani, M., & Jain, A. (2024). Cloud-connected central unit for traffic control: Interfacing sensing units and centralized control for efficient traffic management. *International Journal of Information Technology : an Official Journal of Bharati Vidyapeeth's Institute of Computer Applications and Management*, *16*(2), 841–851. DOI: 10.1007/s41870-023-01527-w

Izhar, M., Naqvi, S. A. A., Ahmed, A., Abdullah, S., Alturki, N., & Jamel, L. (2023). Enhancing healthcare efficacy through IoT-edge fusion: A novel approach for smart health monitoring and diagnosis. *IEEE Access : Practical Innovations, Open Solutions*, *11*, 136456–136467. DOI: 10.1109/ACCESS.2023.3337092

Manjula Devi, C., Sivakarthi, G., Srinivasan, A., Gobinath, A., & Rajeswari, P. (2025). 13 Edge AI in LoRa-Based Health Monitoring. *Edge AI for Industry 5.0 and Healthcare 5.0 Applications*, 185.

Muhoza, A. C., Bergeret, E., Brdys, C., & Gary, F. (2023). Power consumption reduction for IoT devices thanks to Edge-AI: Application to human activity recognition. *Internet of Things : Engineering Cyber Physical Human Systems*, *24*, 100930. DOI: 10.1016/j.iot.2023.100930

Nijhawan, R., & Bhatia, S. (2023). Edge-AI tools and techniques for healthcare. In *Edge-AI in Healthcare* (pp. 17–30). CRC Press. DOI: 10.1201/9781003244592-2

Pandey, A. (2025). Energy-efficient communication in edge computing iot networks. *Metaversalize*, *2*(1), 21–30.

Putra, K. T., Arrayyan, A. Z., Hayati, N., Damarjati, C., Bakar, A., & Chen, H. C. (2024). A review on the application of internet of medical things in wearable personal health monitoring: A cloud-edge artificial intelligence approach. *IEEE Access : Practical Innovations, Open Solutions*, *12*, 21437–21452. DOI: 10.1109/ACCESS.2024.3358827

Santoso, A., & Surya, Y. (2024). Maximizing Decision Efficiency with Edge-Based AI Systems: Advanced Strategies for Real-Time Processing, Scalability, and Autonomous Intelligence in Distributed Environments. *Quarterly Journal of Emerging Technologies and Innovations*, *9*(2), 104–132.

Sathiya, A., Angel, D., Iswarya, M., Poonkodi, R., & Angelo, K. M. (2025, January). IoT Enabled Healthcare Framework Using Edge AI and Advanced Wearable Sensors for Real Time Health Monitoring. In *2025 International Conference on Multi-Agent Systems for Collaborative Intelligence (ICMSCI)* (pp. 384-392). IEEE. DOI: 10.1109/ICMSCI62561.2025.10894492

Sharma, T. P., Solanki, A., Jain, T., Malhotra, A., & Bhutani, M. (2024). Wi-Fi based quadcopter drone with battery monitoring and optimization using crazyflie platform. *International Journal of Information Technology : an Official Journal of Bharati Vidyapeeth's Institute of Computer Applications and Management*, *16*(3), 1887–1898. DOI: 10.1007/s41870-023-01639-3

Shumba, A. T., Montanaro, T., Sergi, I., Fachechi, L., De Vittorio, M., & Patrono, L. (2022). Leveraging IoT-aware technologies and AI techniques for real-time critical healthcare applications. *Sensors (Basel)*, *22*(19), 7675. DOI: 10.3390/s22197675 PMID: 36236773

Yallamelli, A. R. G., & Devarajan, M. V. (2023). Hybrid edge-AI and cloudlet-driven IoT framework for real-time healthcare. *International Journal of Computer Science and Engineering Technology*, *7*(1).

Chapter 10
Overcoming Challenges in Recycling Medical IoT Devices for a Sustainable Future

Pratheesh Manikonda
https://orcid.org/0009-0002-3034-6865
SimplyTurn Technologies, USA

Pavan Kumar Reddy Yellela
Independent Researcher, USA

ABSTRACT

There is a whole range of technical, regulatory, and ethical challenges facing recycling of medical IoT devices. This chapter identifies the barriers to the effective recycling of plastics posing a data security threat, materials for a complexity reason, and risks from contamination and a lack of standard process. It examines current practices of recycling and evaluates their viability in the healthcare application where patient safety and regulatory compliance take precedence. In addition, the chapter considers eco-design, modular architecture, and de/module/safe ways of dismantling that may help device management become more sustainable. Alignment of recycling efforts with the overall green healthcare and circular economy routes is proposed.

DOI: 10.4018/979-8-3373-4134-7.ch010

1. INTRODUCTION

While it is true that the medical IoT devices grew rapidly, and it has contributed significantly to the advancement of the healthcare sector by offering much improved monitoring, diagnosis, and treatment options. Now the range of IoT devices that are essential to patient care goes from wearables like heart rate monitors and insulin pumps, to more sophisticated devices such as insulin pumps and remote diagnostic tools. It is estimated that in coming decades, the global healthcare IoT market will grow immensely, with many more devices being deployed to deliver more personalized, real time, and data driven medical solutions. These devices, which promise much improved patient outcomes and operational efficiency, are becoming very much of an environmental problem. The need for effective recycling solutions for medical IoT devices is growing alongside the increasing number of devices in circulation: millions (Das et al., 2021).

Recycling medical IoT devices is a very complicated process marked with a plenty of technical, regulatory, and ethical challenges. Complicated materials ranging from metals to plastics to circuit boards, to even bio hazardous make recycling some of these devices difficult. First and foremost, however, data security is a big risk since patient data may be sensitive in these devices itself, needlessly complicating safe disposal of the technology. And then the contamination risks associated with recycling medical devices can be high, particularly with materials that have passed through that contact with bodily fluids or something like biohazard environments. In addition, there are no standardized processes for disposing and recycling of IoT devices across the different manufacturers, which further hampers the efficient recycling of IoT devices.

As medical IoT devices become increasingly depended on and the environmental impact from the disposal of these devices stands, the need for sustainable healthcare systems to consider the recycling of these devices is critical for the healthcare and technology industries. However, this is not only for reasons related to the reduction of environmental waste from medical waste, but also because it enables the adoption of a circular economy approach for medical device management, where the devices are designed to be reused, refurbished and recycled in the most efficient way (Singh et al., 2022).

Lessons from the importance of recycling in achieving e-waste sustainability goals: it mitigates the environmental impact of e-waste making it more sustainable and gives way to circular economy promoting the longevity of devices life cycle, reduction in use of new raw materials, resources conservation. Medical IoT device recycling is the key to reducing the banned effects of harmful medical iot devices that are greatly increasing an e waste crisis. Medical device waste is one of the fastest growing waste flows globally and according to reports, e waste growth is one of the fastest growing waste streams worldwide.

Although recycling is well known to be important, the reality of recycling medical IoT devices is not easy. Material complexity; amongst other challenges such as data security, contamination risks and not standardized processes across the industry. In addition to each of these challenges come with their own unique set of obstacles to overcome in order for us to put into practice effective and sustainable recycling processes for medical IoT devices.

In this chapter, we will deal extensively with the problem of these challenges and talk about the breakdown of technical, regulatory, and ethical barriers responsible for poor recycling of medical IoT devices. The chapter identifies the current recycling practice situation and innovations in the field and provides potential solutions to overcome the challenges outlined. These will include streamlining recycling process, improving patient data security as well as promoting eco design practices that can render it possible and affordable to recycle. The chapter will evolve from this exploration of a design solution to provide usable insights for the stakeholders in the medical IoT device supply chain: healthcare providers, device manufacturers, and policymakers to facilitate more sustainable management of medical IoT devices including efforts around their recycling with an ultimate aim of connecting recycling of IoT devices with the aspirations of green healthcare and the circular economy (Bhutani et al., 2022).

On top of that, this chapter applies the challenges to the practical realm of eco design, modular architecture, and safer dismantling practice which will greatly help the recycling process. The chapter will also highlight the importance of bringing all involved parties — manufacturers, healthcare providers, regulators, and recyclers, to work together more closely, fostering an environment where it is easier to manage the lifecycle of devices more sustainably. The bigger picture is to build the foundation of a more sustainable and responsible healthcare ecosystem that can deal with the complete cycle of medical IoT devices and help reduce waste to protect our planet and create a healthier earth.

The rest of the chapter consists of: an exploration of the technical challenges to recycling medical IoT devices from an complexities of materials used viewpoint, and data security and contamination risks respectively. Then it examines current practices of recycling in the healthcare. The chapter then examines innovations in aspects of eco-design and safe dismantling that may overcome recycling barriers. In the next is a discussion about regulatory challenges and the difficulties that privacy of patient laws can create obstacles to compliance with recycling procedures. Finally, this chapter provides set of proposed solutions to overcome these challenges for future medical IoT device recycling in a circular economy framework.

This chapter tries to contribute to the growing dialogue about sustainability in healthcare and the role recycling medical IoT devices may have played to minimise the sector's environmental impact by identifying barriers and proposing concrete solutions. This work intends to inspire further research and collaboration with key stakeholders on innovative solutions to overcome challenges and realize opportunities related to recycling in order to bring sustainable medical device recycling to reality.

2. THE CHALLENGES IN RECYCLING MEDICAL IoT DEVICES

While medical IoT devices have seen a rapid growth, which has led to the improvement of patient care, there are concerns about their effect on the environment as well as the difficulties in recycling their presence. Recycling medical IoT devices is complicated, for everything ranging from the complexity of their materials to risks associated with their data security and contamination, as well as the absence of standardized processes. Responding to these issues is critical for evolving to a more sustainable disposal and recycling of healthcare devices. In the next section, we dive into these challenges in detail, describing each of the hurdles that make the recycling of medical IoT devices difficult in particular.

2.1 Material Complexity

Challenging to recycle the medical IoT devices is the material complexity of the devices themselves. Such devices are usually made of many kinds of materials — metal, plastic, semiconductor, and glass among other things. Because of the variety of materials used, recycling these materials can become more complicated, because they need to be processed differently in order to extract or reuse each option. For example, metals like copper, gold and silver are valuable and can be recovered, but the existence of mixed materials including plastics and composite material makes the separation and recovery process more difficult (Albreem et al., 2023).

Additionally, the dismantling of medical IoT devices consists of small and intricate components, for example, microchips and circuit boards, which are quite a challenge to dismantle manually during the lifetime of the device. Technically, it can be difficult and sometimes expensive to automate the separation of these materials, which discourages effective recycling efforts. While this may suggest that since the materials used are not easily separable and because components are miniaturized and complex, the process of recycling these materials is more expensive and less efficient, this leads to situations where many models might be sent to the landfills rather than being recycled.

Also, most of the medical IoT devices are designed with non recyclable embedded components which also add to the adverse impact on the environment. Such manufacturers also prefer to make the devices compact and durable at the expense of recycling, complicating the implementation of an adequate end of life recycling strategy. However these devices' materials are complex and without proper design modifications recycling efforts will continue to be hampered.

2.2 Data Security Concerns

The second main challenge in the recycling of medical IoT devices has to do with data security of medical IoT devices during the disposal and recycling process. Due to the possession of sensitive patient information such as health records, medical histories and other personally identifiable information, medical devices store sensitive patient information. If not treated properly, these devices could be dangerous and may cause leakage of data, which could ultimately harm the patient privacy or may result in legal liabilities.

For example, in the United States, there are laws like Health Insurance Portability and Accountability Act (HIPAA) while in Europe there is the General Data Protection Regulation (GDPR) that puts a huge regulator on how healthcare related data is handled, stored and disposed of. Under these regulations, the data is required to be completely wiped out before the device can be disposed or recycled. The problem is that incomplete destruction of data remains possible, especially when the devices are heavily used in healthcare and disposal of the devices is not thoroughly examined as per law (Goyal et al., 2024).

Though various devices are designed with inbuilt data encryption or some other security attributes, the recycle course of action generally includes removal or reuse of components that might not have adequate safety against getting the data pulled out. Once data is destroyed there is still the chance that device components will be used again, refurbished, or sold providing access to private data to unauthorized persons. Consequently, patient information security during recycling calls for technical solutions such as secure data wiping technologies and also well laid down procedures ensuring keeping with the privacy regulations.

2.3 Contamination Risks

Medical IoT devices are in contact with bodily fluids, biohazards, as well as infectious materials in certain environments. Therefore, recycling or disposal of these devices that have reached their end of life cycle can cause contamination. Devices left with biohazardous 'waste' on them, such as blood, urine or tissue, can remain on the devices and pose health risks to those dismantling and recycling them.

Take infusion pumps, glucose meters and thermometers: Those and other medical devices might have remnants of bodily fluids that are hard to detect, or impossible to remove, altogether. When these devices are not cleaned thoroughly and properly sterilized before being recycled, they are a significant health risk. In addition, there are risks of crosscontamination —transfer of biohazardous material to other devices or recycling facility parts resulting in broader exposure (Moshkbid et al., 2024).

In order to mitigate these risks, safe dismantling practice and procedures for cleaning medical IoT devices before their recycling are must be developed. Therefore, there are processes which must be adhered to rigorously, to ensure that any device that may have come in to contact with bodily fluids or bio hazards does not end up in the handling of recycling personnel without having been properly sterilized. Besides, facilities handling the recycling of medical devices should be furnished with the necessary specialized tools, protective gear, and containment measures to ensure prevention of exposure to harmful substances in the recycling process.

Table 1. Key challenges in recycling medical IoT devices

Challenge	Description	Implication
Material Complexity	Devices contain diverse materials (metals, plastics, semiconductors) and miniaturized components.	Difficult and costly to separate and recycle; often leads to landfill disposal.
Data Security Concerns	Devices store sensitive patient data that must be securely destroyed before recycling.	Risk of data breaches and non-compliance with privacy laws (e.g., HIPAA, GDPR).
Contamination Risks	Devices may carry biohazards from contact with bodily fluids and infectious materials.	Health risks to recycling personnel; requires rigorous sterilization before handling.
Lack of Standardized Processes	No unified recycling protocols across manufacturers or facilities.	Inconsistent recycling practices; inefficient or improper device disposal.

As shown in Table 1, the combination of the technological, regulatory, and social challenges affect the effective recycling of medical IoT devices to the point where they require coordinated industry standards, secure data disposal methods, and contamination safe recycling protocols.

2.4 Lack of Standardized Processes

Hence, the absence of standardized processes in recycling medical IoT devices creates a key challenge in recycling. While there is more established and widely used protocols for recycling traditional consumer electronics, there are no uniform guidelines or industry-wide standards to recycle in the medical device sector. These absence of standardization creates a gap in the recyclability process, thus making it very difficult for healthcare providers, recyclers and manufacturers to align their efforts.

Device designs are not uniform in any way. Since each manufacturer can use different materials, components, and differing designs, it is difficult for the recyclers to develop efficient ways of disassembly and material recovery. However, due to this variability in device design and in recycling compatibility, recycling operations are not very efficient. In addition, without standardization, healthcare providers face the difficulty of locating the best recycling channels, which in turn increases the unpredictability of recycling rate and the risk of incorrect disposal (Ongaro et al., 2022).

Currently, although some regulatory frameworks and guidelines exist to tackle the medical device disposal (e.g. Waste Electrical and Electronic Equipment Directive in the European Union) aren't precise enough in addressing the special problems associated with medical IoT devices. It is vital to develop clear industry standards and best practices for the recycling of medical IoT devices because they will increase and improve efficiency and help these medical IoT devices be disposed of safely and in an environmentally responsible manner.

3. EXISTING RECYCLING PRACTICES AND THEIR FEASIBILITY IN HEALTHCARE

As previously mentioned, it is a complex task for recycling medical IoT devices. Nonetheless, numerous recycling strategies have been put into use in handling the rising e–waste across the healthcare industry, amid the various challenges littered on its pathway. This section discusses the working of existing recycling practices of medical IoT devices as well as their advantages, setbacks and the role of specialized recycling facilities. Moreover, it discusses the case studies in the healthcare industry that are categorized into real world efforts to deal with these challenges and their outcomes.

Figure 1. Medical IoT device recycling workflow in healthcare settings

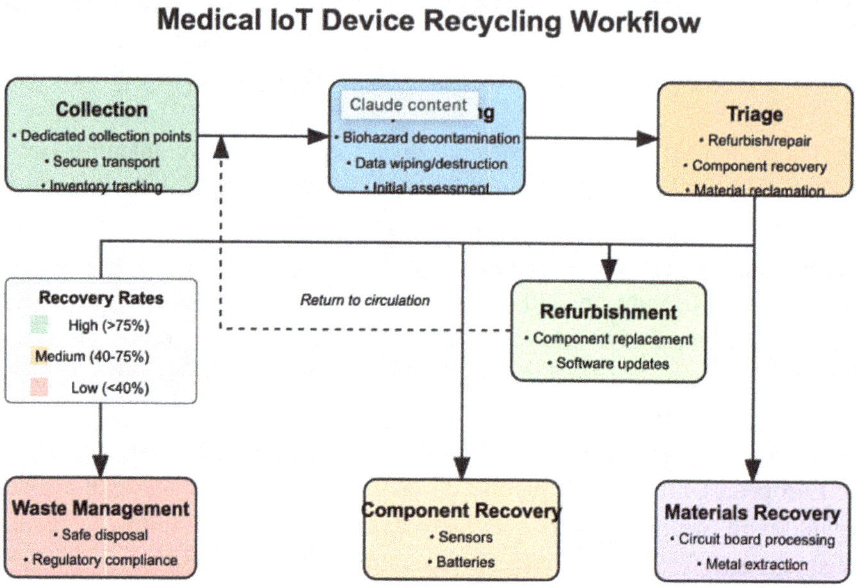

As shown in Figure 1, the medical IoT comprehensive recycling pathway includes collection and preprocessing (e.g., biological hazard decontamination and data destruction); triage decisions to route medical IoT devices to refurbishment, component recovery, material extraction or waste management. It also shows recovery rates of the refurbished devices coming back into circulation as a circular process and illustrates them through approximate recovery rates, based on different paths of the processing. It is based on the workflow of best practices derived from case studies of leading healthcare institutions deploying systematic approaches towards effective management of medical e-waste.

The recycling of medical IoT devices is depicted by a workflow structure (see Figure 1) with the aim to strike a balance between resource recovery and regulatory compliance. A multi pathway approach that recognizes and addresses the idiosyncrasies of healthcare e waste including uniquely addressing special processes for biohazard decontamination, and specifics of data security while maximizing resource recovery. The figure shows how successful recycling programs use a triage based decision framework in order to direct the devices to the appropriate end–of life pathway which is either refurbishment, component harvest, or material reclamation. However, the most notable processes in terms of recovery rates (>75%) are refurbishment processes which result in devices back into circulation, while processes for the specialized recovery of materials can achieve relatively high recovery rates (40 – 75%) from devices that cannot readily be refurbished, but contain valuable elements such as the precious metals from circuit boards or the rare earth components of sensors.

3.1 Overview of Current Recycling Methods Used for Medical IoT Devices

The recycling methods of medical IoT devices depend on the materials that have been used and the processing needed. Typically, the recycling of these devices is achieved through physical dismantling, material sorting, and salvaging of valuable constituent parts. Most recycling facilities have ways of shredding, sorting and separating the raw materials inside the electrical components to extract the metals, including gold, silver and copper. Because gross of these metals a scarce and high demand will also happen to be demanded in developing new devices such as medical IoT systems.

Semiconductors and circuit boards within some medical devices where the processes are mechanical, such as diagnostic and monitoring systems, may be reclaimed. Plastics, glass, and other composites are usually recycled or sent to landfills otherwise. However, these traditional methods are not suitable for medical IoT devices since they have complex design and comprise many different materials (Gupta et al., 2021).

3.2 Advantages and Limitations of Traditional Recycling Practices

Though powerful in regards to some electronic waste recycling, standard recycling techniques are less useful when it comes to recycling medical IoT devices. The methods are one of the main advantages as they recover metals like gold and copper, which can be recycled to make new devices. As a result, virgin materials requirement is decreased and resources are conserved. Furthermore, the traditional recycling practices are very well understood as well as practiced and commonly represents one of the cheaper (low cost) solution by many recyclers.

However, the way this is achieved through traditional recycling methods becomes less clear when it concerned medical IoT devices. For example, taking into account that these devices are made up of different materials—such as mixed plastics, complex circuit boards and specialty metals—the recycling process is more complex. They are also harder to disassemble than many medical IoT devices, ultimately making them difficult to recycle through the automated systems used for this task. In addition, the recycling of metals and certain plastics can go some way to minimising waste, but the non recyclable elements such as certain composite materials are still deposited in landfills, undoing the good work.

Traditional recycling methods also have another huge downside, which is the problem of keeping data safe during the disposal process. As was mentioned earlier, although medical IoT devices typically contain sensitive patient information, yet, shredding or simply incinerating the devices does not ensure that all data from these devices is going to be destroyed completely. The improper disposal of medical devices could result in data breaches and breaches of patient health information confidentiality regulations, namely HIPAA and GDPR, therefore posing a serious risk to the individual's privacy.

3.3 The Role of Specialized Recycling Facilities for Healthcare Equipment

Due to the individuality of recycling medical IoT devices, there has been the development of the dedicated recycling facilities specialized in dealing with equipment specific for the health care. Beyond a doubt, these facilities have advanced technologies and processes that fit with recycling of medical devices including product with complex materials and data security issues. Techniques including manual disassembly are performed by specialized recyclers for more precise sorting of component and materials.

Besides handling medical waste, these facilities also address the contamination risks related with devices utilised in the healthcare environment. As infection control is tight and must be sterile, specialized recycling facilities have state of the art cleaning and decontamination procedures to assure that devices are biohazard free before they are dismantled. In addition, these facilities tend to have the means to dispose of devices that contain sensitive patient data using more secure methods that meet the privacy standards (Chaudhary et al., 2024).

The main purpose of the specialized recycling facilities is to provide a more sustainable and a safer form of recycling, at least compared to others. Regardless of whether facilities adopt stewardship based on ethical, economic, or environmental key officeholder viewpoints, the focus on medical devices allows maximizing the recovery of potential reusable materials to the greatest extent possible, reduce environmental impacts and maintain compliance with the very strict regulations in which healthcare equipment disposal is regulated under.

3.4 Case Studies of Recycling Efforts in the Healthcare Industry and Their Outcomes

There are several healthcare organizations, and recycling initiatives that venture into efforts of developing and implementing effective recycling practices of medical IoT devices. However, there is a case with a hospital or a medical center that has partnered with specialized recycling companies to do away with outdated or non functional medical equipment. The goal of these partnerships was to make sure that devices will be broken down, sterilised and safely recycled in line with the appropriate environmental and regulatory standards.

For example, a US based leading healthcare provider partnered with a specialized recycler that co processes their retired medical devices. The rollout of this partnership has allowed the organization to send almost 90% of its medical waste straight into landfill, and greatly reduced its environmental impact. Advanced recycling technologies in combination with manual disassembly were used by the recycler to reclaim valuable metals and other materials as well as plastics. Furthermore, every piece of sensitive data was wiped off devices safely and thus HIPAA regulations were followed and the risk of data breaches was eliminated.

For another example, a European initiative that is focused on cutting medical device waste and recycling in healthcare is an example. For the past 10 years teams of hospitals, universities, waste management companies have been working on developing a comprehensive medical device recycling program incorporating eco design principles, material recovery and safe disposal practices. This has resulted in the implementation of a circular economy model for medical IoT devices, thereby reusing or refurbishing components to be built into new devices without the need for virgin materials thereby reducing costs to an overall extent. Hospitals have resorted to using components from retired devices instead in order to lower their procurement expenses (Quinto et al., 2025).

Nevertheless, despite these successes, scalability of such programs remains a challenge. Many healthcare facilities, particularly smaller or less resourced ones, still struggle with the logistics and costs of recycling medical IoT devices. For smaller facilities, as there is no way to get larger organizations to provide specialized recycling partnerships or infrastructure. For this reason, further innovation and cooperation by the healthcare sector are key to broader adoption of these recycling efforts.

4. INNOVATIONS IN ECO-DESIGN FOR MEDICAL IoT DEVICES

The healthcare sector is under ever growing load to reduce environmental impact, and eco design emerged as a key approach of modern medical IoT device development. Eco design is about the device design that is resource efficient, environmentally ecological, recyclable and easy to recycle to reduce wastes and achieve sustainability throughout the device lifecycle. In this section, the role of eco designs in solving the problems associated with recycling medical IoT devices is discussed, with special emphasis on the use of recyclable materials, simplification of the devices architecture and the adoption of modular designs. It looks as well at how durable, long lasting design can lead to less frequent disposal of the device, and examples from real world cases where medical device companies have embraced eco design principles.

4.1 The Role of Eco-Design in Reducing Recycling Challenges

Eco-design has a key role to play in reducing the challenges for recycling medical IoT device as it addresses these issues at design stage. Recyclable materials are one of the principles of eco-design. In other words, manufacturers can choose materials that are easier to recycle like certain metals, plastics and bioplastics, so that it will be easier for the recyclers to recover valuable components. For instance, selecting aluminium, which is a metal that is widely recyclable, can greatly minimise the environmental effect compared to other non-recyclable substances.

In addition, eco-design promotes the decrease of device architecture complexity. Typically, medical IoT devices are of a complex nature, crafted with several layers of material that are difficult to divest and recycle. This can be enabled by designing devices containing fewer and more homogenous components, which are simpler to disassemble and recover materials from. For example, having a unified plastic shell, rather than two different plastics together with different metals, makes it more easy to sort and recycle (Bhutani et al., 2023)

Eco-design also motivates to reduce the toxic substances utilised in medical devices. Traditional devices generally contain non biodegradable hazardous materials such as lead, mercury, and cadmium that are harmful if not disposed of properly both to human health and to the environment. Eco-design can make medical IoT devices safer to handle during recycling process, and to this end, it can help to eliminate or reduce the use of these substances.

4.2 Modular Design Approaches that Facilitate Easier Disassembly and Recycling

Another innovative approach in eco design to improve the recyclability of medical IoT devices is the use of Modular Design. Modular design is that in which devices are developed so that they can be easily disassembled, that way separate components or modules can be replaced, fixed, or recycled without the need of dispose of the whole device. By using this approach we come to enhance both the device lifespan and sustainability since the components can be repurposed or restored and prolong the life of the product in general.

Modular design especially suits in the instances of medical IoT device because one can upgrade the part of device i.e. sensors or batter, and doesn't need to throw the entire device. This will decrease the amount of devices that are discarded and lead to reuse of valuable materials. For example, when a medical IoT device has a modular sensor module, the sensors can be replaced when they become past their life, rather than discarding the whole unit. This decreases waste and saves resources, which follows the principles of the circular economy.

In this regard, modular designs are also made with standardized components, which in turn make the recycling process easier. Recyclers can more efficiently recover materials for reuse by providing that parts can easily be separated and identified. This makes recycling process simple and increases overall sustainability of medical IoT devices (Rajagopalan et al., 2024).

4.3 Design for Durability and Longevity to Reduce the Frequency of Device Disposal

This is one of the eco-design key principles because it is about making devices that will last and last. Within the scope of medical IoT devices, this implies the development of products that can endure long term use without the requirement to replace and throw away them often. Manufacturers can improve the environmental impact of the production and disposal of devices in the case of extending the lifespan of medical devices.

A lot of design for durability involves the use of high quality materials and components that will last against the stresses of health care environments. For example, if you consider medical IoT devices, such devices are exposed to often cleaning, sterilizing or rough handling and you need to design such devices to withstand wear and tear. In parallel, units that can be updated or repaired when they become obsolete by improving the software and/or hardware architecture are also possible.

Also, longevity for medical IoT devices contributes to reducing the number of electronic waste generated. In the healthcare area, where budget restraints and the pressures hence accrue from regulation can force premature replacement of devices, longevity design can make significant contribution to waste reduction. This innovation has a potential for reducing healthcare provider's costs, while at the same time promoting environmental sustainability (Matin et al., 2023).

Table 2. Innovations in eco-design for medical IoT devices

Innovation	Description	Impact on Recycling and Sustainability
Use of Recyclable Materials	Materials such as aluminum, bioplastics, and recyclable plastics are prioritized for manufacturing medical IoT devices.	Simplifies recycling processes, reduces environmental impact, and aids material recovery.
Simplified Device Architecture	Devices are designed with fewer, more homogenous components for easier disassembly and recycling.	Facilitates the separation and recovery of materials, improving recycling efficiency.
Reduction of Toxic Substances	Elimination or minimization of hazardous materials (e.g., lead, mercury) in device components.	Reduces environmental and health risks during recycling and disposal.
Modular Design	Devices are built with replaceable modules (e.g., sensors, batteries), enabling easy upgrades and repairs.	Extends device life, reduces waste, and promotes material reuse in alignment with circular economy principles.
Durability and Longevity Design	Devices are designed to withstand prolonged use in healthcare environments, minimizing the need for frequent replacements.	Reduces the frequency of device disposal and e-waste generation, promoting long-term sustainability.

Table 2 illustrates key eco-design innovations that contribute to the sustainability of medical IoT devices, focusing on recyclable materials, simplified designs, and long-lasting product lifecycles.

4.4 Examples of Medical Device Companies Implementing Eco-Design Principles

Fortunately, already some medical device companies have jumped on the eco-design bandwagon with regards to Iot devices and have clearly shown it is possible to develop sustainable healthcare innovation. Such example is Philips Healthcare, using eco-design in its medical devices portfolio by designing energy saving technologies, recyclable materials and modular designs. One of the most important and innovative approaches to eco-design used by the company is to minimize the negative effects on the environment its products have throughout their life, from production to disposal. Philips has also made an effort to draw design for disassembly, so that when the device's life is up, components can easily be recycled (Hernandez et al., 2024).

Another one is Siemens Healthineers, which has incorporated sustainability in its medical device design by choosing energy efficient components and making the material used in its products recyclable. The company strives to limit use of hazardous materials and design for longevity, that its devices can be kept and fixed to increase their life. Moreover, Siemens also promotes the take back programs for disposing devices off and properly recycle them for the healthcare sustainability.

Beyond the above, GE Healthcare has as well put effort in eco design which targets the environmental impact of the imaging devices. The company has adopted energy efficient technologies, is using recyclable materials, and has used design principals that make it easier to disassemble and recycle. GE Healthcare has incorporated sustainability into the design process to reduce environmental costs as well as the economic costs of its medical devices.

These supply the examples of how medical device manufacturers are adopting eco design principles in their products and thus achieving sustainability in healthcare sector. These companies understand the future of medical IoT devices by focusing on recyclable materials, modular design, durability and longevity. Expectations for sustainable healthcare solution are increasing and there is a possibility of more companies adopting eco design practices of medical IoT devices.

5. SAFE DISMANTLING PRACTICES FOR RECYCLING MEDICAL IoT DEVICES

The complexity of these medical IoT devices, the need for patient safety, data security, and compliance dictate the challenge of medical IoT device recycling. Medical IoT devices at end of life cycle are best disposed of or disassembled using safe practices as improper disposal or dismantling can carry with it health risks, contamination, data breach, and environmental damage. The development of safe and efficient dismantling processes, technological innovation that improves the dismantling process, training and certification for healthcare workers in the dismantling process, and collaboration between manufacturers and recyclers are discussed.

Figure 2. Integrated framework for safe dismantling of medical IoT devices

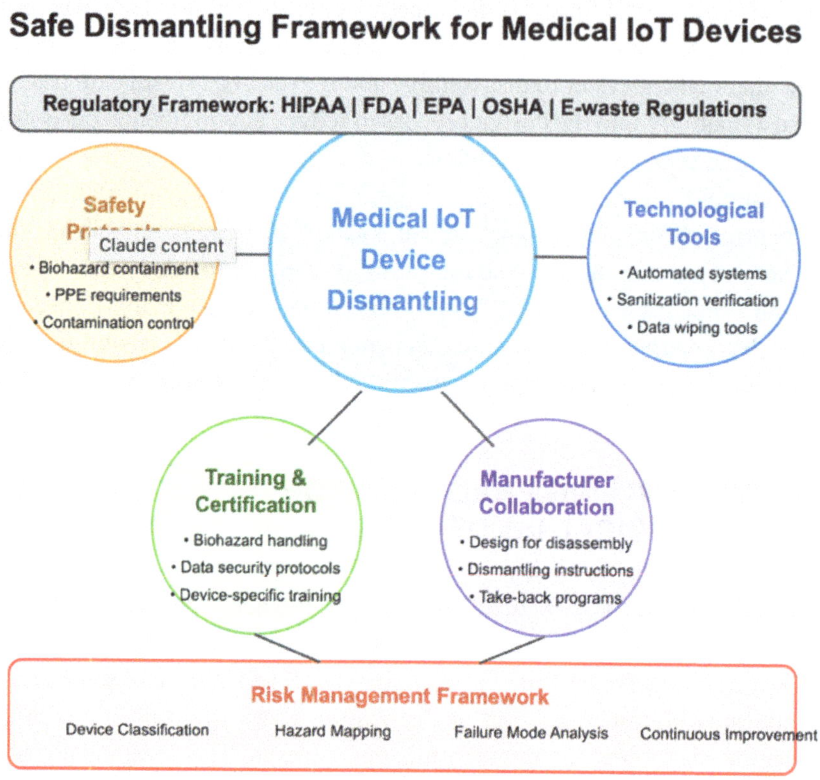

A comprehensive approach to destroy medical IoT devices is necessary to safely dismantle at end-of-life and this is depicted in Figure 2. The four major domains of safety protocol for biohazard management, technological tools in automated and verified processing, training needs of personnel and collaboration with manufacturers revolve around the central focus on the medical device. The application of these elements occurs within an overall regulatory framework and are supported by a basic failure mode based risk management system based on device classification, hazard mapping, failure mode analysis and continuing improvement processes (Selvaraj et al., 2022).

As Figure 2 shows, the safe dismantling of medical IoT devices requires a multi faceted approach where the various technical, procedural and human factors are intertwined. Robust safety protocols need to be integrated with technology that's specialized for medical devices; biohazard containment and data security are unique challenges for medical devices. The training and certification requirements in place provide a basis of competence of the personnel in the handling of such sensitive items, and collaboration with manufacturers enables the more efficient and safer dismantling of such items based on design innovations and technical support. Omitting notes — these elements function in the substantial regulatory envelope comprising of healthcare privacy laws rules, medical device regulations, environmental protection standards or occupational safety criteria. Basing the foundation on a risk management framework guarantees that processes are routinely examined and improved throughout the course of time considering device-specific hazard analyses and through real world dismantling experience. However, safe dismantling practices require simultaneous coordination among multiple stakeholders and disciplines of expertise, and this paper advocates a holistic approach for addressing these issues.

5.1 Development of Safe and Effective Dismantling Processes to Prevent Contamination and Data Breaches

The first step of the medical IoT device recycling is safe and efficient disassembly to avoid contamination and data breaching. Medical devices often contain sensitive patient data, even if mishandled it can allow for the breach of patient privacy laws, most common in the US is HIPAA, and in Europe is GDPR, depending on the patient's nationality. Furthermore, medical devices can carry biohazardous matter like blood or bodily fluids that need to be handled with care to prevent exposure to infectious agents.

To counteract these risks, processes for safe dismantling of the electronic components as well as potentially hazardous materials have to be developed. This means that devices are taken apart in a controlled environment with proper sterilization procedures in order to prevent contamination. For instance, when the device has been exposed to blood or bodily fluids, it should be well cleaned and sterilized prior to being dismantled. The components may first be processed by special equipment (autoclaves) for disinfection, prior to further processing.

In addition, data security is at the top of the dismantling priorities. Most medical IoT devices store sensitive patient information like health records, data, etc. which need to be securely wiped or destroyed before the device is recycled. However, methods of data wiping can be used to ensure that any residual data is removed from devices being dismantled. Physical destruction of hard drives and all other components that store data may also prove necessary to prevent unauthorized access.

5.2 Technological Innovations to Automate or Streamline Dismantling Procedures

As medical IoT device dismantling gets more and more efficient and safer, its role in technological innovation is becoming more critical. Due to its complexity, disassembly of the manufactory takes much time and effort, limiting recycling. Therefore, many companies are investigating automated and semi automated dismantling solutions to make it easier and cut human error.

For instance, robotics and automation have been adopted into the dismantling process to lead delicate and hazardous tasks, including the separation of components or even the extraction of batteries. For instance, specialized tools can be given to robots to unscrew parts apart, separate components apart, and handle potentially hazardous materials. The purpose of these automated systems is to be able to work with multiple types of devices and to not require human input if for example the job is potentially hazardous or time consuming.

The other technological innovation is that advanced sensors and machine learning algorithms help separate materials more efficiently to identify them. Recyclers can use AI powered system to quickly sort the metals, plastics, and circuit boards etc. which will speed up the dismantling process as well as add to the accuracy. Besides enhancing the efficiency, this automation also helps to diminish the possibility of contamination and promotes the entire dismantling process to be safer.

In additional, laser cutting and 3D scanning are being used to very precisely disassemble devices with a minimal chance of destroying sensitive parts or contaminating materials. These technologies enable recyclable components to be separated from waste and recooped as much as possible, and thereby facilitate more environment friendly recycling process.

5.3 Training and Certifications for Healthcare Workers Involved in Device Disposal

The healthcare workers and technicians who are handling the disposing and dismantling of the healthcare IoT devices must be trained on how to deal with the tissue and ions in a safe and effective way. Since recycling of medical devices carries potential hazards such as exposure to biohazardous material and sensitive patient information as well as regulatory issues, guidelines for proper training of staff involved in medical device recycling in terms of the practices involved in medical device disposal and recycling are of utmost importance.

Proper disassembly techniques, sterilization procedures, data security protocols and environmental regulations need to be part of the training programs to be provided. In addition, healthcare workers should be trained about the potential risks from the improper handling of the medical devices including the exposure to infectious materials, chemical hazards or radiation. It is also expected that training should include best practices for maintaining compliance with applicable regulations, e.g. HIPAA, etc (Olabi et al., 2024). The FDA's requirements for medical device disposition (medical device disposal), etc.

In addition, certifications of healthcare workers who dispose of devices may help guarantee that staff have the ability to perform such tasks. Professional organizations or regulatory bodies may offer certification programs that ensure workers have completed standard training and are knowledgeable with the latest industry standards. The certification not only makes the process more safe but also accounts for proper regulatory frameworks during the dismantling and recycling.

5.4 Collaboration Between Manufacturers and Recyclers to Ensure Safe and Efficient Dismantling

Medical device manufacturers and recycling facilities must work closely together in order to dismantle parts of a machine safely, efficiently, and effectively. It is the manufacturers that need to play the key role by designing devices easier to disassemble using materials easier to recycle and supply the recyclers with all detailed information about the components of the device.

For instance, manufacturers may inform recyclers of which parts include hazardous substances or sensitive information and how to safely dismantle products. In addition, manufacturers can partner up with recyclers to design new tools and methods to take apart these intricate devices. This partnership helps guarantee that recycling of those devices always adheres to safety and environmental standards and that all recoverable components are recycled for reuse.

Meanwhile, manufacturers can have shared knowledge passed back from recyclers with valuable feedback on the dismantling process, where improvements in recyclability, or safety can be put in place. The development of new technologies and processes that can be developed in collaboration with this, and that makes recycling of medical IoT devices efficient as well as sustainable, is also possible.

Also, working together among manufacturers, healthcare institutions, and recyclers, can encourage raising awareness and education in safe dismantling practices. Through collaboration, stakeholders can come together and develop a unified way to recycle these medical IoT devices, which would resolve both the environmental and ethical problems related to the disposal of medical devices (Kumar et al., 2023).

6. REGULATORY BARRIERS AND COMPLIANCE IN RECYCLING MEDICAL IoT DEVICES

Recycling medical IoT devices is a complex issue that is surrounded by some regulatory frameworks of device disposal, data security, environmental protection and patient privacy in place. These regulations are very crucial to prevent the recycling process from being unsafe, ineffective and contrary to what is lawful and legally required. Even so, these numerous regulations also intersect in a way that pose great challenges to healthcare institutions, recyclers, and manufacturers. This section looks at how healthcare regulations, environmental laws, and international standards have impacts regarding recycling medical IoT devices from a regulatory barrier and compliance prospective.

6.1 Overview of Regulatory Frameworks Affecting Medical Device Recycling

Several of the regulations governing the recycling of medical IoT devices may vary depending on the jurisdiction, but they are largely meant to safeguard the environment, protect the patients and thwart data security. The Food and Drug Administration (FDA) under the Federal Food, Drug, and Cosmetic Act (FDCA), regulates medical devices in the United States and its guidelines include how medical equipments are disposed to ensure patient safety and adherence to certain regulatory requirements. Whenever the regulations exist, the FDA cover how to safely handle, dispose and destroy medical devices to avoid contamination and not be used.

Besides FDA regulations, Resource Conservation and Recovery Act (RCRA) among many other environmental laws specifies disposal standards of hazardous waste including electronic waste (e-waste). Specific requirements are laid down by these laws on the treatment, storage and disposal of hazardous materials contained in medical IoT devices including batteries, lead, mercury and other hazardous substances. These regulations are put in place to ensure that medical devices are disposed of in a proper way so as to prevent the environment from becoming contaminated and negating human health risks.

In other countries there are such environmental regulations as the European Union's Waste Electrical and Electronic Equipment (WEEE) Directive, that specifies requirements for collection, recycling and recovery of e-waste within member states. The WEEE Directive rests responsibility on manufacturers to make sure the products they bring to market are recyclable and disposed of properly at the end of life cycle. However, as medical IoT devices are recyclable the regulatory standards applied to them in one country may differ from that in another and therefore complicate international trade in recyclable medical IoT devices as it is more challenging to perform cross border recycling (Ghulam & Abushammala, 2023).

6.2 Compliance with Patient Privacy Laws (e.g., HIPAA) and Their Impact on Device Disposal

One of the biggest regulatory hurdles in recycling of medical IoT devices is ensuring that it will abide by patient privacy laws, especially in the US, where the

Health Insurance Portability and Accountability Act (HIPAA) stipulates patient information be strictly protected. Sensitive health data used by many of the medical devices in the market today, including wearable health monitors and diagnostic equipment, need to be securely stored in order to protect patient privacy. If care is not taken, these devices can leak data from Dashboard into the wild (making it vulnerable to data breaches) as well as granting unauthorized access to private personal health information.

To meet the requirements under HIPAA, the law requires that healthcare organizations enforce measures that preserve the confidentiality, integrity, and availability of electronic health information including in data stored on medical devices. Because of this requirement, this affects significantly how medical IoT devices are recycled. To make sure that no residual data is accessible, data protection and data destruction protocols typically must be followed very precisely by recyclers, for example by wiping data from devices or tearing down and physically destroying hard drives and storage components. Non-compliance with HIPAA's requirements to protect data is at best likely to result in substantial financial penalties, lawsuits, and reputational damage from the failure to comply with strict data protection requirements on both the face of healthcare institutions and device manufacturers.

As with the General Data Protection Regulation (GDPR) within the European Union, medical IoT device recycling process is also affected by strict data privacy requirements. With GDPR, that personal data must be securely wiped out when not longer needed, and device recyclers must adhere to this factor too when dealing with medical devices that include personal health information.

The upshots of these laws are that healthcare practitioners and recyclers need to implement secure data wiping technologies and rely on trustworthy hoodlums to take care of away of devices. Therefore, recycling of medical IoTs transcends an environmental and material issue to become a matter of extreme data security (Senarathna et al., 2023).

6.3 International Regulations and the Complexities of Cross-Border Recycling Efforts

One of the biggest regulatory hurdles to the recycling of medical IoT devices is writing to comply with patient privacy laws, especially in the United States where the Health Insurance Portability and Accountability Act (HIPAA) governs with great strings how to correctly protect patient information. Sensitive health data used by many of the medical devices in the market today, including wearable health monitors and diagnostic equipment, need to be securely stored in order to protect patient privacy. If care is not taken, these devices can leak data from Dashboard into the wild (making it vulnerable to data breaches) as well as granting unauthorized access to private personal health information.

To meet the requirements under HIPAA, the law requires that healthcare organizations enforce measures that preserve the confidentiality, integrity, and availability of electronic health information including in data stored on medical devices. Because of this requirement, this affects significantly how medical IoT devices are recycled. To make sure that no residual data is accessible, data protection and data destruction protocols typically must be followed very precisely by recyclers, for example by wiping data from devices or tearing down and physically destroying hard drives and storage components. Non-compliance with HIPAA's requirements to protect data is at best likely to result in substantial financial penalties, lawsuits, and reputational damage from the failure to comply with strict data protection requirements on both the face of healthcare institutions and device manufacturers.

As with the General Data Protection Regulation (GDPR) within the European Union, medical IoT device recycling process is also affected by strict data privacy requirements. With GDPR, that personal data must be securely wiped out when not longer needed, and device recyclers must adhere to this factor too when dealing with medical devices that include personal health information.

The upshots of these laws are that healthcare practitioners and recyclers need to implement secure data wiping technologies and rely on trustworthy hoodlums to take care of away of devices. Therefore, recycling of medical IoTs transcends an environmental and material issue to become a matter of extreme data security.

6.4 The Need for Clearer Standards and Guidelines for Recycling Medical IoT Devices

A major problem in recycling medical IoT devices is the absence of clear and standardized guidelines concerning the recycling process. As such, regulations do exist, but they don't offer detailed standards required for the medical IoT device recycling. It leaves open the chance for interpretation, meaning that devices are not all recycled the same way, nor does each material get recovered. For example, there are few industry wide guidelines similar to disassembly of IoT medical devices, separation of materials, and how to manage de sensitive data during recycling process.

It is very important to develop standardized recycling protocols for medical IoT devices' disposal in a sustainable and compliant way. If implemented, these standards would facilitate recyclers to optimize processes, lower occurrence of data breaches, and also aid in the execution of recycling programs overall. Additionally, this would determine a pathway for designers within healthcare institutions and manufacturers in making decisions that ease recycling and spur the development of more ecologically friendly products.

To implement these standards, this is going to have to be an industry stakeholder, regulatory body and environmental organization collaborative. Such guidelines would provide the tools to harmonize recycling practices across nations and stimulate healthcare sector to adopt principles of circular economy.

7. PROPOSED SOLUTIONS FOR OVERCOMING RECYCLING CHALLENGES

With increasing urgency for sustainability in the efforts of the healthcare sector, healthcare industry must find a way to address the cycle of recycling medical IoT devices. Reduction of the environmental impact, conservation of resources, and promotion of circular economy depend on effective recycling. However, significant technical, regulatory, and ethical barriers impede progress in this area. The solutions in this section propose the steps it believes will help overcome these challenges: 1) developing standardized recycling protocols; 2) improved collaboration between stakeholders; 3) better data security; 4) materials design innovation; and 5) change in regulations.

7.1 Developing Standardized Recycling Protocols for Medical IoT Devices

Recycling of medical IoT devices faces one of the most pressing challenges of having no standardized protocols that provide a path for the entire device disposal lifecycle. Currently, there aren't any well established assembly, recoup, and effergence standards that all industry involve; that causes disarray in the disbursement process. By standardizing recycling protocols, they would all be treated the same, therefore a truly standardized recycling process would be more efficient and environmentally friendly.

This can only be achieved via working together amongst healthcare industry stakeholders, medical device manufacturers and recyclers to devise and utilise a set of guidelines for medical IoThings recycling. All such aspects, from the disassembly of devices to the recovery of metals, plastics and rare earth elements, should be included in these protocols. Standardization would also diminish the risk of data breaches since all the devices would follow the sameistic routine: wiping off sensitive information before recycling to ensure that it adheres to privacy regulations such as HIPAA (Thakur et al., 2024).

This would also standardize practices which recyclers would be able to take best practices for material sorting and waste management. Consistent protocols when applied to the medical industry will ultimately decrease the environmental impact of medical IoT and draw them closer to the point of a circular economy.

7.2 Enhancing Collaboration Between Technology Providers, Recyclers, and Healthcare Institutions

T Therefore, close collaboration between medical IoT technology providers (manufacturers), healthcare institutions and recycling companies is needed in the recycling of medical IoT devices. Each of these parties is imperative in the safe and efficient recycling of medical devices. Healthcare institutions must also make sure they collect such devices, and manufacturers must design devices to be recyclable. The devices have to be processed by recyclers who have the expertise and the facilities required to recycle them in an environmentally acceptable way.

To counter the difficulties in recycling medical IoT devices, it is crucial to boost the collaboration among its stakeholders. Mutual understanding and aligning all parties in their own goals for sustainability can occur through regular communication and partnerships. For instance, device manufacturers can collaborate with healthcare providers to understand the particularities of the medical sector needs and constraints, and recyclers can support then with the info about what materials are used in these devices and how to deal with them.

Partnerships with other companies for take-back and return to manufacturer initiatives could make the process easier. For instance, healthcare institutions could opt to participate in these programs if such devices are retired or refuse to work properly, and would be expected to see to it that these devices are returned into the hands of manufacturers or recycler for the right disposal or refurbishment. These initiatives would provide the incentives for healthcare providers to reduce the burden and assist recycling.

7.3 Improving Patient Data Security During Recycling Through Secure Data Wiping and Destruction Technologies

Recycling medical IoT devices is of course an important topic but the key point for data security of the patient is. Sensitive health information gets stored in many devices that must be completely wiped or destroyed before disposal of the device. Because of this, patient data loss during the recycling process can result in severe data breaches, and, if it is not secured, privacy laws, like HIPAA and GDPR, can be violated.

To resolve this problem, the recycling must include data destruction technologies more advanced than the recycling devices can. Data wiping software that can be applied to already stored information can be used to erase all information that is stored. Physical destruction methods can also be applied to storage components like hard drives or memory chips by shredding or crushing them to prevent that information from being recovered.

A standardized procedure must be adopted by healthcare institutions and recyclers for data destruction: every device should be subject to secure wiping or destruction before it is recycled. Builtin data security features such as secure erasure methods can also be integrated into manufacturers medical IoT devices so that the risks of data breaches are mitigated. Prioritizing patient data security in the recycling process builds a basis for stakeholders to establish trust and confidence in the sustainable recycling practices.

7.4 Innovation in Biodegradable Materials or Reusable Device Components to Reduce Recycling Complexity

Another way to try to overcome recycling issues is through innovative design of medical IoT devices using biodegradable materials or reuseable components. If other eco-friendly materials such as biodegradable plastics, plant based polymers were used, the complexity of recycling medical devices would significantly decrease. These materials are more easily processed and decomposed and thus are less of an environmental burden in their disposal.

Manufacturers can also use biodegradable material itself and also design medical IoT device components that would be reusable and easy to change when worn out and discarded instead of fixing them. An example is modular components with exchangeable parts that can be easier repaired, upgraded, respun to other uses, thereby extending device lifespan and decreasing the reason for recycling.

The design strategies and the use of such materials could potentially decrease the amount of waste produced by medical IoT devices and enhance the recovery efficiency of its waste streams. Collaboration between manufacturers, healthcare providers and environmental organizations would be needed to establish guidelines for adopting and disposing these materials and designs — an onerous process that would encourage adoption of these materials and designs.

7.5 Advocacy for Regulatory Changes to Streamline Recycling Practices and Incentivize Green Innovation

On the other hand, most of the medical IoT devices face regulatory barriers and no concrete framework is available for the recycling of medical IoT devices. A leading way to overcome these barriers is to push for administration changes which will play a vital role in streamlining recycling procedures and enhance green innovation. To address the anomaly of medical device recycling, policymers and regulatory bodies should set clearer and more harmonized recycling standards for medical devices that include the characteristics of IoT devices.

Tax breaks or certifications of eco friendly products could provide incentives to manufacturers to use the recyclable and reusable products. Government can also support research of new technologies and materials for recycling by granting subsidy for green innovation. For manufacturers, these incentives would create the right incentives to prioritize sustainability and, for recyclers, give them an easier task in handling medical IoT waste.

International cooperation over regulatory standards would make the recycling of products that pass across borders easier. The standards of different countries could be harmonized to simplify for cross border recycling and to implement a more sustainable approach for e waste management in global scale.

Table 3. Proposed solutions for overcoming recycling challenges in medical IoT devices

Solution	Description	Impact on Recycling and Sustainability
Developing Standardized Recycling Protocols	Creation of industry-wide guidelines for device disassembly, material recovery, and e-waste handling.	Ensures consistency in recycling practices, improves efficiency, and reduces environmental impact.
Enhanced Collaboration Between Stakeholders	Partnerships between manufacturers, healthcare institutions, and recyclers to ensure effective device recycling.	Fosters mutual understanding and aligned goals, making recycling more efficient and sustainable.
Improving Patient Data Security	Use of secure data wiping and destruction technologies to protect patient information during device recycling.	Reduces risks of data breaches and ensures compliance with privacy laws like HIPAA.
Innovation in Biodegradable Materials and Reusable Components	Adoption of biodegradable materials and design strategies that enable reusable and repairable components.	Reduces recycling complexity and environmental impact, making disposal easier.
Advocacy for Regulatory Changes	Push for clearer and harmonized regulations to streamline recycling practices and incentivize green innovation.	Encourages sustainable practices and facilitates more efficient global recycling efforts.

Table 4 highlights key proposed solutions to address the recycling challenges of medical Iot devices, focusing on standardised protocols, collaboration, data security, material innovation, and regulatory changes to foster a more sustainable recycling process.

8. CONCLUSION

Recycling of medical IoT devices is a complicated but crucial step in healthcare sustainability advancement. With the healthcare industry ever more embracing IoT for better patient care and operation efficacy, the demand in sustainable medical device management has gained greater significance. Regarding the recycling of medical IoT devices, there is great promise of reducing environmental impacts, conserving valuable resources, and minimizing wastes; however, the latter involves several challenges of material complexity, data security, contamination risks, as well as a lack of standardized recycling processes. The barriers to be addressed are big and require a big, coordinated, multifaceted effort that involves collaboration between manufacturer, reporter, recycler and even regulators.

According to key solutions to overcoming these challenges include: development of standard recycling protocols, enhanced collaboration between stakeholders and integration of eco design principles, as well as use of technology for secure data destruction and dismantling. A recycling approach for healthcare IoT can establish a unified recycling system and ensure that medical IoT devices are appropriately treated in the context of promoting circular economy, allowing healthcare sector to minimize its environmental footprint. Additionally, materials science has been advancing and biodegradable or reusable components have been adopted.

9. FUTURE SCOPE

Medical IoT devices offer a major potential in future recycling with the help of the increasing technology advancements, changes in regulations, and growing need of sustainable healthcare practices. As IoT devices in healthcare become more widely used, it will become more necessary to create and implement more effective, scalable, and secure recycling schemes. One area of potential need is to continue developing eco design principles primarily towards medical IoT devices that are easier to disassemble, more modular and made of recyclable materials. New, environmentally friendly materials that are both durable and easier to recycle might help simplify the complexity of device disposal, reducing what they call 'device disposal complexity'.

In addition, the integration of artificial intelligence (AI) and machine learning (ML) in the recycling process presents a very interesting opportunity to further efficiencies in sorting, disassembly, and data destruction. It could be used to optimize material recovery and determine the valuable components, as well as the more efficient separation of hazardous materials. Automation in dismantling processes may reduce labor costs and reduce the risk of contamination so that devices could be dismantled safely and secure. Moreover, blockchain could enhance the recycling chain's transparency and traceability, while leveraging the legal conditions of privacy laws like HIPAA and protects data security across the recycling cycle.

It is expected that better support for sustainable recycling practices will also require regulatory frameworks to grow. This may lead to governments to impose stricter regulations in the design and manufacture of IoT devices with sustainability as an option to stimulate manufactures' initiative. One of the impactful challenges for resolving the global strength of the medical device industry is cross border harmonization of recycling standards to facilitate more efficient and standardised recycling practices across the globe. Finishing, the healthcare sector's stance on circular economy and sustainability in healthcare sector will drive the innovation in the medical IoT device recycling and healthcare waste. Improvements to recycling technologies, materials and practices will present an opportunity to the healthcare industry to bring the environment and patient care a more sustainable future.

REFERENCES

Albreem, M. A., Sheikh, A. M., Bashir, M. J., & El-Saleh, A. A. (2023). Towards green Internet of Things (IoT) for a sustainable future in Gulf Cooperation Council countries: Current practices, challenges and future prospective. *Wireless Networks, 29*(2), 539–567. DOI: 10.1007/s11276-022-03133-3

Bhutani, M., Lall, B., & Agrawal, M. (2022). Optical wireless communications: Research challenges for MAC layer. *IEEE Access : Practical Innovations, Open Solutions, 10*, 126969–126989. DOI: 10.1109/ACCESS.2022.3225913

Bhutani, M., Lall, B., & Agrawal, M. (2023). A novel energy-efficient adaptive superframe structure for OWC-based real-time bio-sensor networks. *Peer-to-Peer Networking and Applications, 16*(6), 3021–3031. DOI: 10.1007/s12083-023-01544-3

Chaudhary, V., Gaur, P., & Rustagi, S. (2024). Sensors, society, and sustainability. *Sustainable Materials and Technologies*, Article e00952.

Das, S. K., Chinnappan, A., Jayathilaka, W. A. D. M., Gosh, R., Baskar, C., & Ramakrishna, S. (2021). Challenges and potential solutions for 100% recycling of medical textiles. *Materials Circular Economy, 3*(1), Article 13.

Ghulam, S. T., & Abushammala, H. (2023). Challenges and opportunities in the management of electronic waste and its impact on human health and environment. *Sustainability (Basel), 15*(3), 1837. DOI: 10.3390/su15031837

Goyal, R., Elawadhi, O., Sharma, A., Bhutani, M., & Jain, A. (2024). Cloud-connected central unit for traffic control: Interfacing sensing units and centralized control for efficient traffic management. *International Journal of Information Technology : an Official Journal of Bharati Vidyapeeth's Institute of Computer Applications and Management, 16*(2), 841–851. DOI: 10.1007/s41870-023-01527-w

Gupta, A., Bhutani, M., Dixit, A., & Lall, B. (2021). FD-OMAC: A novel full-duplex optical media access control for IEEE 802.15.7. *IEEE Access : Practical Innovations, Open Solutions, 9*, 148894–148910. DOI: 10.1109/ACCESS.2021.3124975

Hernandez, D., Pasha, L., Yusuf, D. A., Nurfaizi, R., & Julianingsih, D. (2024). The role of artificial intelligence in sustainable agriculture and waste management: Towards a green future. *International Transactions on Artificial Intelligence, 2*(2), 150–157. DOI: 10.33050/italic.v2i2.552

Kumar, R., Gupta, S. K., Wang, H. C., Kumari, C. S., & Korlam, S. S. V. P. (2023). From efficiency to sustainability: Exploring the potential of 6G for a greener future. *Sustainability (Basel), 15*(23), 16387. DOI: 10.3390/su152316387

Matin, A., Islam, M. R., Wang, X., Huo, H., & Xu, G. (2023). AIoT for sustainable manufacturing: Overview, challenges, and opportunities. *Internet of Things : Engineering Cyber Physical Human Systems*, *24*, 100901. DOI: 10.1016/j.iot.2023.100901

Moshkbid, E., Cree, D. E., Bradford, L., & Zhang, W. (2024). Biodegradable alternatives to plastic in medical equipment: Current state, challenges, and the future. *Journal of Composites Science*, *8*(9), 342. DOI: 10.3390/jcs8090342

Olabi, A. G., Shehata, N., Obaideen, K., Sayed, E. T., Mahmoud, M., AlMallahi, M. N., & Abdelkareem, M. A. (2024). COVID-19: Medical waste management, impact on sustainable development goals, and bibliometric analysis. *Chemical Engineering & Technology*, *47*(1), 4–19. DOI: 10.1002/ceat.202300046

Ongaro, A. E., Ndlovu, Z., Sollier, E., Otieno, C., Ondoa, P., Street, A., & Kersaudy-Kerhoas, M. (2022). Engineering a sustainable future for point-of-care diagnostics and single-use microfluidic devices. *Lab on a Chip*, *22*(17), 3122–3137. DOI: 10.1039/D2LC00380E PMID: 35899603

Quinto, S., Law, N., Fletcher, C., Le, J., Antony Jose, S., & Menezes, P. L. (2025). Exploring the E-waste crisis: Strategies for sustainable recycling and circular economy integration. *Recycling, 10*(2), Article 72.

Rajagopalan, S., McAlister, S., Jay, J., Pham, R. D., Brook, R. D., Nasir, K., & Nieuwenhuijsen, M. J. (2024). Environmental sustainability in cardiovascular practice: Current challenges and future directions. *Nature Reviews. Cardiology*, 1–14. PMID: 39455886

Selvaraj, S., Prasadh, S., Fuloria, S., Subramaniyan, V., Sekar, M., Ahmed, A. M., Bouallegue, B., Hari Kumar, D., Sharma, V. K., Maziz, M. N. H., Sathasivam, K. V., Meenakshi, D. U., & Fuloria, N. K. (2022). COVID-19 biomedical plastics wastes—Challenges and strategies for curbing the environmental disaster. *Sustainability (Basel)*, *14*(11), 6466. DOI: 10.3390/su14116466

Senarathna, W. G. C. M., Sulaksha, L. G. T., Weerarathna, D. M., Jayathma, W. M. V. A., Gamage, D. G. M. L., Thennakoon, T. M. T. N., Hewage, H. T. M., Panagoda, L. P. S. S., Sandunika, D. M. I., & Perera, M. D. V. (2023). Paper recycling for a sustainable future: Global trends. *Journal of Research in Technology and Engineering*, *4*(3), 169–186.

Singh, N., Ogunseitan, O. A., & Tang, Y. (2022). Medical waste: Current challenges and future opportunities for sustainable management. *Critical Reviews in Environmental Science and Technology*, *52*(11), 2000–2022. DOI: 10.1080/10643389.2021.1885325

Thakur, A., Mukhopadhyay, T., & Ahirwar, A. K. (2024). Approaching sustainability in laboratory medicine. [CCLM]. *Clinical Chemistry and Laboratory Medicine*, *62*(9), 1787–1794. DOI: 10.1515/cclm-2023-0973 PMID: 38557335

Chapter 11
Unveiling the Hidden Footprint While Assessing the Environmental Cost of Medical IoT Devices

Vamsi Krishna Reddy Bandaru
https://orcid.org/0009-0001-9396-8250
Cisco, USA

Srinivas Reddy Bandaru
AIIMS, India

Dedeepya Sai Gondi
https://orcid.org/0009-0005-4686-9726
Independent Researcher, USA

ABSTRACT

The prevalent trend of internet of things (IoT) devices adoption in the healthcare sector has created a massive concern about its environmental effect. Although medical IoT devices have huge benefits for patients and overall operational efficiency, they come at a cost to the environment, which could be missed in the supply chain, usage, and disposal stages. This chapter examines the environmental impact of medical IoT devices on the carbon footprint, energy consumption, e waste generation, and also discusses the impact of extraction of raw material. It points out the hidden environmental costs in the development of medical IoT technologies and tries to provide a framework to guide how to design, produce, and dispose more sustainably. The chapter further focuses on whether eco-friendly innovations can

DOI: 10.4018/979-8-3373-4134-7.ch011

be emphasized in order to maximize data usage while minimizing energy and raw material consumption and eventually finding possible measures to reduce the environmental footprint of medical IoT devices to meet the world sustainability goals.

1. INTRODUCTION

The healthcare industry accelerated its ongoing development through Internet of Things (IoT) technology integration which delivered substantial improvements to medical operations as well as patient outcomes and diagnostic solutions. Hospital IoT devices come in multiple formats that include wearable sensors together with connected diagnostic tools and remote monitoring devices along with smart therapeutic equipment. Real-time data collection through these devices lets healthcare professionals deliver evidence-based care and personalized treatments and guide patients' health monitoring irrespective of their global location (Allioui & Mourdi, 2023).

The healthcare industry experienced rapid growth in medical IoT device adoption during the past years because hospitals seek more effective healthcare solutions. Data from market reports predicts medical IoT market growth to explode by surplus billions of dollars during the next ten years. These medical devices provide multiple advantages because they allow remote patient tracking while reducing hospital admission numbers and healthcare expenses and enhancing people's health results. The COVID-19 pandemic boosted the adoption of medical IoT technologies because remote healthcare services became essential at that time.

The benefits of these devices toward better healthcare delivery receive wide recognition yet the environmental implications of these devices gain minimal attention and research. The medical IoT industry continues to expand which heightens the requirement to measure and manage the environmental effects of its entire product lifespan. Medical devices bring instant healthcare advantages despite creating substantial environmental problems throughout their production cycle and operation period and disposal stage.

Medical IoT devices have become increasingly popular but they create various sustainability problems for the environment to handle effectively. Healthcare institutions must address the environmental consequences of digital medical technology adoption when they implement digitalized medical practices. A medical IoT device generates environmental impact starting from its developmental phase through manufacturing and concludes at product waste management. The multiple environmental effects from devices contain their energy requirements together with their material usage and their carbon emissions and their waste generation during their lifecycle (Corigliano, Morrone, & Algieri, 2025).

Finding up-to-date, accurate energy characteristics of these excludes is one of the main pain points because the energy consumption of the devices during their operation is one of the primary concerns. A large amount of energy is required by many medical IoT devices, especially those that need to be used in continuous monitoring and data transmission. However, the energy required by them during their life span leads to greenhouse gas emissions, thereby aggravating the issue of climate change. Moreover, the production processes for these devices are largely dependent on non renewable resources as well as also generate high material wastages that entail high carbon footprints.

Generation of e – waste is another critical issue. Similar to many electronic products, medical IoT devices have short lifecycles because of quick technological advancements as well as necessity to do constant product upgrades. As a result, e – waste is accumulated, when these devices, having reached the end of their useful life, are discarded. Improper disposal of medical IoT can contaminate the environment since many medical IOTs contains hazardous materials such as heavy metals and toxic chemicals that can pollute water, soil and air.

In addition, the medical IoT devices typically involve intricate supply chains comprised of several countries and industries. Different levels of environmental regulations and practices can make these supply chains obscure the true environmental cost of the devices. Furthermore, raw materials extraction including rare earth metals required for sensors and batteries also leads to very severe ecological impacts from habitat destruction to pollution and natural resources depletion. Considering these complexities, it is obvious that a full understanding of the environmental impact of medical IoT devices is essential to guide future attractive and helpful innovations and policies to foster a sustainable future (Sallam, Mohamed & Mohamed, 2023).

Environmental footprint of medical IoT devices is hidden in the material selection of various design, manufacturing and material disposal processes that make it difficult for consumers and healthcare providers to fully understand extent of its ecological impact. Usually this hidden footprint is not taken into account in the process of development and deployment of medical IoT technologies; they are mostly concerned about performance, cost effectiveness, and patient outcomes. However, in its very success, and therefore, in its future sustainability, the devices for remote healthcare are prone to potentially long term negative environmental consequences which are the critical aspect that can actually undermine the benefits they give to the healthcare system.

Many medical IoT device are developed in the design phase without considering their environmental impact. Often the secondary concerns are with what the materials are, how energy is consumed, and how easily they can be recycled compared with functionality and cost. For example, the use of non-recyclable plastics, toxic chemicals, and non-sustainable raw materials in the production of medical devices exacerbates their environmental impact. Although some manufacturers are starting to add sustainability to their designs, these attempts are small compared to the increasing necessity for hospitals to employ greener technologies.

The manufacturing process itself also adds the environmental footprint. It is energy and water-consuming, and it generates waste and emissions. Manufactures sometimes get these components from various suppliers in different countries which increases the carbon footprint arising from transportation and locating these plants. However, it is rarely accounted for in estimating how much overall the device costs the environment.

In the use phase there is additional environmental burden imposed by the constant need of energy to power medical IoT devices. A lot of devices need to be plugged to power sources or frequently recharged, and consequently, contribute to ongoing energy demand. Similar to the case with the other devices, also these devices can consume large amounts of energy that in turn translate to significant greenhouse gas emissions, particularly in the regions where electricity is mainly produced from fossil fuels when scaled across healthcare systems (Kumar, 2025).

One of the major environmental risks is still the final phase, that is the disposal of materials. The growing issue of e-waste is due to medical IoT devices meant for disposals prematurely or wrongly. This means that, since many devices bear batteries as well as sensor and other electronic components, disposal of these would result in these harmful chemicals leaching into the environment as well. Medical electronics often cannot be recycled and repurposed at the end of its life cycle because it is too complex and costly to recycle.

The novel contributions in this chapter are as follows:

- In this chapter, I provide a detailed and holistic quantitative evaluation of environmental impact of medical IoT devices throughout their whole product lifecycle – from design and manufacturing to operation and disposal, and so on. Most studies concentrate on individual stages, but this chapter delineates the role of each stage in relation to the total ecological footprint. On a more comprehensive note, this approach identifies key areas of improvement to help improve a more sustainable design and usage of the medical IoT sector.
- The main contribution of the chapter is to identify and explore 'hidden' environmental costs of medical IoT devices. However, these costs are often underestimated or ignored because in addition to the ecological impacts of raw material extraction and e-waste disposal, the device's operation is also very energy intensive. The chapter brings into attention these hidden factors as it motivates towards more sustainable decision-making and policy formulation for sustainability of medical IoT technologies.

- The chapter also provides an actionable framework for embedding sustainability at every stage of design and development of medical IoT devices. Specific strategies are mentioned regarding the use of eco-friendly materials, energy efficient components, recyclable designs and others to minimize the environmental footprint of medical IoT devices. This piece closes the gap in between a reasonable healthcare innovation and a reasonable environmental responsibility and delivers practical answers for medical IoT device producer, healthcare provider and policymakers to construct a extra sustainable medical IoT ecosystem.
- The chapter then brings forward future policy and regulatory recommendations to tackle environmental related problems raised by medical IoT devices. It focuses on the necessity for global standards and regulations that would promote sustainable manufacturing, usage and end of life practices. The chapter also contributes to the ongoing global conversation on how to adhere technological progress with environmental stewardship in the healthcare work environment.

The entire chapter will be divided into several sections for providing a structured and holistic analysis of the environmental footprint of the medical IoT devices. Firstly, it includes an Introduction that presents the ascent of medical IoT technologies and the need to measure their ecological result. The Lifecycle of Medical IoT Devices is then analyzed in detail through design, manufacturing, operations and disposal stages. The focus of the chapter then shifts towards critical evaluation of Environmental Cost in terms of carbon footprint, energy consumption and e waste generation. Next comes the Environmental Implications which attempts to probe into broader ecological impacts e.g. pollution and resource depletion. Real world medical IoT device and its environmental cost are illustrated by including Case Studies. Then the chapter goes on to introduce Innovative Solutions for decreasing the environmental footprint including sustainable design methods and energy efficient technological methods. The paper concludes by discussing Policy and Regulatory Frameworks that can encourage sustainable practices and then closes with a short Conclusion with some of the main findings and recommendations. This organization guarantees a graceful flow of thought from foundational concepts to actionable solutions, and furnishes theory as well as practice.

2. UNDERSTANDING THE LIFECYCLE OF MEDICAL IoT DEVICES

Medical IoT device lifecycle includes several stages, where each of them affect its environmental footprint. The stages from initial design and development to its end-of-life disposal together determine the sustainability of the device. This helps to understand each phase, so that improvements can be made at points where harm to the environment can be reduced. In this section, we look into the key stages of a medical IoT device's lifecycle, study their environmental costs and explore how medical IoT could diminish its ecological footprint (Aminizadeh et al., 2024).

2.1 Design and Development Phase

Any medical IoT device can only achieve any functionality, cost and environmental impact starting from this design and development phase. In this stage important decisions are to be done about which materials, components and technologies will be used in the device. For instance, many of the medical IoT devices, like wearable health trackers, smart implants, as well as some diagnostic tools, incorporate advanced sensors, microchips and batteries. The environmental footprint of the device is greatly impacted by the materials used, for instance plastics, metals, and electronics. It turns out that a high cost of resource extraction or a low recyclability can lead to a larger ecological impact during the entire life of the device.

Also, this phase dictates the device's energy efficiency as well as durability. To select such a device, designers usually have to choose between maximizing the device's performance features like its processing power, connectivity capabilities, and small form factors, and ignoring the environmental costs both in the form of their device's energy consumption and waste generation. Some of the eco-design principles like the use of recyclable materials, low power consumption components and modular designs may not be at the forefront. There are some manufacturers that have adopted some sustainable design practices in the design, but the focus is still on the design that yields the most innovative design, performance, and least cost (Cappellini & Schirru, 2024).

2.2 Manufacturing and Supply Chain Considerations

After the design phase, the medical IoT device goes in to production. One phase that has a lot of environmental impact is this phase because it entails a very large and complex sourcing of raw materials, component production, and final assembly of the product. The assortment of these specialized devices typically includes the sensors, the circuit board, the battery and the display, which are almost always sourced from multiple suppliers from around the world. The carbon footprint also added on by transporting these components as their supply chain stretches across different continents.

Another key fact is the energy consumption during manufacturing. Most of the medical IoT devices are manufactured in facilities that require energy intensive processes, usually with energy that is drawn from non renewable sources. Also, the extraction and processing of raw materials including lithium for batteries, rare earth metals for electronic components for example, affect significant environmental degradation. These devices also incur environmental costs—habitat destruction, soil erosion, pollution—because of mining operations. In addition, packaging and defective components produced during manufacturing contribute to environmental impact. Although some manufacturers are starting to change their production processes, including reducing waste and converting to renewable energy sources, scaling effort across the world is difficult to overcome on a global scale (Regona, Yigitcanlar, Hon, & Teo, 2024).

2.3 Usage and Operational Impact

Medical IoT devices' use in hospitals, clinics, or at home by an individual patient runs into the operational phase of such devices. During this phase, these devices consume energy to do their jobs, and the energy consumption smart device is function specific and depends on the smart device's device complexity and connectivity requirements. Many of these devices are expected to persistently monitor patient data or transmit real time patient information to healthcare providers, or even provide therapeutic interventions, requiring a constant power supply, or continually recharging the device. This phase consumes the energy which results in overall environmental footprint of the product, particularly the devices are quite huge in healthcare systems.

Medical IoT devices, in addition to their energy use, also have a set of environmental impacts through the use of wireless connectivity. Wi-Fi, Bluetooth, or cellular networks are used by many devices to send data, while critical for real time monitoring and diagnostics, therefore also constitute a source of energy consumption. Further, the demand for data center to process and store the transmitted information enhances the energy footprint. These networks can also greatly improve the efficiency of these networks—and enables the devices to run using as little power as possible—can considerably reduce the impact of their operations. With higher dependence on IoT technologies in healthcare systems, it is imperative to investigate possible measures to minimize energy usage while devices age, so that the value of these devices does not suffer the consequences of environmental costs (Sabet, 2025).

2.4 End-of-Life and Disposal

The problem, however, is when medical IoT devices reach the end of their useful life, they are a problem with serious environmental repercussions. Like most electronical products, med devices are consisted of a number of hazardous materials which can cause harm to the environment when not disposed of properly. It is for this reason that many medical IoT devices contain batteries, printed circuit boards (PCB), and other parts that, when discarded in landfills, will release toxic chemicals. In addition to containing hazardous materials, these medical IoT devices include plastics that are not easy to recycle, which accumulate to general waste in general.

The process of recycling medical IoT devices has some challenges. Very often there are many of small parts, and disassembly is not easy, making acquisition of precious material such as gold, copper and silver on board more difficult. Moreover, certain healthcare laws may prevent recycling of medical tools since several of these devices are considered medical waste and need to be thrown away by specific protocols. Then the large number of these devices end as e waste, a prospect that becomes a grave environmental threat in countries with poor waste management systems.

To combat these challenges, however, some manufacturers are testing take back programms or collaborating with recycling firm to offer the right way to get rid of and recycle medical IoT devices. Another demand is for devices that can be easily re cycle and can be composed of non harmful material to environment. Nevertheless, these measures are yet in incipience and when these practices are adopted by more medical IoT device manufacturers, they will be able to effectively minimize the environmental impact caused by the disposal of their products (Arega & Sharma, 2023).

So, the lifecycle of medical IoT products is such, and all this results in the total environmental footprint. From design and fabricating these technologies, to use and finally disposal, there are opportunities to minimize their ecological impact. However, by understanding the environmental costs to each phase of production, distribution, and the disposal of products, healthcare organizations, companies producing these products and government policy makers can, collectively, reduce environmental costs of the medical IoT industry and encourage more sustainable practices.

Table 1. Comparative environmental impact of different categories of medical IoT devices

Device Category	Typical Use Case	Power Requirements	Estimated Lifecycle (Years)	Primary Environmental Concerns
Wearable Health Monitors	Vital signs tracking (heart rate, SpO_2)	Low (battery-powered)	2–3	Battery waste, short lifecycle, plastic casings
Implantable IoT Devices	Pacemakers, neurostimulators	Moderate (long-life batteries)	5–10	Non-recyclability, hazardous disposal, invasive extraction
Remote Diagnostic Tools	Glucose monitors, smart thermometers	Moderate	3–5	Sensor calibration waste, packaging, periodic electronic updates
IoT-Connected Imaging Devices	MRI, CT scan integration for real-time data	High (continuous power use)	7–10	High energy consumption, complex e-waste, embedded rare earth materials
Smart Infusion Pumps	Automated medication delivery	Moderate	4–6	Electronic waste, energy use, biohazard potential
IoT-Based Environmental Sensors	Monitor hospital air quality, temperature	Low	5–7	Battery and sensor degradation, data transmission energy use

Table 1 compares the environmental impact of different categories of medical IoT devices with regards to power requirements, life cycle duration and main environmental problems. This then shows how the usage context and functioning of the device impacts their ecological footprint; for instance, implantable devices, while having a long lifetime, have large deposition problems since it cannot be recycled and can be dangerous. In contrast, wearable health monitors are low in energy consumption, yet the rapid obsolescence and consequent replacement of the devices increases waste. Imaging devices with high power stand out for high energy consumption and the use of rare earth in terms of environmental impact. This table helps in grouping devices according to their characteristics regarding operation and their environment, and therefore, stresses the importance of customized sustainability strategy for medical IoT devices considering the specific ecological consequences for each type of device in the medical IoT ecosystem.

3. ASSESSING THE ENVIRONMENTAL COST

However, medical IoT device's contribution to its environmental costs is usually underestimated due to prioritization of the device's healthcare benefits rather than its ecological footprint. Nevertheless, with increasing expansion in the use of these devices, it is essential to evaluate the actual environmental expenses related to their production, operations, and disposal. This section discusses the key elements of environmental impact of medical IoT devices with specific reference to carbon emission, energy usage, e waste generation and the environmental effects of raw material extraction (D'Alessandro et al., 2024).

3.1 Carbon Footprint of Medical IoT Devices

Medical IoT devices' carbon footprint is the total amount of greenhouse gases (GHGs) produced during the lifecycle of the products, from cradle to grave. Energy usage in the production phase can be especially energy intensive, and the footprint of the carbon is typically greatest here. Energy sources that emit carbon dioxide and other GHGs at high levels are used in many manufacturing processes including component fabrication, assembly, and testing. Besides direct emissions at the manufacturing stage, transportation of components or finished devices within the global supply chains also contribute to the carbon footprint, especially in case of long-distance shipping.

The carbon footprint of medical IoT devices continues to grow during the operational phase — especially for devices that have to transmit data or otherwise remain in constant connectivity. Consider wearable health device is an example where it uses wireless networks like Wi-fi, Bluetooth or cellular networks to continuously watch the vital sign and send data to health care providers (Towhiduzzaman et al., 2017). Moreover, with all this information transmitted, their energy consumption contributes even more to the overall carbon emissions of these networks and their data centers which process the sent information. As the use of IoT devices and technology increases in the healthcare sector, there is also a significant concern over its cumulative carbon footprint when it comes to achieving zero emission and promoting greener technologies.

3.2 Energy Consumption During Operation

The most prominent environmental costs of medical IOT devices being energy consumption during its use. These devices are meant to run continuously, gathering and sending data often in a real time fashion to healthcare professionals. For instance, wearable health monitors, smart thermometers, and glucose sensors have to be powered all the time in order to collect patient data and to provide timely feedback to the device users or to caregivers on charting the health progress of patients. Also, processing of information on other (mobile) devices is often with small batteries fueling them and ongoing charging or vibrant replacement is the norm, hence elevated energy consumption.

Medical IoT devices energy consumption during the operation can be different depending on the complexity, functionality and requirements for the connectivity. Vital signs may be monitored by relatively simple devices with low energy demands, whereas more complex devices such as connected imaging devices or those able to perform advanced diagnostics would consume much more power. Furthermore, several of these devices depend upon wireless networks to transport data, which further adds to energy use in maintaining the networks. Furthermore, the data centers storing and processing the information from these devices use substantial power run the facilities. Due to the increasing number of connected devices in healthcare, a total energy demand is expected by these devices, motivating the design of energy efficient solutions (Mallo et al., 2024).

There is much room for improvement in terms of energy efficiency in the case of medical IoT devices. However, manufacturers can optimize communication protocols and wireless protocols, incorporate energy saving technologies and by prioritizing low power components will reduce the operational energy consumption of these devices. Further, the advancement of renewable energy sources of data transmission and storage implies still more reduction of the environmental impact of energy use of the medical IoT devices.

3.3 E-Waste Generation and Management

Electronic devices and components that are discarded because they are no longer useful are referred to as e–waste or electronic waste. There is an ever increasing problem of e-waste and a large contributor is in medical IoT devices. DUE TO QUITE FAST TECHNOLOGICAL ADVANCEMENTS, MANY OF THESE DEVICES HAVE A RELATIVELY SHORT PRODUCTS LIFE SPAN AND NEW MODELS KEEP REPLACING THE OLD ONES. This is why there are large volumes of discarded medical IoT wearables each year, which in turn create waste in large quantities. The problem with e-waste starts because they can contaminate soil and water with toxic heavy metals (mercury, lead, cadmium), chemicals and nonrecyclable plastics.

The health care regulations that classify them as a medical waste also complicate the disposal of medical IoT devices. Suchly is an attempt to classify these devices through an ontology so these are not recycled with other electronics, as they tend not to. But the programs and initiatives are in place already to recycle the medical devices, so there is no need for new legislation for them; the problem is, IoT devices are not covered because they're too complicated, and do not conform to the standard protocols for waste management. Moreover, numerous devices are very small with intricate components which are hard to disassemble and recycle. This makes recovery of the valuable materials, such as gold, copper and silver used in the manufacture of the devices more difficult.

Manufacturers and healthcare providers can look into take back programs in order to lessen the environmental impact of ewaste, and medical IoT devices can be properly recycled and disposed when possible. Moreover, designing devices that are conscious of material recycling, reusable materials, and fewer complexities in components will reduce the waste production as well as make the recycling process more efficient. Still, a global concerted effort is needed to solve the e waste issue and build systems conducive to responsible recycling and reuse of medical IoT devices.

Figure 1. Environmental impact lifecycle of medical IoT devices

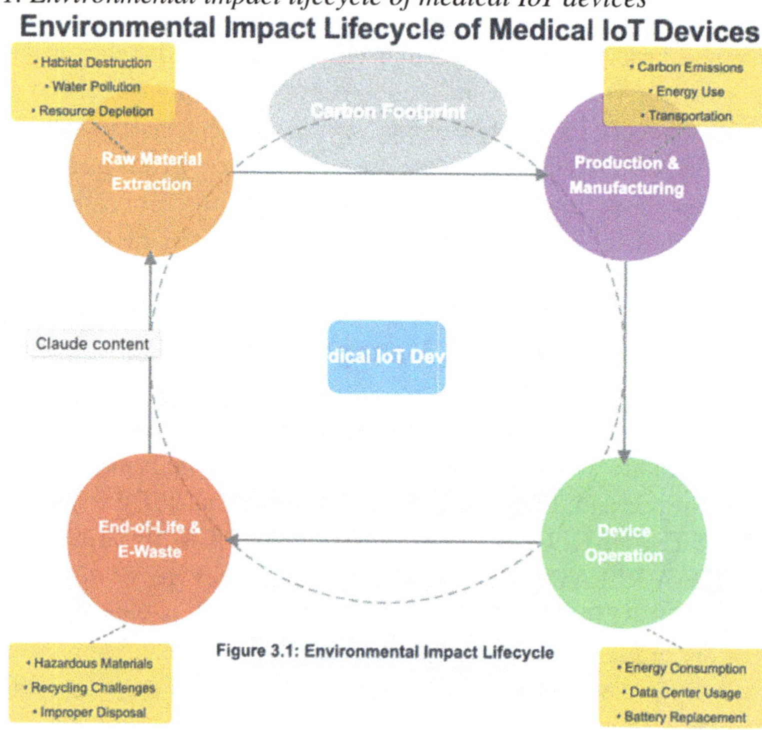

In Figure 1, this would be the four routes of lifecycle of environmental impact of medical IoT device starting from raw material extraction, production & manufacturing, device operation & end of life & e-waste. The representation of each phase is a circular node with the associated environmental impacts being depicted in adjacent boxes. The cyclical arrangement of the diagram is carefully chosen to reflect continuous connecting of the environmental impacts within the lifecycle of medical IoT device. The diagram particularly bring forth problems like habitat destruction and resource depletion during extraction, carbon emissions during production, energy consumption during operation, and hazardous materials management during disposal.

3.4 Raw Material Extraction and Environmental Impact

The environments are negatively affected by the extraction of raw materials for making medical IoT devices as habitats are destroyed, water is polluted and natural resources are depleted. Additionally, many medical IoT devices utilize materials including rare earth metals, lithium, cobalt which are utilized for sensors, batteries, circuit boards and etc. However, the mining of these material is a very environmentally damaging process. For instance, lithium mining (used in most rechargeable batteries) will have adverse implications for areas where the mining process uses massive amounts of water, leaving water scarce. Soil and water become contaminated from cobalt mining in places where environmental regulations are low (Tariq, Ahmed, Khan, & Bashir, 2025).

Besides the direct environmental impacts of extracting raw materials, mining process can indirectly affect social and ethical aspects (poor working condition, abuse and deprivation of local communities). As the medical IoT device demand increases, this increases the pressure on the supply chain of these critical materials. Given this, mining continues to contribute to an exacerbation of the negative environmental consequences because more and more land is deemed necessary for mining extraction, and more and more resources are depleted.

Thereby, the raw material extraction of the medical IoT industry should realize the alternatives to mitigate the environmental and social effects, such as using recyclable materials, increasing the resource efficiency, and selecting raw materials from the ethically responsible suppliers. Moreover, innovation in the field of sustainable material science, specifically in terms of developing biodegradable or recyclable components, can greatly mitigate the need for virgin materials and reduce medical IoT devices' environmental footprint.

Finally, medical IoT devices have multiple types of environmental costs as they generate many carbon emissions, consume a lot of energy, generate e waste, and extract raw materials. However, to allow for the benefits to not be cancelled out by negative environmental impact of medical IoT technologies, manufacturers, healthcare providers and policymakers should collaborate to address these issues and encourage sustainable practices across the lifecycle of these devices.

4. ENVIRONMENTAL IMPLICATIONS OF MEDICAL IoT DEVICES

With the growing trend for adoption of medical IoT devices across the globe, the environmental effects of manufacturing, using and disposing of these devices are increasing. While such devices undoubtedly provide a range of benefits in the enhancement of healthcare outcomes, it is equally important to appreciate the wider environmental risks that they introduce. This section then discusses potential medical IoT device impacts on ecosystems and natural resource, pollution and toxicity risks, and the world view regarding how sustainable is the healthcare technology.

4.1 Impact on Ecosystems and Natural Resources

Manufacturing and Operating medical IoT poses huge ecological and natural resource consequences. The production of many of these devices is done by extracting raw materials that are sometimes taken with activities such as disrupting natural habitats and depleting valuable resources. For instance, metals such as lithium, cobalt, and rare earth elements, which are necessary for batteries and electronics, are mined, causing such impacts like land degradation, deforestation and water pollution. These environmental disturbances can have an effect on the local ecosystem, biodiversity and the health of the local communities (Shoomal, Jahanbakht, Componation, & Ozay, 2024).

The growing demand for medical IoT devices has further been supporting the pressure on natural resources as healthcare systems across the world rely deeply on such technologies for monitoring, diagnosis as well as treatment. While there is an escalating need for such materials as plastics, metals, and minerals used in these devices to further strain the planet's finite resources. Moreover, the manufacturing of these devices (including energy demands) and energy consumption during their operations account for the depletion of non renewable energy sources that aggravate global energy crisis.

With the increasing use of medical IoT devices, practices for more sustainable management of resources need to be introduced. It involves the development of lesser dependant on scarce natural resources as well as the use of the materials that are more abundant, recyclable and more environmentally friendly than the current ones. Moving towards a circular economy, where the materials are used and reused without going to the point of recycling, can greatly minimize the pressure on ecosystems and on natural resources.

4.2 Pollution and Toxicity Risks

The main source of such pollution and toxicity risks of medical IoT devices are its materials of construction and waste. These devices contain many hazardous substances including heavy metals, toxic chemicals and flame retardants which can be an environmental and human health risk should these devices are not properly managed. At the end of their lifecycle, when disposed of improperly, these devices can release these substances into soil, water and air.

Medical IoT devices also can pollute through the improper disposal that releases toxic substances, such as mercury, lead, cadmium, and brominated flame retardants. In these chemicals leach into the environment, poisoning water sources and estrogentic chemicals harm wildlife and can even be absorbed into the food chain. The environmental risks of e – waste are large in countries that have inadequate waste management infrastructure and practice improper ways of disposal, like open air burning and landfilling. The environmental pollution that results from medical IoT devices is made more severe by these practices, and threatens to harm both humankind and the environment for the long term.

Furthermore, while manufacturing medical IoT devices, chemicals are usually employed which are pollutants in air and water in the production process. For instance, circuit boards, batteries or other electronic components are fabricated using solvents and other chemical that poses a threat to both the workers and the environment. If not properly controlled, these processes can lead to toxic emission that cause further damage to ecosystems.

The manufacturers of medical IoT devices should therefore opt for non toxic, eco friendly materials during the making of this class of devices to combat the pollution and toxicity risks they pose. Furthermore, better recycling systems for medical IoT devices will lead to a reduction of the pollution caused by e waste. So devices can be easy to disassemble and recycle, as well as promote responsible device disposal and recycling programs, which can also go far to reduce the environmental toxicity.

4.3 Global Perspective on Healthcare Technology Sustainability

From a global scale, the sustainability of all areas of Healthcare technology, including medical IoT devices is a cause of concern. Nowadays, since more and more IoT devices are being used by the healthcare systems in the world, it becomes extremely important to balance the pros and the cons of these systems keeping the environment in mind. When it comes to the developing countries, the implementation of medical IoT devices poses a big challenge in itself as there is no availability of proper infrastructure through which e-waste management can be taken care of and sustainable manufacturing practices can be practised. Technology is changing and advancing at a rapid pace and devices are discarded at a high turnover rate to exacerbate the problem of e – waste especially in countries which have shortage of waste management facilities.

On the flipside, other developed countries have the advantage of needing to be more concerned with the environmental ramifications of the medical IoT devices. There is an ever increasing number of healthcare providers and policymakers looking for ways to decrease environmental impact. Unfortunately, the global nature of medical IoT supply chain means that environmental issues are often ignored in regions where there is weak regulatory environment and thus uneven progress in sustainability occurs across the world (Martínez-Ramón, Calvo-Rodríguez, Iribarren, & Dufour, 2024).

To view the healthcare technology sustainability with a global perspective, international collaboration in establishing universal standards and guidelines for the design, production and disposal of medical IoT devices is required. Creating policies that foster environmentally friendly practices by manufactures as well as have the healthcare systems prepared to deal with the environmental ramifications of the technology, countries need to collaborate as a unit. Furthermore, international cooperation, mainly by developed countries, is necessary to manage the e-waste disposal problems especially in developing countries, through the provision of necessary waste management infrastructure and recycling industry.

Furthermore, the idea of sustainability in the context of healthcare technology is gaining much importance, thus creating an opportunity to utilize a greener alternative. Renewable energy sources, biodegradable materials, and low impact manufacturing processes for making medical IoT devices are some innovations that can greatly decrease medical IoT devices' environmental footprint. Additionally, the circular economy principles like product take back schemes and device reuse can be included to reduce the environmental impact and create favourable circularity around healthcare technology.

Figure 2. Environmental implications of medical IoT devices

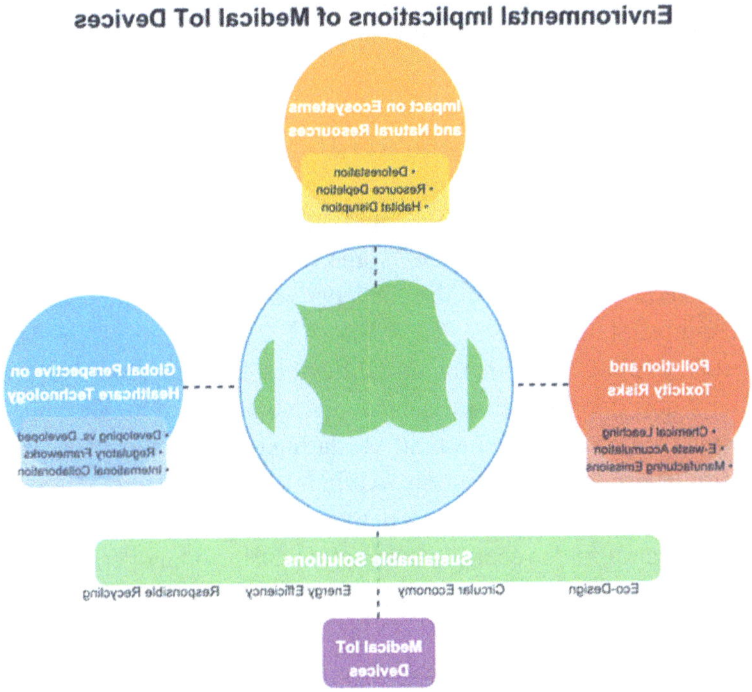

This global perspective of the environmental implications of medical IoT devices is presented in Figure 2. For worldwide impact, the central Earth is depicted surrounded by three key environmental implications categories: Impact on Ecosystems and Natural Resources on top, Pollution and Toxicity Risks on the left, Global perspective on Healthcare technology on the right. Every one of these categories includes specific concerns such as deforestation, e waste accumulation, and disparity of regulatory framework. These environmental challenges are shown to have their source in the medical IoT devices at the bottom whereas the 'Sustainable Solutions' banner gives the possible mitigation strategies such as eco-design, circular economy, energy efficiency and responsible recycling.

5. CASE STUDIES: ENVIRONMENTAL IMPACT IN REAL-WORLD SCENARIOS

The use of medical IoT devices is growing; consequently, it is important to understand how they impact the real world environment. In this section, we evaluate environmental footprints of popular medical IoT devices with their own case studies. Additionally, we assess lessons from the industry leaders that have embraced sustainable strategies in the design, manufacturing, and disposal of these devices to aid the healthcare sector in minimising their environmental impact and advantages these devices provide.

5.1 Analysis of Popular Medical IoT Devices

Likewise, the medical IoT device environmental impact is variable, based on its design, functionality and lifecycle. Some of the most widely used medical IoT devices are used as a showcase of the challenges and opportunities for dealing with their environmental footprint.

Therefore, wearable health monitors, devices like smartwatches and fitness trackers have now become indispensable for personal health monitoring. These are devices which are on and are constantly collating data on vital signs such as heart rate, activity levels and sleep patterns. Although health benefits due to the use of these devices are important, the impact of them on the environment is mostly related to battery production and to energy consumption during their operation. Due to their high frequency of recharging, energy demands are high and lithium-ion batteries – the main batteries in many of these devices – are synonymous with resource depletion and toxicity. In addition, since these devices become obsolete or are replaced with newer ones, they tend to add to the increasing e-waste problem.

Smart infusion pumps have been routinely used in the hospital settings to provide specified amounts of fluids and medicines to the patient. Basically, these are devices that have sensors, which are connected to the hospital networks in order to convey the real-time data concerning the infusion rates and about how the patients respond. Environmentally, these pumps are difficult because their parts are complicated plastics, metals, and batteries. Moreover, the pumps are frequently used for a number of years before being disposed of or replaced, thus creating e-waste. As they are critical in healthcare settings, these devices bring forth the need to design more durable, recyclable and energy efficient medical devices that can be used for a longer time before being disposed of.

Medical Imaging Devices: Due to advancing IoT technologies, MRIs (magnetic resonance imaging), etc. are getting more integrated with other medical imaging devices to improve the accuracy of their diagnostics and to remain operationally smooth. These devices often require large energy inputs on operation and for the cooling systems required to maintain its best working condition. While connected imaging devices make use of rare earth elements for their sensors, as well as being difficult to recycle, this adds to their environmental cost. However, these devices involve a very energy intensive production process and can thus only be produced within a limited carbon budget, and they also have hazardous materials embedded in components that pose a significant problem at the end of their life.

By way of case studies, we describe the diversity of the environmental problems presented by medical IoT devices. The environmental footprint of some of the devices, for instance wearable health monitors, is low whereas the devices with a higher footprint, such as connected imaging systems, have considerably higher requirements. All of these devices operate by the generation of e-waste, waste energy, and are reliant on nonrenewable resources in their own manufacture (Chauhan, Bahad, & Jain, 2024).

5.2 Lessons from Industry Leaders in Sustainable Practices

There are several companies and organization who provide to the medical IoT space and have taken steps to reduce environmental impact of there product. Industry experts literally open the minds offering their combined knowledge on utilizing sustainability in designing, manufacturing, and disposing off medical IoT devices.

Philips Healthcare: Philips has successfully decreased the ecological footprint of the medical devices offered, including IoT-enabled items. The company has opted for a circular economy model where it is trying to reduce waste and increase the lifecycle of its devices. Philips has made sure that their products are designed to be recyclable, making use of materials that can easily be taken apart for recycling. In addition, the company has improved the energy efficiency and added low power components in the effort to reduce energy consumption when running the company's devices. For instance, Philips' patient monitoring systems are developed to be energy efficient and therefore contribute to lowering the operational carbon footprint of hospitals.

Additionally, Philips has taken the initiative of implementing takeback programs, whereby used medical devices are returned for correct waste disposal. Through this initiative e – waste generated by their products is reduced and also that valuable materials are extracted and reused. Philips' commitment to sustainability goes beyond the boundary of the organization; the company collaborates with its suppliers to guarantee that raw materials are sourced sustainably and environmental and social requirements met.

This includes Medtronic, a World leader of health care products and services, who has decided to introduce and incorporate sustainability into its corporate strategy at reducing its environmental impact. Hence, the company has gone a step further to reduce the carbon footprint of their products. Another one of Medtronic's key sustainability efforts is the design of energy efficient pacemakers and insulin pumps that use minimal energy while in operation. Medtronic has been able to reduce the total environmental impact of these devices on efficiency.

Secondly, Medtronic has devoted to minimizing e-waste in products by extending product life and designing modules. A company also developed products that are upgradeable on few small waste and can continue to remain in the use without the need of frequent replacement. In addition, Medtronic has also committed to reach net zero carbon emissions across its own operations by 2030, while exerting efforts to reducing emissions from the use of its products in healthcare facilities.

GE Healthcare: It took initiatives to enhance the sustainability aspect of medical IoT devices by maintaining energy efficiency and also reducing the environmental impact of products construction. Green design principles, including energy saving technologies, are now a fact of life for such things as imaging systems, and patient monitoring equipment are now part of the company's mantra. For example, GE Healthcare's imaging systems are built to use less energy than previous versions in order to help reduce carbon footprint as these systems have grown in popularity. Furthermore, GE has incorporated sustainable materials and, in some cases, sustainable and recycled materials into its devices.

Sustainability also exists within GE Healthcare's product supply chain reduction of environmental impact. In the act of ensuring that raw materials are responsibly sourced and that it minimizes waste throughout the manufacturing process, the company depends heavily on cooperation with its suppliers. At the end of the life cycle, GE has also been involved in seeking solutions for proper disposal and recycling of medical devices which will recover valuable materials and manage hazardous substances safely.

How Elsevier, Philips, Medtronic, and GE Healthcare integrate sustainability into the design, production, and disposal of medical IoT devices give us a feel of what needs to be done by other companies in the field. Adoption of the circular economy principles, connecting energy efficiency, responsible sourcing and recycling can serve as an example to others in the healthcare industry.

The environmental challenges and opportunities in the sector can also be demonstrated through the case studies of popular medical IoT devices and lessons learned from the industry leaders. Medical IoT devices definitely provide great value to the healthcare field, but they also come with an environmental cost. Learning from other industry leaders when it comes to sustainability practices, manufacturers and healthcare providers can abide by more eco friendly solutions, decrease e waste and also reduce energy consumption. Following the trend as the healthcare sector adapts to IoT, this will be necessary to avoid designing and using these devices at the expense of the environment.

6. INNOVATIVE SOLUTIONS FOR REDUCING ENVIRONMENTAL FOOTPRINT

As the environmental impact of medical IoT devices becomes a growing concern, it is crucial to explore innovative solutions that can significantly reduce their environmental footprint. From eco-friendly design approaches to sustainable manufacturing processes, energy-efficient devices, and circular economy initiatives, there are numerous opportunities to make medical IoT devices more sustainable. This section discusses these innovative solutions and their potential to mitigate the negative environmental effects of medical IoT devices.

6.1 Eco-Friendly Design Approaches

Eco friendly design principles are one of the best means of lowering the environmental impact of medical IoT devices, and should be adopted right from product development process. Eco design considers the environmental aspects of the whole life a device, from material selection and manufacture to use and end use. The objective is a device which is functional, efficient and environmental responsible.

The choice of sustainable materials can begin an eco-friendly design. They can also choose biodegradable materials, or recyclable plastics and plant-based alternatives for instance instead of nonrecyclable plastics or hazardous chemicals. It also lowers the production and disposal waste as well, as having more components means more waste. Simplified designs are also simpler to disassemble at the end of a product's life, improving the ability to recycle and diminishing e-waste. Medical devices, like wearable health monitors, for example, can be constructed with easy to replace or upgrade modular parts that do not require disposal of the entire device.

Secondly, there is a requirement of longevity and durability of the device. When devices are built to last longer, replacement circuits with fewer parts susceptible to wearing out are less needed, and less waste ends up being generated. Designing for longevity allows manufacturers to be part of addressing the environmental burden of often turning over products.

6.2 Energy-Efficient Devices and Technologies

The environmental footprint of medical IoT devices is highly dependent on the energy consumption. Given that more and more IoT based solutions are being adopted by healthcare systems, the summed up energy usage by these devices can be significant. Therefore, developing energy-efficient devices is required in order to minimize the environmental impact of medical IoT devices.

A way to get energy efficient is the device power management optimization. There are many medical IoT devices that transmit data to cloud systems continuously, for example, wearable health monitors, infusion pumps, and so on. And as this connectivity becomes a constant, it can easily become high energy consuming if the devices are not designed with energy efficiency in mind. These devices can be made to use much less energy through the use of low power processors, energy harvesting techniques and smart sleep modes. For instance, piezoelectric generators and solar cells can be examples of energy harvesting technologies that power certain medical IoT devices for enable them to work autonomously without dependency on batteries.

In addition, the energy consumption of medical IoT devices for data transmission can be reduced through the integration of the wireless communication protocols such as low power Bluetooth (Bluetooth Low Energy or BLE). With energy efficient wireless protocols used by manufacturers, overall operational energy demand can be reduced leading to reduction in carbon footprint of the healthcare systems.

Additionally, the need for fewer recharges or battery lifespans can reduce the number of batteries that need to be replaced and thereby reduce the waste, waste byproduct, and resource consumption of these devices. Manufacturers may also design devices that can be used with recyclable or environmentally friendly batteries thereby lessening the impact of disposing of used batteries on the environment.

6.3 Sustainable Manufacturing Processes

To reduce the environmental impact of medical IoT devices at production stage, sustainable manufacturing processes are necessary. The traditional manufacturing techniques usually involve energy intensive processes, cause big wastes and also entail the use of hazardous chemicals. Since some businesses are slowly adopting more environmentally friendly manufacturing methods, they are able to lower their carbon footprint as well as the environmental footprint.

Additive manufacturing, or 3D printing, is one such method of maintaining the sustainability of manufacturing. 3D printing, which builds devices layer by layer using only the materials necessary, allows for more precise choice of material use and minimal waste of materials, in addition it allows to produce the lightweight and complex designs that otherwise would be difficult to create using traditional methods, thus reducing the energy and the material needs in the production. Additionally, 3D printing can enable the production of medical IoT devices tailored to be customized devices instead of sculpted large mass manufactured ones which have tremendous levels of waste involved in them.

One other practice in sustainable manufacturing is that energy efficient production practices are used. Manufacturing facilities that use renewable energy sources, i.e. solar or wind power for their energy needs help cut down on carbon emissions from production. Likewise, energy recovery systems can be adopted by companies in order to capture and reuse waste energy, thereby decreasing the environmental impact of the companies' operations.

Lean production methods can also be adopted by manufacturers to eliminate waste and maximize efficiency in the production process. Therefore, lean manufacturing practices can significantly reduce the environmental footprint of medical IoT devices, by eliminating the material use, which are not really needed, reducing the energy consumption and streamlining the workflows. In addition, environmentally conscious companies can apply for certifications such as ISO 14001 in order to receive a certification that guarantees their manufacturing processes are at high levels of 'environmental management'.

6.4 Circular Economy and Recycling Initiatives

A circular economy is a promising solution to reduce environmental footprint of medical IoT devices. A circular model for an economy is focused on, like reuse, repair, refurbishment and recycling rather than linear production and disposal. Medical IoT devices in a circular economy should be designed to be reused or recycled upon completion of their life cycle; avoid waste; and save on valuable resources.

A key component of medical IoT in a circular economy is to setup specific take back and recycling programs. Old or outdated medical devices can be collected, refurbished, and recycled at the manufacturer and providers levels for components or valuable materials, such as metals, plastics that can be used in new medical devices. It requires less virgin materials and promotes the flush of e-waste from landfills.

Designing the devices to be easy to disassemble and recycle may further support recycling initiatives. Manufacturers can do this by using recyclable materials for their product and using modular designs so that their products are better suited for recycling at its end of life. Furthermore, the development of closed loop supply chains, where the materials in old items are reused to manufacture new devices, can decrease the apparent requirements for raw goods and lower total environmental effect.

Additionally, the adoption of repair and refurbishment programs can lengthen the service life of redundant medical IoT devices as well as decrease frequency of replacements. Just like factory farming is less waste producing, but offers more 'intents' from a given land mass, these programs can focus on repairing or upgrading devices with minimal waste, so healthcare providers can get a greater bang for their equipment buck. This method doesn't only reduce the e waste but also contributes to making healthcare technology affordable and accessible.

It is key to find innovative ways of decreasing the medical IoT device environmental footprints, so that healthcare technology can continue improving patient outcomes because environmental footprints are not compromised. Designing medical IoT devices in an environmentally friendly approach, having energy efficient devices, sustainable manufacturing processes and circular economy initiatives are all crucial roles in making medical IoT devices more sustainable. To reduce the environmental impact of medical IoT devices while maintaining its effectiveness, reliability, and accessibility for healthcare sector, manufacturers, healthcare providers and policymakers should embrace such solutions. Instead, if the medical IoT device demand continues to grow, the adoption of these innovative practices will be crucial in achieving a balanced amount of the benefits of these technologies with the environmental responsibility needed.

Table 2. Innovative strategies for reducing the environmental footprint of medical IoT devices

Sustainability Strategy	Target Lifecycle Stage	Primary Goal	Potential Benefits
Eco-Friendly Product Design	Design and Development	Minimize resource use and facilitate recyclability	Reduces material waste, enhances product longevity, simplifies end-of-life processing
Energy-Efficient Microcontrollers	Usage and Operation	Lower energy consumption during device runtime	Decreases carbon footprint, extends battery life, reduces hospital power loads
Modular Hardware Architecture	Design and Manufacturing	Enable component reuse and easier repairs	Lowers replacement frequency, minimizes e-waste, supports upgradeability
Green Manufacturing Techniques	Manufacturing and Supply Chain	Reduce emissions and resource consumption	Limits industrial pollution, improves factory energy efficiency
Use of Biodegradable Materials	All Stages (esp. casing/disposables)	Limit long-term environmental contamination	Promotes safer disposal, reduces landfill accumulation
Recycling and Circular Economy Models	End-of-Life	Maximize recovery of materials and reuse	Conserves raw materials, supports a closed-loop lifecycle
Renewable Energy Integration	Manufacturing and Operation	Power production and usage with green energy	Reduces dependence on fossil fuels, supports clean energy initiatives

Table 2 presents some innovative means of minimizing environmental footprint of medical IoT throughout their lifecycle with some innovative means. The table helps construct a structured view of how environmental impact of a device can be lowered in its all stages (design, manufacturing, usage, and disposal) by aligning each sustainability initiatives to a right stage. For example, following eco friendly design principles and module hardware do a lot to reduce materials waste, make future upgrades or repairs easier, energy efficient microcontroller make operations green. Green manufacturing techniques which draw from renewable energy sources in totality supports the transition to more sustainable production as well as operation. Further, circular economy models and biodegradable materials help in dealing with the critical matter of e – waste. The results of this table are highlighting the fact that a lifecycle approach sensitive to multifacet is needed for the sustainable innovation of medical IoT technologies.

7. POLICY AND REGULATORY FRAMEWORK FOR SUSTAINABLE MEDICAL IoT

Growing fast in the healthcare sector are the medical IoT devices, whose sustainability requires the development of policies and regulations for their growth. Given that these devices are becoming central to healthcare systems across the world it is important to consider their environmental impact in policy decisions and regulatory frameworks. These policies are often shaped by governments, international organizations, and healthcare institutions and there the adoption of the sustainable practices in the medical IoT space. In this section, we shall look at global standard and regulations that govern sustainability with regards to medical IoT, role of healthcare organizations in driving sustainability, future trends and regulatory trends that shape future sustainable medical IoT to be.

7.1 Global Standards and Regulations

The global standards and medical IoT sustainability regulations are shifting and functioning bodies on the international level are understanding that the issue of medical technology impact on environment needs to be addressed. A number of organisations have developed guidelines and regulations tied to sustainability principles for the lifecycle of medical devices, including those with an IoT element.

The International Organization for Standardization (ISO) 14001 is one of the important global standards which focuses on the Environmental Management Systems. This standard offers an organizational framework for managing an organization's environmental responsibilities to prevent its activities from harming the bio environment. This standard can be adopted by medical device manufacturers to set their own environmentally conscious practices within their own production processes for medical IoT devices thereby reducing carbon footprint and resource consumption associated with medical IoT devices.

The Waste Electrical and Electronic Equipment (WEEE) Directive is another relevant regulatory framework that regulates the disposal of electronic devices, even including medical IoT ones, in European Union. The requirement of this regulation is that manufacturers should be responsible for the collection, recycling and correct disposal of e-waste, recovering the valuable materials and preventing the release of hazardous substances into the environment. In terms such as medical IoT, similar regulations like the Restriction of Hazardous Substances (RoHS) directive also control the utilization of destructive materials in electronics which encourage the utilization of better, more maintainable segments for medical IoT gadgets.

Besides these existing standards, regulatory enforcement relating with the environmental aspects of medical devices are more seriously beginning to be enforced in the European Union, the United States, and Asia regions. For instance, the EU's Circular Economy Action Plan seeks to increase the sustainability of electronic products (e.g. medical devices) to promote circularity, reduce waste and enhance resource efficiency. Along with sustainability becoming a growing concern around the world, it is anticipated that there will be wider standards and regulations, which provide increasingly clear direction for manufacturers and healthcare providers alike to follow.

7.2 Role of Healthcare Organizations in Sustainability

Hospitals, clinics, and medical networks are among healthcare organizations that are in efforts of promoting the sustainability of medical IoT devices. Healthcare institutions are in the position to influence design, production and disposal of medical devices as end users of these technologies. Yet, healthcare organizations can reduce the environmental footprint of devices they use by adopting sustainable practices, and, in turn, showcase an example for the whole healthcare industry.

Healthcare organizations also contribute to sustainability from the procurement of medically relevant and environment friendly medical IoT devices. Sustainability principles, eco design, energy efficiency and recyclability can be incorporated into products by manufacturers that collaborate with the healthcare institutions and this can aid in reducing the environmental footprint of the operations. Furthermore, healthcare organizations can voice support for policies that will foster the creation of sustainable medical technologies and have a share in the medical IoT sustainability discourse about the future.

In addition, medical devices at the point of care can be provided by healthcare providers and the provision of these devices and other clinical operations can follow sustainability best practices to reduce the environmental impact of medical devices used. Hospitals and clinics can set up recycling programs for utilized medical devices at their end of life cycle and recycle their processed components and dispose of hazardous materials in the most effective manner. Additionally, hospitals can introduce energy efficient infrastructure to reduce the carbon footprint of the hospital's operation, including using of renewable energy to supply electricity in the hospital's premises and energy saving technologies in medical equipment.

Another avenue of contribution to sustainability for healthcare organizations is for them to participate in the increasing trend of medical device take-back programs. These programs consist of collecting used or obsolete medical devices for sound recycle and disposal. The promotion of such initiatives by healthcare institutions can assist in e waste reduction and also in pushing for the reuse of valuable materials in the loop of the lifecycle of medical IoT devices.

8. CONCLUSION

As medical IoT devices become more an essential part of yesterday's healthcare systems, the environmental footprint of the device becomes a pressing issue. Although medical IoT devices are capable of transforming patient care, monitoring and treatment, it is crucial to recognize and mitigate their concealed environmental footprint. In this chapter, the life cycle of medical IoT devices has been analyzed from design and development through to end of life disposal wherein there are many stages where environmental impacts happen. In addition, it has considered significant factors like energy consumption, carbon footprint, e waste generation and the extraction of raw materials, and stressed on the application of sustainable practices at all stages of lifecycle of such devices. Eco friendly design approaches, energy efficient technologies and initiatives of circular economy are promising ways to reduce the environmental footprint of medical IoT devices. For healthcare to remain sustainable, healthcare organizations and policymakers must actively use environmental responsible practices and promote legislation support of more environmentally friendly technologies. To reach an environmentally sustainable future of medical IoT devices is a team effort, an innovation, an ongoing commitment to balance an advance in healthcare with care for the planet.

9. FUTURE DIRECTIONS

Innovations in things such as medical IoT devices will continue to advance the future of sustainable medical IoT devices, alongside strengthened regulatory frameworks. With increasing environmental issues, more attention will be paid to the development of green materials and energy conservation technologies on medical IoT devices. Manufacturing companies will probably define further working on the goods, which will not only be highly functional but will also be eco friendly, across a long time and easy to recycle. In addition, renewable energy sources will be integrated in to medical IoT devices like energy harvesting technologies, to reduce dependency on conventional power sources. Moreover, the foreseeable future regulations will encourage a more transparent and accountable way of sustainability in the manufacturing of medical devices. This could include requirements on manufacturers to reveal the environmental effects of the products they manufacture and also to obey more strict e–waste management, energy efficiency, and sustainable resources standards. The importance of sustainability in medical IoT development is of no less significance, especially with the healthcare industry growing in size at such an alarming rate as the devices we create must always be advancing within the context of a healthier future, not one where the future of the planet is at risk.

REFERENCES

Allioui, H., & Mourdi, Y. (2023). Exploring the full potentials of IoT for better financial growth and stability: A comprehensive survey. *Sensors (Basel)*, *23*(19), 8015. DOI: 10.3390/s23198015 PMID: 37836845

Aminizadeh, S., Heidari, A., Dehghan, M., Toumaj, S., Rezaei, M., Navimipour, N. J., & Unal, M. (2024). Opportunities and challenges of artificial intelligence and distributed systems to improve the quality of healthcare service. *Artificial Intelligence in Medicine*, *149*, 102779. DOI: 10.1016/j.artmed.2024.102779 PMID: 38462281

Arega, A., & Sharma, D. P. (2023). Towards smart and green features of cloud computing in healthcare services: A systematic literature review. *Journal of Information Systems Engineering & Business Intelligence*, *9*(2), 143–158. DOI: 10.20473/jisebi.9.2.161-180

Bhutani, M., Lall, B., & Agrawal, M. (2023). An efficient and adaptive superframe structure for IEEE 802.15.7-based real-time sensor networks. [No journal information available]

Bhutani, M., Lall, B., & Agrawal, M. (2023). A novel energy-efficient adaptive superframe structure for OWC-based real-time bio-sensor networks. *Peer-to-Peer Networking and Applications*, *16*(6), 3021–3031. DOI: 10.1007/s12083-023-01544-3

Cappellini, I., & Schirru, E. (2024). Bridging anesthesia and sustainability: A special article for a path towards eco-conscious practice. *Anesthesia Research*, *1*(3), 168–179. DOI: 10.3390/anesthres1030016

Chauhan, D., Bahad, P., & Jain, J. K. (2024). Sustainable AI: Environmental implications, challenges, and opportunities. In Explainable AI (XAI) for Sustainable Development (pp. 1-15). Springer.

Corigliano, O., Morrone, P., & Algieri, A. (2025). Navigating the challenges of sustainability in the food processing chain: Insights into energy interventions to reduce footprint. *Energies*, *18*(2), 296. DOI: 10.3390/en18020296

D'Alessandro, C., Szopik-Depczyńska, K., Tarczyńska-Łuniewska, M., Silvestri, C., & Ioppolo, G. (2024). Exploring circular economy practices in the healthcare sector: A systematic review and bibliometric analysis. *Sustainability (Basel)*, *16*(1), 401. DOI: 10.3390/su16010401

Kaur, G., Gupta, V., Settia, N., Duggal, S., & Bhutani, M. (n.d.). Reilluminating wireless horizons: Exploring complete information transmission through visible light. [No journal or source information available]

Kumar, R. (2025). A comprehensive review of MCDM methods, applications, and emerging trends. *Decision Making Advances, 3*(1), 185–199. DOI: 10.31181/dma31202569

Mallo, S. F., Abdulqader, D. M., Abdullah, R. M., Ismael, H. R., Rashid, Z. N., & Sami, T. M. G. (2024). A review on feasibility of web technology and cloud computing for sustainable ES: Leveraging AI, IoT, and security for green operations. *Journal of Information Technology and Informatics, 3*(2), 89–104.

Martínez-Ramón, N., Calvo-Rodríguez, F., Iribarren, D., & Dufour, J. (2024). Frameworks for the application of machine learning in life cycle assessment for process modeling. *Cleaner Environmental Systems, 14*, 100221. DOI: 10.1016/j.cesys.2024.100221

Regona, M., Yigitcanlar, T., Hon, C., & Teo, M. (2024). Artificial intelligence and sustainable development goals: Systematic literature review of the construction industry. *Sustainable Cities and Society, 108*, 105499. DOI: 10.1016/j.scs.2024.105499

Sabet, M. (2025). Sustainable halogen-free polymer composites for next-generation flexible electronics. *Journal of Nanoparticle Research, 27*(2), 33. DOI: 10.1007/s11051-025-06234-0

Sallam, K., Mohamed, M., & Mohamed, A. W. (2023). Internet of Things (IoT) in supply chain management: Challenges, opportunities, and best practices. *Sustainable Machine Intelligence Journal, 2*, 3–1. DOI: 10.61185/SMIJ.2023.22103

Shoomal, A., Jahanbakht, M., Componation, P. J., & Ozay, D. (2024). Enhancing supply chain resilience and efficiency through Internet of Things integration: Challenges and opportunities. *Internet of Things : Engineering Cyber Physical Human Systems, 27*, 101324. DOI: 10.1016/j.iot.2024.101324

Tariq, U., Ahmed, I., Khan, M. A., & Bashir, A. K. (2025). Bridging biosciences and deep learning for revolutionary discoveries: A comprehensive review. IAES International Journal of Artificial Intelligence (IJ-AI), 14(2), 867-883.

Chapter 12
Blockchain for Ethical Supply Chains:
Transparency in Medical IoT Manufacturing

Reeta Parmar
https://orcid.org/0009-0003-7189-1226
SAITM, India

Abhishek Singh
SAITM, India

Puneet Garg
https://orcid.org/0000-0002-4635-546X
KIET Group of Institutions, Delhi, India

Tripti Sharma
https://orcid.org/0000-0001-8692-261X
MSIT, India

Placido Rogerio Pinheiro
https://orcid.org/0000-0002-1718-1712
University of Fortaleza, Brazil

ABSTRACT

The remarkable rise of Medical Internet of Things (IoT) equipment has significantly changed healthcare services, providing real-time monitoring, diagnostics, as well as data-driven treatments. But, as with any mass-produced machine, the universally outsourced nature of the production of these devices has brought up huge ethical and transparency issues on where the raw materials are sourced from, the way in

DOI: 10.4018/979-8-3373-4134-7.ch012

which labour is conducted, counterfeit components, and compliance with local regulations. In this paper, the authors are presenting the usage of the blockchain technology to set up an ethical and tamperproof transparent supply chain for the Medical IoT manufacturing. With based on the decentralised and immutable environment provided by blockchain, the framework guarantees full traceability of each part, from raw material to device assembly. Smart contracts facilitate verification of compliance and supplier audit Sn, and the consensus mechanisms grant data integrity requiring no central control.

1. INTRODUCTION TO MEDICAL IOT AND ETHICAL SUPPLY CHAINS

Medical IoT is poised to revolutionize health care by integrating embedded sensors, connectivity, and data analytics to support continuous real-time monitoring, diagnostics, and even therapeutic responses (Raj et al., 2023). Wearable heart monitors, connected insulin pumps, and smart diagnostic systems are just a few of the IoT applications in medical devices that are not only improving patient quality of life, but also helping to reduce hospital readmissions and driving the growth of personalized medicine (Garg et al., 2021).

However, the increased availability of these products also comes with an increasing reliance on intricate, global supply chains. Medical IoT equipment is made up of a number of layers of component production ranging from the component production plant and the manufacturer to the chips, sensors, circuit boards, cases, software module producers from all kinds of regions (Gautam et al., 2024). This complex network of sourcing has huge implications in making sure that everything in the life cycle of production remains ethical, regulatory and safe.

Ethical issues raised supply chains are the use of conflict minerals, forced or child labor, counterfeits or sub-standardized parts, and a general reluctance to embrace environmentally friendly practices. Further, lack of transparency in the provenance of components and their handling can not only undermine the integrity of medical products, but can also impact the trust patients and regulators place in healthcare technologies (Khan et al., 2024).

In this environment ethical supply chains that are transparent, and can be verified, and are compliant with human rights and safety practises comes to the fore. By extension most existing supply chain systems are centralized and utilize physical audits that take a long time to complete, and as such these existing supply chain systems do not provide real-time visibility, immutable tracking of a product and secure transparency throughout the life cycle of a Medical IoT device (Bennett and L. 2024).

In order to fill these holes, blockchain technology arose as a hopeful tech. Allowing for decentralized, immutable records of transactions and material movements, blockchain offers an infrastructure where each constituent whether that's a raw material provider or a hospital can ascertain the origin and ethical nature of ingredients with security and transparency(Bai et al., 2020). When paired with IoT sensors and auto-executable smart contracts, blockchain can help to guarantee that devices are ethically produced, shipped in the right conditions, and that a package was not tampered with on its way to you (Saberi et al., 2019).

2. KEY COMPONENTS OF BLOCKCHAIN TECHNOLOGY

Blockchain is a disruptive decentralized ledger technology that changes how data is recorded, validated and shared. The four main components of this system: decentralization, immutability, consensus mechanisms, and smart contracts provide building blocks for secure and transparent trustworthy systems. Each part is essential to maintain the integrity of the application and the ethical operation of any application such as supply chain management in IoMT for medical devices manufacturing (Bennett and L. 2024).

Figure 1. Components of blockchain technology

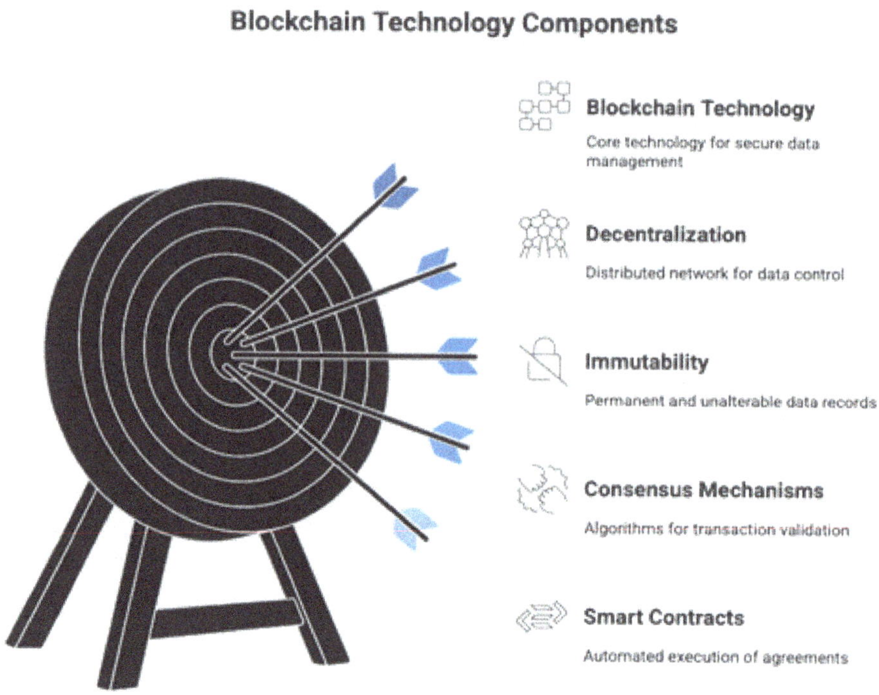

2.1 Decentralization

Conventional systems use centralized databases which is under the control of a single entity. In contrast, blockchain is a distributed network, in which multiple nodes (also referred as participants) keep updating the same ledger. This eliminates single points of failure, minimizes the risk of data tampering, and makes it so that every stakeholder manufacturers, suppliers, regulators, and healthcare providers has the same level of access to supply chain records (Chauhan et al., 2021). According to MedicalIOT protocol, the decentralization will achieve transparency and cooperation among trusted parties for the MedicalIoT supply chain.

2.2 Immutability

Any information that is recorded on a blockchain has to satisfy a consensus mechanism in order to be altered or deleted. This irreversibility ensures that all events of the supply chain visible on the blockchain, such as production dates, quality measures, transportation information, and certifications, cannot be corrupted or lost. Any alteration in the history of records is revealed, dissuading fraudulent activities and ensuring integrity throughout the Medical IoT manufacturing cycle (Garg et al., 2021).

2.3 Consensus Mechanisms

Consensus mechanisms is protocols which ensure that every node in the network agrees on the validity of a transaction. Popular mechanisms include:

- **Proof of Work (PoW):** Secure but costly in energy
- **Proof of Stake (PoS):** Efficient and scalable.
- **Proof of Authority (PoA):** Suitable for permissioned networks where authorised nodes (such as verified manufacturers or regulators) validate transactions

For the Medical IoT industry, we often want a mechanism more like PoA as it serves high performance, while being compliant with regulations (Nanda et al., 2023).

2.4 Smart Contracts

Smart contracts are autonomous computer programs that carry out a series of instructions based on agreed or predefined conditions stored in the blockchain. Within an ethical supply chain, smart contracts are crucial for transparency and accountability(Khan et al., 2024). They can verify supplier certifications and ensure that only certified, compliant suppliers participate in the supply chain. They can also schedule payment releases based on confirmed deliveries, thereby minimizing delays and disputes. Sustainable standards, like environmental and labor compliance standards, can also be enforced through smart contracts by embedding these conditions into the transaction logic itself (Wang et al., 2023). In addition, they can generate alerts if the level of quality level or the due date is violated and the problem can be addressed in due time. Through reducing human intervention requirements, the contracts serve to maximize operational efficiency and maintain regular, verifiable ethical compliance.

3. ROLE OF BLOCKCHAIN IN ETHICAL COMPLIANCE AND TRACEABILITY

Applications integrating blockchain into a supply chain, especially for vital industries, such as Medical IoT (Internet of Things) production has greatly increased the potential for ethical compliance and traceability. These two aspects are critical not only for compliance purposes, but also for safeguarding consumer confidence, brand equity, and social responsibility throughout the world's supply networks (Mann et al., 2018).

3.1 Ethical Compliance through Transparency

Ethical compliance such as raw material originating responsibly, fair labor practices and meeting of environmental standards can be hard to trace in the traditional supply chain, with its complexity and lack of transparency through multiple supply tiers (Khan et al., 2024). Blockchain solves this by establishing a transparent and immutable ledger in which every single transaction, event or certification is logged and visible to all authorized participants. For instance, should a part used in a medical device be from a conflict area, be made with child labor, blockchain can record and alert the producer of the point of origin (Nanda et al., 2023). Smart contract can be programmed to automatically ensure compliance if a supplier has not uploaded a valid certification (e.g., ISO, FDA), transaction can be n then blocked or have a flag to go through an audit. This ethical check automation enables companies to establish compliance without depending on tedious manual audits, which are expensive, infrequent, and easily circumvented.

3.2 End-to-End Traceability

Thanks to blockchain, the full traceability of the product, from extraction of raw materials and through logistics and final use in hospitals or healthcare facilities is traced by recording every product movement and transformation in the supply chain (Mann et al., 2018). Every movement of the product's lifecycle is recorded on the blockchain in combination with timestamps, geolocations, and the responsible entity for that task, forming a chronological digital twin of the tangible product. In the case of a device failure, recall, or compliance investigation, owners can readily track how any Medical IoT device has been used and its history so that they can pinpoint where the trouble began (Garg et al., 2021).

This traceability is particularly important for high-risk medical devices, as it allows even minor issues to potentially have a major impact on patient safety. It also guarantees regulatory bodies (FDA, CE, ISO auditors etc.) have access to accurate, timely and immutable data for inspections and certifications (Nanda et al., 2023).

3.2 Enhancing Accountability across Stakeholders

Because blockchain is immutable and distributed, there is accountability from all entities involved in the supply chain. Once a transaction is checked and entered, it can't be replaced or reversed, discouraging bad behaviour like marking of items, unauthorized materials substitution, or record doctoring. For instance, manufacturers cannot retroactively issue certificates, distributors cannot tamper with temperature logs, and suppliers cannot forge origin allegations. In addition, clients or auditors can confirm the ethical origins of the product to the claims of the producer by reading the blockchain, which creates a trustworthy relationship between consumers and the brand (Shi et al., 2019).

3.3 Automating Compliance with Smart Contracts

Smart contracts self-executing computer code sitting on the blockchain – can be programmed to enforce moral code and regulations. For example, a smart contract might confirm that all raw materials are certified as conflict free before allowing them to be manufactured into a medical device. If a supplier doesn't adhere to specific labor or safety standards, deliveries can be automatically rejected or alerts sent out to compliance officers. This rules-based automation, in the terms of reduction of manual work, this makes the procedure more efficient and less prone to human error or fraud.

4. SMART CONTRACTS FOR SUPPLY CHAIN AUTOMATION

Smart contracts are self-executive code that is deployed to a blockchain and can enforce rules, making contracts more transparent, secure and autonomous. Furthermore, in supply chain automation, smart contracts minimize the role of intermediaries, lessening manual intervention, having real-time application of business terms; which makes smart contracts an interesting prospects for the Medical IoT based supply chain of the future (Bennett and L. 2024).

4.1 Automation of Supply Chain Transactions

With smart contracts, different supply-chain processes can be automated, helping to bring efficiency in the work and eliminating errors or delays. Principal operations include validation of purchase orders, tracking and confirmation of shipments, payments triggered by receipt of goods, as well as compliance checks. For example, for a medical sensor, a smart contract can release payment automatically once a supplier has shipped a batch and uploaded the details to the blockchain, it verifies that the shipment has occurred and that quality compliance data has been uploaded to the blockchain. Such automation serves to increase the transparency and the trust in transactions amongst the parties, as well as ensure that transactions are promptly effected and performed according to pre-agreed contractual conditions (Kutybayeva et al., 2025).

4.2 Enforcing Predefined Conditions

Smart contracts are executed automatically only upon the arrival of certain conditions that are verifiable to prevent disputes and make all participants fair. In the case of Medical IoT equipment, this functionality is especially crucial because the devices in use are critically important (Nanwal et al., 2021). For instance, it could require a supplier to provide a valid compliance certificate, such as ISO 13485, before allowing a shipment to be approved, ensuring regulatory requirements are addressed at the time of shipment initiation. What's more, if the cold-chain temperature of a device is off during shipping, the smart contract can automatically pause the delivery or offer a quality alert. The automation and conditional enforcement ensures heightened safety and accountability in medical supply chain.

4.3 Enhancing Transparency and Accountability

Since all activities that are regulated by smart contracts are securely recorded on the blockchain, all companies in the supply chain ranging from component manufacturers to final distributors face the music. Such inherent transparency makes fraud costly and guarantees full auditability in real time, instant resolution of disputes and permanent records of contract execution (Khan et al., 2024). Every action and event is logged alongside the timestamp, which makes tampering impossible without detection. For instance, one could easily verify when a Medical IoT device was built and to what specifications and under what conditions it was distributed. This level of traceability builds trust between partners and is a system facilitating compliance with regulatory and quality requirements (Panda et al., 2024).

4.4 Improving Efficiency and Reducing Costs

Smart contracts mitigate the necessity for a multilayer process of manual supervision and third\-party confirmation services, therefore optimizing supply chain processes and improving overall efficiency. They automate critical processes, drastically lowering administrative, procurement and shipping costs, and lowering the likelihood of error or miscommunication between staff. This automation is especially important in heavily regulated areas like healthcare, where even small instincts or documentation errors can put patient lives at risk or result in them paying heavily in non-compliance fees (Chauhan et al., 2021). Once these smart contracts are in place, organizations can keep a tight grip on workflows all the way through and, at the same time, observe velocity, precision and compliance with regulatory rules.

5. DECENTRALIZED IDENTITY AND SUPPLIER VERIFICATION

Manual Supply Chains: In legacy supply chains, and especially in fields such as Medical IoT Manufacturing is the process of identifying and authenticating suppliers slow, non-transparent and subject to fraud (Nanwal et al., 2021). This is even more so in ethical supply chains which rely on trust, compliance and traceability. Decentralized identity (DID)-based systems running on blockchain technology offer disruptive change, offering secure, verifiable, and non-forgable digital identity for manufacturer, suppliers, and even stakeholders (Bai et al., 2020).

5.1 Decentralized Identity (DID)

Decentralized identity (DID) is an identity framework built on blockchain technology that enables people, organizations, or things to own and control their digital identities without relying on a central entity (Bai et al., 2020).These identities are hidden and cryptographically attested on a decentralized ledger, where integrity, tamper-resistance, and non-repudiation are guaranteed. In the case of supply chain (and particularly supply chains for, for example, the vital medical manufacturing sector), each supplier, each manufacturer, might have a distinct DID (Khanfar et al., 2021). This DID is securely connected to important information and credentials such as compliance certification (e.g., ISO13485 certification for medical devices), transaction history, the geographical location of manufacturing facilities, license, audit results, and ESG (Environmental, Social, Governance) performance scores. Bringing about transparency and tamper-proofing digital identity, DIDs increase accountability, traceability and trust at all stages of the supply chain (Kutybayeva et al., 2025).

5.2 Supplier Verification Using Blockchain

In a blockchain-based supply chain, supplier onboarding and validation is now much faster and more reliable. Each new supplier onboarding is recorded with a blockchain-based Decentralized Identity (DID) solution, which forms the company's digital identity. Certificates of compliance, licenses, and partner endorsements verification credentials are digitally signed by trusted certifying bodies and written to the blockchain in a way that they cannot be tampered with. By simply being able to check out the supplier's credentials, certifications, and transaction history directly on the blockchain, once a transaction or partnership is in the process of being formed, any other party can quickly verify who the supplier was, without having to do their own fact-checks or depend on intermediaries (Shi et al., 2019). This frictionless process significantly reduces onboarding time, creates trust among network participants and reduces the risk of bad actors (counterfeiters and non-compliant and unethical, non-vetted suppliers) creeping into the network.

5.3 Real-Time Trust Anchors

By tying verified identity data to individual supply chain events, blockchain provides real-time trust anchors against which the veracity of data and the trustworthiness of participants can be validated. For instance, a trusted medical sensor producer can push its production logs on the blockchain directly. Every log entry is attributed with their Decentralized Identifier (DID), and the time it was created, providing irrefutable proof of provenance. This enables distributors, hospitals and regulators to immediately verify that a medical device originated from a certified, compliant manufacturer. This real-time verification facilitates end-to-end traceability and accountability - particularly as the global supply chain remains characterized by fragmented oversight and high exposure to counterfeit or substandard products (Mann et al., 2018). The fact that blockchain can anchor identity to data and events and therefore provide trust at the core of all participants of the network.

5.4 Benefits for Medical IoT Manufacturing

- **Removes Counterfeit Risk:** Real suppliers can't fake identities Verified on the blockchain.
- **Faster Compliance:** It automates compliance checks of a supplier to regulations of a medical-grade products.
- **Supports ethical auditing**: Buyers select which suppliers to work with if the latter are known to have ethical traceable data (Panda et al., 2024).

- **Facilitates Cross-border Trust:** Blockchain allows a single source of truth across borders.

6. REAL-TIME MONITORING WITH IOT AND BLOCKCHAIN INTEGRATION

The combination of IoT and blockchain is influencing real-time monitoring in supply chains, particularly in medical IoT manufacturing, where the integrity and security of products is important (Khanfar et al., 2021). Using both sensor-based data capture and cryptographically secure distributed ledgers, it enables verification of product conditions, from temperature and humidity, through to the identification of shock events and location on a single, permanent record for the entire supply chain.

6.1 Role of IoT in Real-Time Condition Monitoring

The packaging, transport vehicles or storage units can have IoT sensors installed in them which can become invaluable from the moment of manufacture to the time at which these products reach their final destination (Upadhyay et al., 2023). This sensors are used to control important environmental parameters permanently and with it sensitive goods can be properly handled or transported. In a medical space that could mean temperature used for devices that ship with cold chain logistics, humidity for corrosion or damage prevention and vibration for sensitive parts that are easily damaged in transit (Pournader et al., 2020). What's more, GPS tracking gives you all the live up-to-date positions on shipment locations to adhere to route integrity and ward off theft and hijacking. The multiple sensor nodes of these IoT devices report real-time data, usually to an edge gateway or cloud service, providing instantaneous analysis and decision-making. This consolidation of IoT and supply chain logistics improves visibility, compliance and mitigates the risk of damage and regulatory discord.

6.2 Blockchain as an Immutable Audit Trail

Data retrieved by the IoT devices is hashed and stored on the blockchain, creating a secure and traceable log of the environmental parameters seen on the product throughout the supply chain. The data can no longer be edited or deleted after it has been logged. Every sensor reading also is time-stamped, and signed cryptographically with a key to guarantee its authenticity and source. All the involved actors from the manufacturer, through the logistics provider, to the regulator and the end user, have access to the same verified data at the same time and that gives transparency and trust (Pournader et al., 2020). By preventing forged and tampered transport logs and environmental conditions, the blockchain-IoT solution closes critical loopholes in standard supply chains, increasing compliance and enforcement – especially for sensitive medical devices and equipment.

6.3 Benefits in Medical IoT Supply Chains

- **Regulatory Compliance**– Regulating authorities (FDA and EMA) demand evidence that medical IoT devices were transported in a certified manner. Blockchain provides verifiable records.
- **Tamper Detection:** Any kind of irregularity, including, but not limited to, insulation leak to a higher temperature outside of the safe range, is written into an immutable log allowing immediate remediation.
- **Alerts as the World Changes:** Smart contract-governed alerts will be issued or informed through things when IoT readings cross set points.
- **Chain of Custody Transparency:** Every transfer from manufacturer to distributor to hospital is recorded and tied to sensor readings in a complete visibility chain.

7. CONSENSUS ALGORITHMS FOR MEDICAL IOT SUPPLY CHAINS

This is even truer when it comes to blockchain-enabled supply chains (i.e., the manufacturing of medical IoT) where it is important to select the most appropriate consensus algorithm. Consensus algorithms decide on how participants in a network of distributed nodes can come to an agreement on the legitimacy of transactions, promoting data consistency over a blockchain (Khanfar et al., 2021). Healthcare sector, as well, requires a high level of trust, secure and efficient systems, but also works in permissioned networks with participants being trusted entities (authorized suppliers, producers, and regulators) (Prakash et al., 2023). This section provides a comparative analysis of the three prominent consensus algorithms, PoW, PoS and PoA and explains why PoA is suitable for medical IoT supply chain.

7.1 Proof of Work (PoW)

Proof of Work mining (PoW) is a process where miners are solving cryptographic puzzles in order to validate transactions and add them to the blockchain. While that's mathematically quite secure in itself, since it's so difficult to compute, and has worked well for networks like Bitcoin, this method has its downsides (Garg et al., 2023). PoW is very costly in terms of energy and its ecological aspects and it is not a preferred choice for sustainable applications (Sharma et al., 2024). There is also low transaction per second and confirmation lag, which does not fit the real-time, high speed needs of health care systems. In a Medical IoT setting both which need off-line validation of data and have strict energy constrains, PoW is regarded unreasonable for use for a consensus of blockchain. For modern healthcare logistics, this is not in line with the speed, scalability and environmental sustainability that is desired and required (Venkatesh et al., 2020).

7.2 Proof of Stake (PoS)

Proof of Stake (PoS) is the consensus mechanism in which validators are voted in to validate transactions based on the crypto currency to which they hold and are willing to stake as security. This approach is much more energy saving compared to the Proof of Work consensus, and enables faster transaction processing, which offers a great appeal to many blockchain applications(Sharma et al., 2024). However, PoS has notable limitations. It can bring centralization because the rich individuals have more power over the network (Garg et al., 2023). This economic validator selection may not represent reliability and adherence to regulation and compliance. Nobody is transparent about who is doing what in a PoS system, where things are particularly obscure with regard to how validators store accountability given specifics of their industry. In the medical IoT context, of patient safety, compliance and verifiable trust, may be inefficient for PoS It does not automatically grant the institutional oversight and trust assurances required for handling sensitive health data and supply chains (Zhang & Chan2020).

7.3 Proof of Authority (PoA)

Security model: It's a Proof of Authority consensus where a limited set of validators curator-selected validators approve transactions. On the contrary of PoW and PoS, PoA is based on well-known validators who are trustable instead of computational power or assets at stake (Wang et al., 2023). Advantages of this model include high transaction throughput and low latency for real-time supply chain operations, low energy consumption to promote sustainability, and an identity-based trust model, where validators, such as licensed manufacturers, regulatory bodies, or accredited hospitals, are held responsible for the on-chain events (Venkatesh et al., 2020). In addition, PoA offers strong auditability, since all transactions and validator actions can be traced. But it is more centralised than PoW or PoS in the sense that it relies that the mixed network will not have enough stake to outrace the stakers and its more centralized than PoW since you are trusting that validators are honest rather than the decentralisation being a feature. It is relevant to Medical IoT as PoA is a permissioned blockchain types. It permits only accredited and permissioned entities, such as compliant medical vendors or healthcare providers, to run validator nodes and facilitate full regulatory compliance, while keeping data integrity and system operations smooth and efficient in the medical supply chain(Chauhan et al., 2021).

7.4 Justification for PoA in Medical IoT Supply Chains

To effectively manage the supply chain in a medical IoT you need to balance the wishes of transparency, trust, and control so that patients are safe, regulations are met, and operations run smoothly. Proof of Authority (PoA) does a great job filling this gap by being a consensus model for permissioned, high-trust environments. In those cases it ensures that regulation and ethical bodies can be integrated in a natural manner - doing enforcement and auditing. PoA allows for real-time tracking by validating quick, continuous data streams from IoT sensors (e.g., temperature, humidity and geolocation) vital for the quality assurance in delicate medical instruments(Zhang & Chan 2020). And as validators in a PoA network are vetted trusted parties, accountability can be embedded into the system for gaudy residue and supply chain failure scenarios, there is someone to blame. Its architecture also provides operational scalability, which is ideal for large-scale deployments by the enterprises, and consortiums of accredited stakeholders, including hospitals, logistics companies, and regulatory bodies(Pournader et al., 2020). Perhaps this is the reason why PoA has a great potential to be a practical solution for securing and enhancing supply chains of Medical IoT due to its trade-off between speed, compliance and accountability.

8. DATA PRIVACY AND SECURITY CONSIDERATIONS

In the Medical IoT supply chain, it is of utmost importance to protect sensitive assets, such as patient-related device data, proprietary component specifications, and regulatory compliance documentation. The immutability and the transparency of blockchain are attractive for data sharing, however privacy issues should be taking into consideration such as to make sure ethical and safe use of data.

Figure 2. Data Privacy And Security

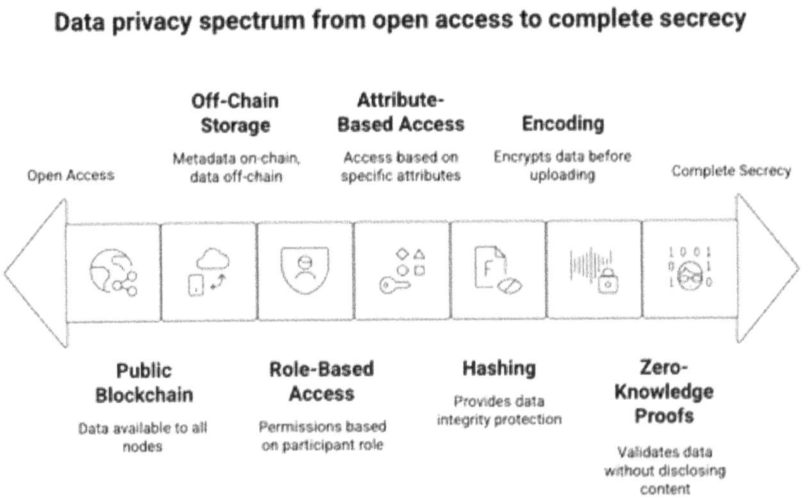

8.1 Handling Sensitive Data in a Decentralized Environment

Blockchain networks, are usually spread across several parties such as manufacturers, regulators, logistics companies, and hospitals, especially in the supply chain domain. this increases the threat of unauthorized access to sensitive details where security policy is not established. Public Blockchain make all data available to all nodes by design, which is not practical for sensitive medical information. Therefore, Medical IoT networks should be permissioned to only allow verified and authorized parties to access data..

8.2 Off-Chain Data Storage

The architecture is frequently hybrid, due to the fact that the blockchain is not well-suited for large amounts of data storage (e.g., logs and diagnostics files). Only crucial metadata (e.g., file hashes, timestamps, and digital signatures) is published on-chain, with data stored in off-chain secure storage solutions like IPFS, cloud databases, or institutional data servers (Subramanian et al., 2021). This method ensures the integrity of blockchain while achieving storage and privacy compliance at ratio achievable.

8.3 Cryptographic Techniques for Data Protection

- **Hashing**: Provides integrity protection by creating a unique fingerprint for each data file. Any modification will cause the hash to be different, and it indicate an inconsistency.
- **Encoding:** Data is E2E encrypted prior to upload of chain. Consumer access is restricted to those with missed call decryption keys.
- **Zero-Knowledge Proofs (ZKPs:** They make it possible to validate contents (e.g. authenticity, compliance) of data without disclosing the content itself. This is extremely useful for creating privacy during audits.

8.4 Role-Based and Attribute-Based Access Control

In a blockchain-based Medical IoT supply chain, sensitive data are controlled using smart contracts which enforce fine-grained access control policies. These smart contracts can use Role-Based Access Control (RBAC) where permissions are assigned based on a participant's role for the network, like a manufacturer, distributor, or regulator. For instance, a regulator may be given permission to see all compliance logs, but a distributor can only see the data of the progress of shipment. Also Attribute-Based Access Control (ABAC) can be used where access decisions are based on certain attributes, e.g. certification level, geographic location, and or operational status of a device. This flexible, automated method prevents unauthorized access to, or manipulation of, specific data, adding increased privacy, and compliance, and trust throughout the supply chain's lifecycle.

8.5 Regulatory and Compliance Alignment

For Medical IoT the blockchain solution should comply with international privacy regulation in order to guarantee the safe and ethical handling of sensitive health data. Important standards include: HIPAA in the US, which emphasizes the strong privacy and security requirements for health data; the GDPR in the EU, which makes data-focused rights such as data minimization and the 'right to be forgotten' enforceable; and ISO/IEC 27001, which specifies the requirements for information security management systems. This regulatory requirement can be fulfilled for by programming smart contracts to enforce automatic compliance. For example, provide time-based access to sensitive information, revoke permissions after an event or period of time, or invoke automated deletion or anonymization processes when policy dictates. By integrating compliance logic at the infrastructure level of the blockchain, the Medical IoT systems can enforce privacy, security, and legal guarantees without relying only manual supervision (Beniwal et al., 2021).

9. CASE STUDY: TRANSPARENT MANUFACTURING OF IMPLANTABLE DEVICES

In the heavily regulated field of healthcare, medical implants (e.g., pacemaker, cochlear implant, insulin pump or orthopedic screw) are subject to the strictest safety, traceability and ethical requirements (Prakash et al., 2023). Even the slightest malfunction or imitation of such devices could have fatal outcomes. In this case study, the application of blockchain technology was adopted, in order to build trust and ethical responsibility on the manufacturing and distribution of an implantable device(Bai et al., 2020).

9.1 Background and Problem Statement

The standard supply chain of medical devices is from raw material suppliers to component manufacturers, to assemblers, to logistics provider, to hospitals, and finally to the hospital patient. Key challenges include:

- No clear view of orders from quote on the shelf.
- Counterfeit components entering unnoticed.
- Paper or loose digital files that are easily manipulated.
- Could be held responsible to a limited extent for delays, QC and Sourcing compliance.

These flaws seriously threaten both patient safety and compliance.

9.2 Blockchain Implementation Framework

A permission-based blockchain network was implemented with all legitimate stakeholders being certified raw material providers, ISO-certified manufacturers, government regulators, logistics and the hospital purchase department. Key Features of the Implementation.

9.2.1 Unique Device Identity (UDI) on Blockchain

Each of the implantable issued were attributed a cryptographically secure Unique Device Identity (UDI) it was written forever to the blockchain. The UDI took into account key information such as date of manufacturing, lot number, source of the raw material and quality assurance test results, resulting in a collection of tamper-resistant, traceable records on the origin and quality of the device.

9.2.2 Smart Contract-Driven Compliance Checks

Foreground During production, the inspection itself was also enforced by smart contracts at a few key compliance points. They allowed for automatic verification of supplier certifications, verification of ISO standards prior to assembly, and checkpoints for approvals at all stages of production. This contributed to consistent quality and regulation compliance, but it did not depend on manual oversight.

9.2.3 Integration with IoT Devices

Real-time IoT sensors tracked ambient humidity and temperatures for implantable devices while stored or transported. The real-time information from these sensors was also hashed and stored on the blockchain, enabling a permanent audit trail that could also be leveraged to support compliance verification and quality control.

9.2.4 Regulatory and Ethical Tracking

Blockchain made sure that the raw materials were sourced from ethically approved mine and reliable suppliers – removing the conflict minerals and unethical practices.

9.2.5 Hospital Access and Verification

Prior to implant, for example hospitals and surgeons would be able to scan the UDI QR code to check the full provenance of the device on the secure blockchain dashboard to determine authenticity, compliance and safety.

9.3 Outcomes and Benefits

- **Increased Transparency:** Everyone had access to tamper-proof, shared data.
- **Enhancement of Patient Safety:** No counterfeit implants were found after implementation.
- **Audit Efficiency:** Compliance audits of regulation went from weeks to hours with instant access to the blockchain.
- **Supply Chain Traceability:** Real-time tracking back to culprit for any delays or failures.
- **Cost Saving:** No paper work and manual verification reduced the time and operational cost.

10. BENEFITS AND LIMITATIONS OF BLOCKCHAIN-BASED SUPPLY CHAINS

Blockchain is a game changing technology in how we secure and protect supply chain information with its ability to provide transparency, reinforce ethical processes, and alleviate fraud. However, just like any technology, it comes with great advantages and some cons that need to be thoroughly weighed before wide-scale implementation.

The blockchain system provides an innovative means for managing supply chains and helps to achieve traceability, ethical compliance and reduction on fraud (Mann et al., 2018). But, just as with any technology, it has some very attractive advantages, along with the usual set of limitations to make well-thought-out decisions prior to widespread adoption.

10.1 Benefits

10.1.1 Enhanced Transparency and Traceability

The unique tamper-proof, unchangeable ledger that is made available by blockchain, in which every transaction, every movement of goods is recorded and sealed with a time-stamp. Such transparency allows direct tracking from raw material to end user and full traceability. It is also an important mechanism for combatting counterfeiting for high-risk sectors such as medical IoT manufacturing and aids companies to meet regulatory compliance more cost effectively and effectively.

10.1.2 Ethical Accountability

Supply chains are notoriously criticized for getting materials from dubious or illegal sources. Blockchain provides a solution to this by allowing verifiable ethical sourcing, based on recorded and unalterable supplier certifications (Upadhyay, Prashant, and Pradeep Tomar). It also enables the tracking of labor and environmental conditions up and down the supply chain. By letting in only the authorized accounts (verified vendors) and absolutely not letting in people who aren't, blockchain ultimately helps maintain more accountability and ethical and legal legality.

10.1.3 Cost Reduction through Automation

By automating such critical supply chain processes as payments, quality checks, and customs clearances, smart contracts can ease transactions while minimizing delays. They help reduce the cost of operations and speed of doing business across the network due to less intermediaries and use of lesser manual paperwork and that will be very much beneficial to all.

10.1.4 Improved Security

The cryptographic protocols of the blockchain ensure that data is protected from tampering and unauthorized access, thus maintaining integrity and confidentiality of all recorded information. This level of security is particularly important in medical supply chains, where safeguarding sensitive health data and ensuring medical device integrity are central to patient safety and regulatory compliance within.

10.1.5 Real-Time Data Sharing and Collaboration

Transparency and Consistency Members of the blockchain network can see the same real-time data, which runs parallel to guarantee trust and accountability within most supply chain ecosystems. This shared view of information increases alignment, reinforces trust, and enables more rapid decisions to be made with an increased level of knowledge across a globally distributed base of users. All participants in a blockchain network share the same, up-to-the-minute information in the system, so there's never any mystery about what's happening throughout the supply chain.

10.2 Limitations

10.2.1 Scalability Issues

Public blockchains have known issues supporting vast amounts of transactions which causes the network to become congested and process transactions slowly. Even in permissioned blockchains, performance could be degraded through addition of significant participants by the application of MapReduce (or similar batching mechanism) whose behaviour is unpredictably hurting both responsiveness and scalability.

10.2.2 Integration Complexity

It can be technically challenging and expensive to integrate blockchain with current legacy systems, ERPs, and IoT systems, as it often requires highly customized and upgraded setups. Calibration between different systems would also be an important factor to address, ideally through standardization and inter-operability of different systems, which can complicate matters in multi-stakeholder situations.

10.2.3 Data Privacy

With its immutability, the blockchain itself is a good place to store data. But storing some sensitive data can sometimes breach privacy regulations like GDPR and HIPAA. Therefore, a hybrid model can be the solution that stores just metadata or hashes to the chain but keeps personally sensitive data off-chain (Kumar et al., 2025). This approach facilitates compliance, but also introduce a complexity regarding the architecture of the system and the management of the data.

10.2.4 Adoption and Trust Challenges

It might be difficult to get all supply chain members to adopt a blockchain network. Smaller suppliers or suppliers with global reach may not have the technical know-how, infrastructure, or resources to participate, resulting in inconsistent participation and capping the true potential of a connected, transparent, blockchain-powered supply chain.

10.2.5 Energy Consumption (in Some Models)

Energy consuming consensus mechanisms, such as Proof of Work (PoW), have very low sustainability and not suitable for supply chains. So, more effective algorithms like PoA or PBFT are used in most cases (Venkatesh et al., 2020). Yet, although they achieve better speed and power efficiency, they can decrease the degree of decentralization and network trust assumption.

11. OVERCOMING TECHNICAL AND ADOPTION CHALLENGES IN BLOCKCHAIN DEPLOYMENT

Although the use of blockchain technology could have revolutionized transparency and ethical behaviour in Medical IoT production, it is still suffering from several technical and up- take deficiencies (Upadhyay et al., 2022). Seamless enterprises and their ecosystems need to be scalable, secure and inclusive in the implementation. These challenges need strategic solutions.

Apart from serving as a disruptive technology that could improve transparency and morality in the manufacture of Medical IoT, implementation of blockchain technology is faced by challenges of both technical and adoption nature.

11.1 Integration with Legacy Systems

One of the biggest technical challenges is the adoption of blockchain technology within an existing legacy ecosystem: ERP systems, inventory management software, and IoT infrastructures. These systems were not originally created for decentralized ecosystems, thus the interoperability is complicated and resource-heavy(Bai et al., 2020).This process can be bridged if blockchain networks can be integrated with traditional IT systems w/o impacting the day to day operations by development standardized APIs, middleware, and secure gateways where data can be easily exchanged between and blockchain network and traditional IT environments.

11.2 Scalability and Performance Constraints

Public Blockchain, in particular those relying on consensus such as Proof of Work (PoW) consensus suffers from limited transactional throughput and delay. Even in permissioned Blockchain, the performance bottleneck will be challenged with large numbers of entities. In the case of healthcare supply chains, where time-critical decisions have to be made, the impact of latency can be critical to responsiveness and patient safety (Garg, 2025). This can be addressed by using scalable consensus algorithms such as PoA or PBFT, which provide higher throughput and lower power consumption, especially in permissioned, high-trust networks.

11.3 Data Privacy and Regulatory Compliance

Immutability of blockchain can be an advantage, but it may be in conflict with privacy-related requirements like GDPR (right to forget) and HIPAA (health-related data protection). Due to these legal restrictions, storing the medical record information directly on the chain is not permissible (Kumar et al., 2025). A hybrid model of hashes or metadata stored on-chain and the actual data stored off-chain in secure, compliant storage is often required. But it complicates data architecture and requires strong encryption and access control.

11.4 Lack of Standardization and Interoperability

Medical supply chains can traverse multiple entities across different nations, and communications are made on different blockchain platforms (e.g., Hyperledger, Ethereum, Corda). The unavailability of common data formats and interoperability standards constitutes a major obstacle to end-to-end transparency (Wan et al., 2020). For this reason, creating common standards for cross-chain communication and blockchain-agnostic APIs is important to ensure seamless sharing of information and collaboration within the heterogeneous infrastructure landscape.

11.5 Limited Technical Expertise

Blockchain is a relatively new, niche field, which means it represents a barrier of entry for many organizations, particularly small suppliers or underserved hospitals in growing markets that do not have the technical knowledge or financial means to implement and maintain a blockchain-based infrastructure. It can be counteracted by investment in capacity development, easy to use platforms and open training programs to lower entry thresholds and stimulate wider uptake.

11.6 High Initial Costs and ROI Uncertainty

The deployment of blockchain solutions requires high initial investments in terms of software, hardware, and workforce. The return on investment (ROI) is not immediately obvious for many organizations, especially smaller stakeholders." To overcome this, parties should begin with smaller scale proving ground initiatives to leverage high-impact use cases, such as anti-counterfeiting, compliance clarity or UDI management, and outline building blocks that demonstrate tangible value for the participants, before it is rolled out across the full supply chain (Upadhyay et al., 2024).

11.7 Resistance to Change

Adoption also faces resistance at an organizational level, as blockchain threatens the traditional centralized models of data ownership and trust (Wan et al., 2020). Trust is the biggest barrier Stakeholders' concerns about sharing confidential data, losing control and trusting in distributed infrastructure and partners are the main barriers to a successful DT. Trust can be built by transparent governance systems, legal frameworks and incentives (e.g., low costs of compliance, faster payments) to continue this collaboration and uptake.

11.8 Validator Governance and Accountability

In the permissioned PoA or PBFT networks, validators should be pre-approved actors like producers, regulators, or the logistics companies. But building out governance infrastructure that monitors the actions of validators and serves as an accountability mechanism is a challenge. Well-defined roles, consequences for non-compliance and third party monitoring are a must for keeping the network integrity and trust of the stakeholders (Upadhyay et al., 2023).

12. FUTURE TRENDS AND RECOMMENDATIONS

As blockchain evolves, it is anticipated to play a bigger part in ensuring ethical and transparent supply chains, especially in high-stakes settings such as medical IoT manufacturing. Multiple innovations and technologies are driving the future of blockchain in the supply chain as well as important policy and regulatory considerations. This section considers important future directions and provides strategic advice for stakeholders.

12.1 AI Integration for Predictive Analytics and Smart Decision-Making

Combining AI with blockchain in the context of supply chains can add a lot of intelligence, automation and even adaptability (Garg, P., Sharma, N., & Sonal, B. S). AI-based applications such as predictive maintenance and demand forecasting drive actionable insights that, when logged and verified on-chain, are capable of reducing downtime, lowering waste and enabling better resource planning (Wan et al., 2020). AI agents can also work together with IoT sensor data and perform on-the-fly actions, based on the sensors that execute smart contracts and authorize certain things, such as a lawnmower or a mobile working robot (Garg2024)." This means automatic responses to anomalies, such as variations in temperature, in real-time without humans needing to get involved. With enhanced AI and blockchain blending over time, we create self-correcting systems that compare future supply-chain activity to known past behaviours, constantly improving operations while reinforcing ethical standards and regulatory compliance.

12.2 Zero-Knowledge Proofs (ZKPs) for Privacy-Preserving Transparency

Introduction to Zero-Knowledge Proofs Zero-Knowledge Proofs are a family of cryptographic protocols that enable one party (the prover) to prove to another party (the verifier) that some statement is true, without revealing any information about the statement (or "witness") except that it is true.

- There are scenarios where ZKPs could provide for the confidential validation of adherence of a supplier to a standard, the provenance of a good, the quality

of a good and others without revealing trade secrets, proprietary information, or harmful information to the validator.
- This is particularly important in medical device manufacturing, where privacy and compliance are paramount.

ZKPs are able guarantee cryptographic acceptability while satisfying the requirements of both industrial and legal standards; that is, the audit requirements.

12.3 Cross-Chain Interoperability

"Blockchain is moving beyond being just for the financial industry and becoming more intermingled among industries; so you're going to need interoperability between the blockchain and the different Blockchain that are used for supply chain-based solutions too (Wan et al., 2020)." Supply chains generally are composed of multiple companies, potentially with different platforms, i.e., Hyperledger, Ethereum or private consortium chains. In order to facilitate widespread cooperation, the need for cross-chain bridges and protocols that open up the ability to share data securely and quickly between networks is absolutely essential. This synergy enables traceability, promises more transparency and offers real-time coordination, despite their global, multi- and cross-chain supply chain nodes. At the end of the day, it enables end-to-end visibility and traceability for the entire supply ecosystem.

12.4 Development of Global Policy and Regulatory Frameworks

For blockchain applications to transform medical and ethical supply chains, however, clear legal and governance mechanisms are needed:

- Only then can international norms on the sharing of data, interoperability and reporting be set.
- Governments and regulators need to cooperate on the creation of reliable digital identities, licensing regimes and legality of smart contracts.
- Policy frameworks should also promote the development of green blockchain protocols, particularly in energy-constrained industries.

13. CONCLUSION

In this chapter, the disruptive potential of blockchain technology is examined concerning its impact on ethical considerations and transparency in the supply chain of Medical IoT (Internet of Things) manufacturing. The fabrication of counterfeit components, unscrupulous sourcing, divided and disjected inspections, and non-transparent processes are also emerging issues together with the demand for connected healthcare systems. Blockchain provides a decentralized, tamper-proof, and verifiable structure that can efficiently tackle these problems. The role of specific blockchain elements, e.g., smart contracts, decentralized identity and consensus algorithms (in particular, Proof of Authority) to help automate compliance, provider vetting and real-time tracking are considered. The layering of IoT sensors on blockchain enables ongoing monitoring of device parameters (such as temperature and humidity), which is critical in maintaining product integrity during manufacturing and transport.

The data privacy concerns is the topic analysed also in the chapter, that focused the attention on the off-chain storage, cryptographic constructions and the possibility of using zero-knowledge proofs (ZKPs) in order to have transparency preserving privacy. A concrete use case, where one specifies the challenges around implantable medical devices (IMD), adds the industry experiences and benefits as traceability, ethical validation and regulatory convenience of the method. In conclusion, we discussed the future trend for AI incorporated in smart supply chain decision-making, and cross-chain for multi-network cooperation, and strong global policy frameworks for blockchain adoption in medical industries. Although there are still some outstanding challenges such as user acceptance, integration with legacy systems, and keeping validator trust, the general advantages of blockchain in transparent, secure, and ethical operations in the medical-IOT context are sizeable and promising.

REFERENCES

Bai, C. A., Cordeiro, J., & Sarkis, J. (2020). Blockchain technology: Business, strategy, the environment, and sustainability. *Business Strategy and the Environment*, *29*(1), 321–322. DOI: 10.1002/bse.2431

Beniwal, S., Saini, U., Garg, P., & Joon, R. K. (2021). Improving performance during camera surveillance by integration of edge detection in IoT system. [IJEHMC]. *International Journal of E-Health and Medical Communications*, *12*(5), 84–96. DOI: 10.4018/IJEHMC.20210901.oa6

Bennett, L. (2024). The Role of Blockchain Technology in Enhancing Supply Chain Transparency in Italy.

Chauhan, S., Singh, M., & Garg, P. (2021). Rapid forecasting of pandemic outbreak using machine learning. *Enabling Healthcare 4.0 for Pandemics: A Roadmap Using AI, Machine Learning, IoT and Cognitive Technologies*, 59-73.

Garg, P. (2024). Prediction of female pregnancy complication using artificial intelligence. In *Artificial Intelligence and Machine Learning for Women's Health Issues* (pp. 17–35). Academic Press.

Garg, P. (2025). Explainable AI & Model Interpretability in Healthcare: Challenges & Future Directions. *EKSPLORIUM-BULETIN PUSAT TEKNOLOGI BAHAN GALIAN NUKLIR*, *46*(1), 104–133.

Garg, P., Dixit, A., Sethi, P., & Pruthi, J. (2023, December). Strengthening Smart City with Opportunistic Networks: An Insight. In *2023 International Conference on Advanced Computing & Communication Technologies (ICACCTech)* (pp. 700-707). IEEE. DOI: 10.1109/ICACCTech61146.2023.00117

Garg, P., Pranav, S., & Prerna, A. (2021). Green internet of things (G-IoT): A solution for sustainable technological development. In *Green Internet of Things for Smart Cities* (pp. 23-46). CRC Press. Garg, P., Sharma, N., &Sonal, B. S. INTELLIGENT SYSTEMS AND APPLICATIONS IN ENGINEERING.

Garg, P., Srivastava, A. K., Anas, A., Gupta, B., & Mishra, C. (2023). Pneumonia Detection Through X-Ray Images Using Convolution Neural Network. In *Advancements in Bio-Medical Image Processing and Authentication in Telemedicine* (pp. 201–218). IGI Global. DOI: 10.4018/978-1-6684-6957-6.ch011

Gautam, V. K., Gupta, S., & Garg, P. (2024, March). Automatic Irrigation System using IoT. In *2024 International Conference on Automation and Computation (AUTOCOM)* (pp. 100-103). IEEE.

Khan, N., Hossain, S., Khadka, U., & Sarkar, S. (2024). Blockchain in Supply Chain Management: Enhancing Transparency, Efficiency, and Trust. AIJMR-Advanced International Journal of Multidisciplinary Research, 2(5).

Khanfar, A. A., Iranmanesh, M., Ghobakhloo, M., Senali, M. G., & Fathi, M. (2021). Applications of blockchain technology in sustainable manufacturing and supply chain management: A systematic review. *Sustainability (Basel)*, *13*(14), 7870. DOI: 10.3390/su13147870

Kumar, B., Garg, V., Ahmed, K., Garg, P., Choudhary, S., & Baniya, P. (2025, March). Enhancing Healthcare with Blockchain: Innovations in Data Privacy, Security, and Interoperability. In *2025 3rd International Conference on Disruptive Technologies (ICDT)* (pp. 932-938). IEEE.

Kutybayeva, K., Razaque, A., & Rai, H. M. (2025). Enhancing Pharmaceutical Supply Chain Transparency and Security with Blockchain and Big Data Integration. *Procedia Computer Science*, *259*, 1511–1522. DOI: 10.1016/j.procs.2025.04.106

Mann, S., Potdar, V., Gajavilli, R. S., & Chandan, A. (2018, December). Blockchain technology for supply chain traceability, transparency and data provenance. In *Proceedings of the 2018 international conference on blockchain technology and application* (pp. 22-26). DOI: 10.1145/3301403.3301408

Nanda, S. K., Panda, S. K., & Dash, M. (2023). Medical supply chain integrated with blockchain and IoT to track the logistics of medical products. *Multimedia Tools and Applications*, *82*(21), 32917–32939. DOI: 10.1007/s11042-023-14846-8 PMID: 37362711

Nanwal, J., Garg, P., Sethi, P., & Dixit, A. (2021). Green IoT and Big Data: Succeeding towards Building Smart Cities. In *Green Internet of Things for Smart Cities* (pp. 83-98). CRC Press.

Panda, S. K., & Satapathy, S. C. (2024). Drug traceability and transparency in medical supply chain using blockchain for easing the process and creating trust between stakeholders and consumers. *Personal and Ubiquitous Computing*, *28*(1), 75–91. DOI: 10.1007/s00779-021-01588-3

Pournader, M., Shi, Y., Seuring, S., & Koh, S. L. (2020). Blockchain applications in supply chains, transport and logistics: A systematic review of the literature. *International Journal of Production Research*, *58*(7), 2063–2081. DOI: 10.1080/00207543.2019.1650976

Prakash, A., Avasthi, S., Kumari, P., Rawat, M., & Garg, P. (2023). 18 Modern healthcare system: unveiling the possibility of quantum computing in medical and biomedical zones.

Raj, G., Verma, A., Dalal, P., Shukla, A. K., & Garg, P. (2023). Performance comparison of several LPWAN technologies for energy constrained IOT network. *International Journal of Intelligent Systems and Applications in Engineering*, *11*(1s), 150–158.

Saberi, S., Kouhizadeh, M., Sarkis, J., & Shen, L. (2019). Blockchain technology and its relationships to sustainable supply chain management. *International Journal of Production Research*, *57*(7), 2117–2135. DOI: 10.1080/00207543.2018.1533261

Upadhyay, D., Garg, P., Aldossary, S. M., Shafi, J., & Kumar, S. (2023). A linear quadratic regression-based synchronised health monitoring system (SHMS) for IoT applications. *Electronics (Basel)*, *12*(2), 309. DOI: 10.3390/electronics12020309

Upadhyay, P., & Tomar, P. "Alzheimer's disease Classification using various machine learning approaches: A Review." In 2022 4th International Conference on Advances in Computing, Communication Control and Networking (ICAC3N), pp. 232-237. IEEE, 2022. DOI: 10.1109/ICAC3N56670.2022.10074339

Upadhyay, P., & Tomar, P. "Alzheimer's Disease Classification Using Ensemble Methods." In *International Conference on Signal*

Upadhyay, P., Tomar, P., & Yadav, S. P. (2024). Comprehensive Systematic Computation on Alzheimer's Disease Classification. *Archives of Computational Methods in Engineering*, *31*(8), 1–32. DOI: 10.1007/s11831-024-10120-8

Venkatesh, V. G., Kang, K., Wang, B., Zhong, R. Y., & Zhang, A. (2020). System architecture for blockchain based transparency of supply chain social sustainability. *Robotics and Computer-integrated Manufacturing*, *63*, 101896. DOI: 10.1016/j.rcim.2019.101896

Wan, P. K., Huang, L., & Holtskog, H. (2020). Blockchain-enabled information sharing within a supply chain: A systematic literature review. *IEEE Access : Practical Innovations, Open Solutions*, *8*, 49645–49656. DOI: 10.1109/ACCESS.2020.2980142

Wang, B., Lin, Z., Wang, M., Wang, F., Xiangli, P., & Li, Z. (2023). Applying blockchain technology to ensure compliance with sustainability standards in the PPE multi-tier supply chain. *International Journal of Production Research*, *61*(14), 4934–4950. DOI: 10.1080/00207543.2022.2025944

Zhang, R., & Chan, W. K. V. (2020, July). Evaluation of energy consumption in block-chains with proof of work and proof of stake. *Journal of Physics: Conference Series*, *1584*(1), 012023. DOI: 10.1088/1742-6596/1584/1/012023

Chapter 13
AI and IoT in Global Health:
Ethical Lessons From Pandemic Response

Leelawati Pokhrel
https://orcid.org/0009-0005-2675-1720
SAITM, India

Ajay Kumar
https://orcid.org/0000-0003-4739-8127
Raffles University, India

Puneet Garg
https://orcid.org/0000-0002-4635-546X
KIET Group of Institutions, Delhi, India

Neetu Anand
https://orcid.org/0000-0003-1514-2282
MSI, Delhi, India

Narinderjit Singh
INTI Universitry, Malaysia

ABSTRACT

COVID-19 epidemics catalysed the outstanding distribution of artificial intelligence (AI) and Internet of Things (IoT) technologies in global health systems, originally changed how social illness monitoring, contact tracking, diagnostics and delivery of the health care system. While these technologies demonstrated remarkable ability to rapid epidemic response from the-AI diagnostic image system, which helped the

DOI: 10.4018/979-8-3373-4134-7.ch013

radiologists overwhelmed in the contact applications made in the IT anesthesia, which tracked the transfer of disease, also emphasized important moral challenges on secrecy security, Healthcare, Algorithm. This chapter offers a comprehensive study of the moral implications of the implementation of AI and IoT during the epidemic, analyses both remarkable successes and important errors in different health systems and national contexts.

1. INTRODUCTION

The COVID-19 epidemic represented a watershed moment for digital health technologies, which originally turns into artificial intelligence (AI) and Internet of Things (IoT) to global health systems. Typically, the decades of gradual implementation would have been taken as governments, health institutions and technology companies, which were designed to distribute digital solutions on a unique scale and speed (WhitaLov et al., 2020). This extraordinary acceleration created a unique historical moment, where the promises and crises from digital health technologies were felt together, both offer remarkable successes and calm errors that shape our understanding of the distribution of moral technology in the health care system.

In the first months of the epidemic, traditional health monitoring and insufficient inadequacy of reaction systems were detected. Manual contact tracking proved to be insufficient to the scale and speed of viral transmission, overwhelmed by demand for clinical capacity, and the health care system struggled to allocate resource allocation and patient flow. In this context, AI and IoT technologies emerged as the necessary tools to meet these challenges. The AI-operated diagnostic system promised to increase the radiologist overwhelmed by breast imagination, while the machine learning algorithm offered the ability to predict the outbreak of the disease, optimize resource allocation and identify effective treatments. In addition, IoT unit smartphones provide unique opportunities for health monitoring at the publication level, contact transport and symptomatic monitoring (Upadhyay et al., 2024) .

However, this technical optimism implementation was angry with the reality of challenges and unexpected results. The rapid distribution of these systems often occurs without adequate views on moral implications, leading to a waterfall of privacy violations, algorithm bias, digital monitoring of overpine and existing health inequalities (Morley et al, 2020). Covid-19-— the most population for older individuals, racial and ethnic minorities and people with low socio-economic conditions are the most impaired often they are able to arrive or benefit from minimum digital health interventions, creating new forms of health inequalities.

The chapter is from these real world experiences and receives action -rich lessons for future global health initiatives, which suggest moral contours extended to traditional biomedical morals to remove unique challenges with digital health technologies. We check how algorithm can guide the importance of responsible technology by maximizing the principle of disability, privacy in design and maximizing public health benefits and reducing damage to the weaker population.

The intersection of AI, IoT and Global Health offers unique moral challenges that extend far beyond the structure of traditional biomedical ethics. While classical bioticians mainly focus on individual relationships between patient and supplier and autonomy, profits, non-man and justice principles, digital health technologies work on individual privacy rights, collective public health benefits, population monitoring and complex concerts of social justice ideas. These technologies produce large amounts of data, which can be collected, analysed and performed on the parameters of the population, can individually change the relationship between personal health information and collective health results (Upadhyay et al., 2022) ..

Figure 1. Ethical challenges in digital health technologies

Epidemic response provided a unique real -world laboratory to understand how these competition values and interests play in behaviour. Unlike controlled research environment or gradual technology rolling, epidemics created a situation where digital health technologies were rapid, scope and under the conditions of significant uncertainty and urgent. This natural experiment provided valuable insight into practical challenges to balance personal rights with collective benefits, perform the real performance of AI systems under stress, and social and political dynamics that shape the technology and resistance.

The global nature of the epidemic also emphasized the importance of cultural, political and economic references in the design of distribution and acceptance of technology. Countries with different political systems, privacy traditions and technical infrastructure approached clearly digital health solutions in different ways. South Korea's extensive digital monitoring system, credit card posts, mobile phone placement and CCTV recording, and show the potential effectiveness of aggressive digital contact tracking by increasing intensive privacy and civilian questions. On the other hand, European countries such as Germany and Switzerland adopted the Privacy Preservation Method, which preferred personal data security, but faced challenges in using prices and technical implementation. These non -conformity approaches provided valuable comparative insight into trade ties between privacy and public health efficiency.

The epidemic also revealed the complex mutation between technical abilities and social acceptance. Even the most sophisticated AI and IoT system proved to be ineffective when responsible for human behavior, cultural norms and social mobility. Contact the contact applications, regardless of their technical sophistication, can only be effective when adopted by important parts of the population. AI diagnostic tools, regardless of how accurately the laboratory is in settings, proved to be problematic when different patients used to be used in real health services for the world, with population, image processing equipment and clinical workflows compared to people used in development (Upadhyay et al., 2022).

In addition, epidemics highlighted the important importance of data quality, algorithm bias and representation in the AI system. Many AI diagnostic tools were trained on datasets that did not represent the diversity of the global population adequately, affecting systematic prejudices that affected clinically accurate and treatment recommendations. The discovery that the AI-oriented pulse oximetry system systematically reduced the oxygen level in the blood in patients with dark skin colours, provided how historical prejudices in medical data can be eliminated and enforced by algorithms, with potential life-drank consequences for minority population (Sozoding et al. 2020).

The IoT dimension of epidemic response raised equally complex moral questions. Location trekking, close to the distribution of proximity sensing and physical monitoring created new forms of digital monitoring that expanded beyond the surroundings of traditional health care system. Surveillance systems in the workplace were quickly expanded and tracked the status of employee movements, interactions and health in ways that were unthinkable before the epidemic. While these systems were justified by legitimate public health problems, they established new criteria for digital monitoring, which remained outside the immediate health emergency.

Epidemic response also emphasized the global dimensions of equity. While countries with high meters quickly deployed sophisticated AI and IoT systems, low and moderate-or-in often have digital infrastructure, technical capacity and lack of financial resources to implement similar solutions. This digital division not only affected the immediate epidemic response skills, but also threatened to increase global health differences in the long term. The concentration of AI development and deployment in rich countries and regions meant that these systems were often designed and tested in contexts that do not reflect global population disease patterns, health care and population characteristics (Upadhyay et al., 2023).

The speed of epidemic response also created a system of outstanding collaboration and data sharing, which challenged the traditional models for intellectual property, data ownership and research regime. Rapid sharing of viral genetic sequences, clinical data and research results enabled rapid vaccine development and detection of treatment, but also raised questions about data sovereignty, profit sharing and commercialization of health data gathered for public health purposes.

As we move beyond the intense phase of the epidemic, the lessons from this extraordinary period of technical deployment quickly become important for the development of future health system and guiding emergency preparation. The epidemic demonstrated both AI and IoT technologies in global health contexts both the ability and significant risk. This revealed the need for new moral contours that can solve unique challenges with digital health technologies, while preserving basic human rights and promoting equity.

The purpose of this chapter is to synthesize these lessons under actionable guidance for decision makers, health services, technologists and moralists working at the intersection of AI, IoT and Global Health. By examining both the successes and the errors in epidemic response, we can develop a stronger approach to the distribution of moral technology that maximizes public health benefits by protecting personal rights and promoting social justice. The effort to correct this balance is ahead of the current epidemic, as digital health technologies are ready to play a rapid central role in global health systems in the coming decades.

2. TECHNOLOGICAL LANDSCAPE: AI AND IoT IN PANDEMIC RESPONSE

2.1 AI Applications in COVID-19 Response

Artificial intelligence technology was distributed in several domains during epidemics, each presented moral ideas and reveals both AIS transformative capacity and underlying limits in health services.

Diagnostic AI systems: Machine learning algorithms were quickly developed and the breast X beams, CT scans and other image models were distributed for Covid -19 diagnosis. Companies such as Zebra Medical Vision, AIDOC and Quure.AI developed AI equipment to help radiologists identify potential COVID-19 cases, with some systems requiring accuracy rates more than 95%. However, many of these systems were trained on limited data sets from specific hospitals or areas, increasing serious concerns about generality and prejudice in different populations (Ball, 2021).

Figure 2. Diagnostic AI systems overview

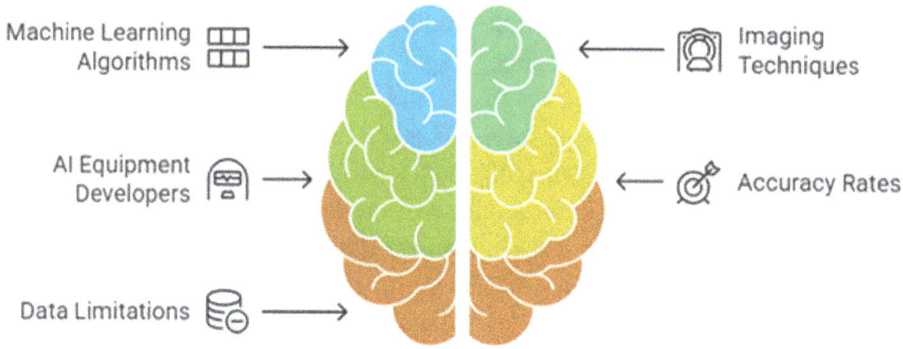

AI COVID-19 Detection Challenge showed that more than 400 AI models were developed for the diagnosis of COVID-19, but some rigid clinical verifications or real settings were demonstrated. Most of the discovery was that many AI systems performed poorly when tested on data on imaging equipment compared to people used in different hospitals, patient population or training. For example, some models unconsciously learned to identify specific imaging tools or hospital protocols instead of real disease patterns, causing incorrect confidence in clinical accuracy. The public of the distribution of these systems often traditional regulatory inspections, some units are used in clinical practice without FDA approval or equivalent regulatory withdrawal (Roberts et al., 2021).

- **Epidemiological Modelling and Prediction**: AI-powered epidemiological models became crucial tools for predicting disease spread and informing policy decisions worldwide. The Institute for Health Metrics and Evaluation (IHME) developed influential models that guided lockdown decisions across multiple countries, while Imperial College London's mathematical models provided critical early warnings about the potential severity of the pandemic. Google's COVID-19 forecasting models and Facebook's symptom mapping surveys demonstrated how tech companies could contribute to public health surveillance. However, these models often contained assumptions and biases that disproportionately affected certain communities. Many models were trained primarily on data from high-income countries with robust healthcare systems, limiting their applicability to low- and middle-income settings. The models also frequently failed to account for social determinants of health, cultural factors affecting disease transmission, and the differential impacts of interventions across diverse populations. Some models systematically underestimated disease burden in minority communities or overestimated the effectiveness of interventions in areas with limited healthcare infrastructure (Ferguson et al., 2020).

- **Medication Detection and Vaccine Development:** AI demonstrated remarkable success in accelerating the identity of potential medical science and vaccine candidates. Deepmind's Alphold system significantly helped to understand SARS-COV-2 protein structures, providing insight to indicate drug design efforts. Benevolentai identified existing medicines, which can be renovated for COVID-19 treatment, while various pharmaceutical companies used AI to speed up medical screening processes. MRNA vaccine development of modern to adapt pig protein design and predict immune reactions was quite sharp by AI-powered approaches. However, the benefits of the discovery of AI-touched drugs were not evenly distributed globally. Most AI-operated drug search efforts were concentrated in rich countries and multinational companies, with limited access to resulting innovations for low-income population. Computer resources required for advanced AI drug discovery created obstacles to researchers in resource-limiting environments potentially increased global health inequalities (Jumper et al., 2021).
- **Clinical Decision Support:** The AI system was deployed to help health professionals in clinical decision -making, including patient trips, treatment recommendations and resource allocation. The purpose of these systems was to optimize rare resources such as fan and ICU beds while improved patient results. However, several clinical decision support systems incorporated prejudice that reflects existing inequalities in health services, and potentially eliminates uneven treatment patterns in different demographic groups.

2.2 IoT Technologies in Health Monitoring

IoT devices played increasingly critical and complementary roles in pandemic response, particularly in monitoring, surveillance, and health data collection, fundamentally transforming how health systems approached population-level health management and individual patient care.

Figure 3. IoT device roles

2.2.1 Contact Tracing Applications

Countries worldwide deployed smartphone-based contact tracing apps using Bluetooth Low Energy (BLE) proximity sensing technology, representing one of the largest IoT health deployments in history. Apple and Google's joint Exposure Notification API became the foundation for numerous national contact tracing apps, including Corona-Warn-App in Germany, Swiss Covid in Switzerland, and various state-level applications across the United States. These systems used smartphones as IoT devices to automatically detect and log proximity encounters between individuals, creating decentralized networks of health monitoring devices. However, the effectiveness of these systems varied dramatically based on adoption rates, technical implementation, and population compliance. Germany's Corona-Warn-App achieved approximately 28 million downloads but faced challenges with battery optimization and Bluetooth reliability across different device types. The decentralized approach, while privacy-preserving, created technical complexities that limited real-world effectiveness. Many apps suffered from high false positive rates due to Bluetooth signal interference, detection through walls and barriers, and inability to distinguish between safe and risky encounters. Furthermore, the digital divide significantly impacted the equity of contact tracing systems, as populations without smartphones or with older devices were systematically excluded from these digital health interventions (Troncoso et al., 2020).

2.2.2 Wearable Health Monitor

Consumer portable equipment such as smartwatch, training trackers and special health monitors were quickly renovated to monitor the COVID-19 symptoms and early detection. Covid-19 portable study operated by Stanford Medicine showed that Apple watches, fitbits and Garmin devices that equipment can detect physical changes associated with infections several days before the beginning. These devices monitored heart rate variability, relaxing heart rate, sleep patterns, activity levels and skin temperature to identify potential infections. Researchers found that changes in heart rate variability and increased relaxing heart rate may indicate COVID-19 infection with proper accuracy. However, the placement of the portable health monitor increased significantly confidentiality and equity problems. Many users were not aware of the limit for health data collection and sharing through their devices. The data collected by these wearable were often stored in cloud services, divided by third parties, or used for commercial purposes beyond immediate health monitoring. In addition, the tiring units were mainly used by young, rich and more technical literary populations, and produced prejudice in the sample as a limited generality of conclusions to a wider population. The cost of these units also created obstacles to individuals and families with low incomes, and expanded inequalities in Digital Health (Mishra et al., 2020).

Figure 4. Wearable data usages range from personal to commercial

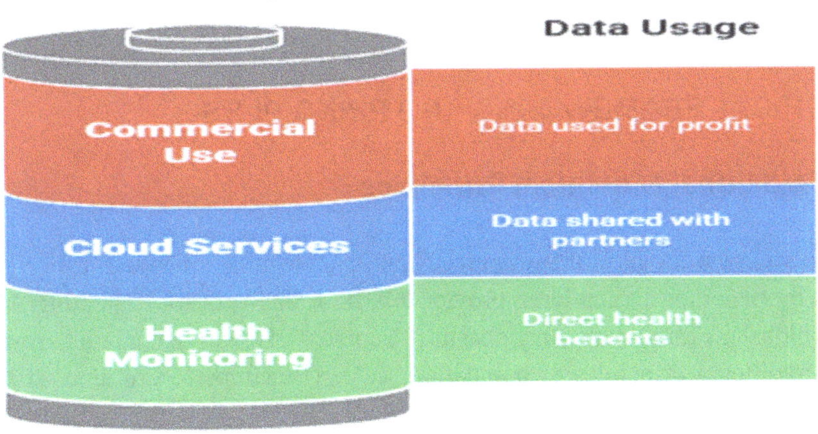

2.2.3 Environmental Monitoring Systems

The IoT sensor was placed widely to monitor air quality, coating level, temperature, humidity and ventilation in public places, workplaces and health services to reduce viral transmission risk. The integrated network of environmental sensors to adapt the smart building system HVAC system, monitors the level of CO2 as a power of attorney for ventilation efficiency and track the coating to implement social external measures. Universities, offices and retail sites implemented IoT-based occupant monitoring systems that used anonymous sensors to track foot traffic and density. In these systems, many sensor types are often combined to create extensive environmental monitoring networks including infrared Peepal counters, Wi-Fi device detection and data viewing system. However, environmental monitoring systems expressed concern about generalization of monitoring in public places and the ability to creep beyond public health goals.

2.2.4 Remote Patient Monitoring

IoT units enabled long-distance monitoring, so that health professionals can allow significant signals, drug farming and symptomatic progress to COVID-19 patients who recovered. Pulse oximetry, blood pressure monitor, thermometer and other connected health equipment allowed continuous monitoring of high-risk patients, which reduced the stress and exposure risk for the hospital's capacity for health professionals.

3. ETHICAL FRAMEWORKS AND PRINCIPLES

3.1 Traditional Bioethical Principles

Classic bio principles of autonomy, benefits, non-milk zones and justice established by Benamps and children provide a basic framework for evaluating AI and IoT in global health contexts, in global health contexts. However, epidemics revealed these traditional principles of population scales (Benamps & Children, 2019) to digital health technologies.

Autonomous and personal self -determination: The autonomy principle traditionally emphasizes personal self -determination and the right to make informed decisions about one's own health care. When it comes to AI and IoT distribution during the epidemic, this theory faced outstanding challenges when individual alternatives directly affected collective health results. Contact tracking of applications understood this stress most clearly - these systems must be adopted broadly to achieve epidemiology -efficiency, but meaningfully personal consent is necessary for moral distribution. The term "informed consent" became particularly complicated when handling the AI system, whose decision -making process was opaque or when IoT units collected data in background processes. Many individuals lacked technical knowledge to understand how their data was processed, divided or used by the AI algorithm, was actually reduced to the basis for informed consent. In addition, the emergency of epidemic response led to the implementation of the "Opt-Out" instead of the "Opt-in" system, where individuals were automatically nominated under Digital Health Authority until they chose to retire actively. This approach, although potentially more effective for public health goals, raises serious questions about such consent

3.1.1 Beneficence and the Duty to Do Good

The principle of beneficence requires that healthcare interventions actively promote patient well-being and maximize benefits. AI and IoT technologies demonstrated significant potential for beneficence during the pandemic through improved diagnostic accuracy, enhanced disease surveillance, and more efficient resource allocation. AI diagnostic systems promised to augment healthcare provider capabilities and reduce diagnostic errors, while IoT monitoring systems offered the potential for early detection of health problems and improved patient outcomes. However, determining what constitutes "benefit" became complex when considering population-level interventions. The benefits of digital contact tracing, for example, were primarily collective rather than individual, requiring a broader interpretation of beneficence that encompassed community and population health outcomes. This expansion of beneficence to include collective benefits raised questions about how to balance individual and societal goods, particularly when the benefits were uncertain or unevenly distributed across different populations.

3.1.2 Non-Maleficence and the Imperative to "Do No Harm"

The principle of non-melolphics, traditionally interpreted as avoiding direct losses for patients, required significant revival when it comes to AI and IoT distribution. Digital health technologies presented new forms of potential injuries that expanded beyond traditional clinical risks. Algorithm bias in the AI system can cause systematic discrimination against some demographic groups, resulting in the inequality of health consequences that violate the principle of non-man. The distribution of AI diagnostic tools without sufficient confirmation in different populations leads to the risk of reducing or reinforcing existing differences in the health care system. Privacy violations through IoT units caused psychological injury and violated personal dignity, while digital surveillance systems posed the risk of civilian freedom and democratic rule. The injury concept also includes collective losses, such as the erosion of public confidence in the health care system or generalization of future surveillance technologies can be abused.

Figure 5. Privacy violations & digital surveillance

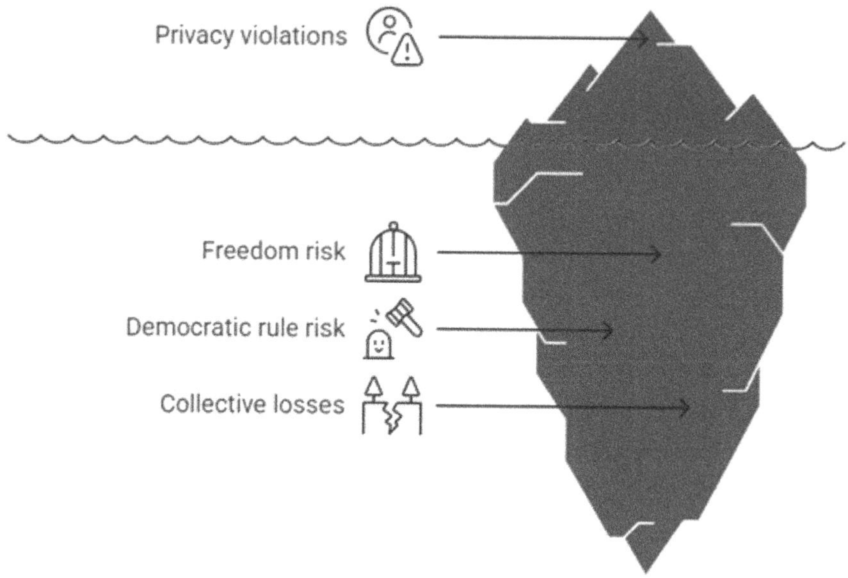

3.1.3 Justice and Fair Distribution

The principle of justice requires fair distribution of benefits and burdens in healthcare delivery. The pandemic starkly revealed how AI and IoT technologies could either promote or undermine health equity, depending on their design and implementation. Digital health interventions often exacerbated existing inequalities by requiring access to smartphones, reliable internet connectivity, and digital literacy skills that were unequally distributed across populations. Elderly individuals, racial and ethnic minorities, and those with lower socioeconomic status were systematically excluded from many digital health benefits, creating new forms of healthcare disparities. The development and deployment of AI systems primarily in wealthy countries and institutions meant that the benefits of these technologies were concentrated among already privileged populations, while their risks were often borne by marginalized communities who had little voice in their development or governance.

3.2 Emerging Ethical Frameworks

3.2.1 Emerging Ethical Frameworks

The epidemic response required the development of a particularly new moral framework to address the unique challenges generated by digital health technologies, as traditional biomedical principles proved to be inadequate to operate AI and IoT system working with population scales with outstanding data collection and algorithm's.

Algorithm justice and prejudice Bold: The term algorithm's fair AI system emerged as an important moral structure to ensure that it works equally in different demographic groups, geographical areas and clinical contexts. Unlike traditional medical morality that focuses on individual relationships between patients and suppliers, the righteousness of the algorithm requires systematic evaluation of how AI systems affect different population. Framework contains several dimensions of justice, such as individual habituation (similar individuals should receive equal treatment), justice in the group (different demographic groups should experience equal results) and detailed justice (decisions should be the same in a delegate world where the person belonged to their own demographic group). During the epidemic, the implementation of Algorithm Fairness was discovered significant challenges in defining and measuring justice in diverse population. For example, the AI diagnosis system is mainly trained on data from a demographic group, and often performs poorly on others, leading to systematic clinical errors affecting minority population. The structure of algorithm justice also emphasized the tension between different justice norms - adaptation of one type of justice often compromised others, which requires a clear value decision on what justice as justice as justice

3.2.2 Privacy by Design and Data Minimization

The principle of design privacy, originally developed by N. Cavocian, was required for IoT Purinogenin health contexts during the epidemic. This structure emphasizes entering privacy safety from the beginning of system design, instead of adding the same. The approach contains seven basic principles: Integrity instead of reactive measures, privacy as standard settings, complete functionality (no unnecessary trade tape), end-to-end security, visibility and transparency and honour for the user's privacy. During the epidemic, the successful implementation of privacy of the Design Apple-Google Exposure Notification API, which used cryptographic techniques, included contact tracking while holding personal contact data decentralized and encrypted. However, many epidemic response systems were unable to use adequate confidentiality according to design principles, instead collected broader personal information without obvious requirements or propatation. Frameworks also emphasized data minimization --able minimum data required for the prescribed purpose is a data collected for public health purposes, was not presented with commercial or surveillance activities. These principles became especially important because many organizations tried to maintain and reproduce the health data collected during the emergency period (Cavocian, 2009).

3.2.3 Solidarity and Reciprocity in Digital Health

The new disposition that emphasized collective responsibility and mutual help emerged to address the natural social nature of digital health interventions under pandemics. The concept of solidarity developed by philosophers such as Barbara Pincec and Elena Bix believes that individual health results are linked and collective action is necessary to meet shared health challenges. This structure is outside traditional individual centric morals, which is to include social obligations and shared responsibility for health results. In the context of AI and IoT Purinogen, the structure of solidarity emphasized that individuals have both rights and responsibilities in digital health systems. For example, participating in contact tracking of applications represents both a personal benefit and contribution to social health. However, the structure of solidarity also emphasized the need for reciprocity - as people who contribute to collective health articles should receive the right benefits in return. This principle challenged the systems that retrieved data from the weak population that focused the benefits of the privileged groups. The structure of solidarity also emphasized the importance of building and maintaining public belief in digital health systems, and identifying that effective epidemic response requires extensive voluntary participation that can only maintain through transparent, responsible and fair system design (Prainack and Buyx, 2017).

Figure 6. Cycle of solidarity in digital health.

3.2.4 Responsible Innovation and Anticipatory Governance

The epidemic reinforced the development of the innovation structure that is responsible that insists on estimating and addressing the possible negative consequences of new technologies before the damage caused the loss. These frameworks require ongoing stakeholders to engagement, transparent decision -making and adaptive governance structures that can respond to new challenges.

4. CASE STUDIES: ETHICAL CHALLENGES IN PRACTICE

4.1 Contact Tracing: Privacy vs. Public Health

The distribution of digital contact tracking systems worldwide provided clear examples of moral tensions between privacy and public health efficiency.

4.1.1 South Korea's Extensive Approach

South -Korea used an aggressive digital monitoring system in combination with credit card items, mobile phone space data and CCTV recording for contact tracking. This approach is extremely effective in controlling the spread of the virus, and aroused significant privacy and would be unacceptable in many democratic communities (Park et al., 2020).

4.1.2 European Privacy Protection Model

European countries such as Germany and Switzerland adopted the privacy savings method using decentralized systems that had contact data on individual equipment. Despite being more morally acceptable, these systems achieved low adoption rates and faced technical challenges that limit their efficiency (Wish et al., 2021).

4.1.3 Implications of Digital Divisions

Contact tradition apps require ownership and digital reading skills, effectively except for older population, low income communities and those without reliable internet access. It created a two -level system where the weakest population was the least possible benefit from digital health intervention (Yadav et al.,2022).

4.2 AI Bias in Diagnostic Systems

The rapid distribution of ai clinical equipment during epidemic emphasized important problems with algorithm bias and generality.

4.2.1 Consultonic Bias in Pulse Oximetry

AI -directed pulse oximetry system, which is widely used for COVID -19 monitoring, was found to systematically reduce the oxygen level in the blood in the blood in patients with dark skin. This prejudice located in historical data set limits led to delay or inadequate treatment for minority patients (Sjoding et.al., 2020).

4.2.2 Socio-Economic Bias in Risk Exception

AI models to predict COVID-19 results often include socio-economic factors such as authority for health conditions, which may eliminate health inequalities. These algorithms can systematically reduce the risk of unworthy population or reduce the risk in rich societies (Rajkomar et al., 2018).

4.2.3 Geographical Bias in Imaging AI

Many AI diagnostic tools were mainly trained on datasets of high-or-II countries, which limits their efficiency in low and medium.-sized or medium settings, where disease presentation and imaging equipment can vary considerably.

4.2.4 IoT Privacy Violations

The extensive distribution of IoT equipment for health monitoring during the epidemic led to several privacy violations, which emphasized the immediate need for a strong data security structure in digital health implementation (Soni et al., 2022).

Location tracking Overs: Many countries implemented aggressive mobile phone tracking systems that preceded the necessary public health measures. In Israel, the government authorized the domestic security agency Shin Bet to use ROMdata, originally gathered for terrorist purposes, such as to track Covid-19 patients and take quarantine measures. This resurrection of monitoring infrastructure aroused serious concerns about the function of the function and normalization of digital monitoring. Similarly, countries such as Taiwan and South Korea joint cell tower data, GPS information and credit card transaction items in wide range to create wide movement profiles of individuals. These systems are effective for contact transport, and often lack clear legal frameworks, sunset part or meaningful consent mechanisms. Many of these tracking systems continued to work beyond the immediate epidemic period, where the data was maintained for an extended period without a clear justification or deletion period (Ienca & Vayena, 2020).

4.2.5 Workplace Surveillance Expansion

The epidemic accelerated the distribution of the IoT competition's surveillance systems, which fundamentally replace the expectations of employee-employer relationships and privacy. Companies quickly implemented thermal cameras, proximity sensors and health monitoring equipment to track employees' movements, interactions and health. These systems are often more monitored than necessary for the prevention of COVID-19 prevention, creating broad surveillance networks that track productivity, social interactions and personal habits. Many employees needed to use IoT units that monitored their location in buildings, and track their match with social deformed measures and recorded their talks with colleagues. The data collected by these systems were often stored indefinitely, divided by third parties, or used for purposes beyond the Covid -19 prevention, such as performance evaluation and disciplinary functions. Workers often had a limited ability to get out of these surveillance systems without jeopardizing the employment, which caused a huge situation to reduce meaningful consent (Ball, 2021).

4.2.6 Health Data Commercialization

Epidemic created outstanding opportunities for commercial institutions to collect and MIL's health data through IoT units. Portable unit producers, smartphone papers and health technology companies collected large amounts of sensitive health information under the cover of public health requirements. Many users were unaware that their data was shared with commercial partners, used for marketing purposes or sold to third parties. Exercise tracks and smart watches collected detailed physical data, including heartbeat, sleep patterns, activity levels and location, often without clear disclosure of how this data will be used beyond immediate health monitoring. Some companies used the epidemic health data collection to create a detailed consumer profile, which was then used for targeted advertising, insurance and other commercial purposes. The emergency of the epidemic was often used to justify data collection practices that were unacceptable under normal conditions, with limited inspection or place of responsibility (Raj et al., 2023).

Figure 7. Data collection during epidemics

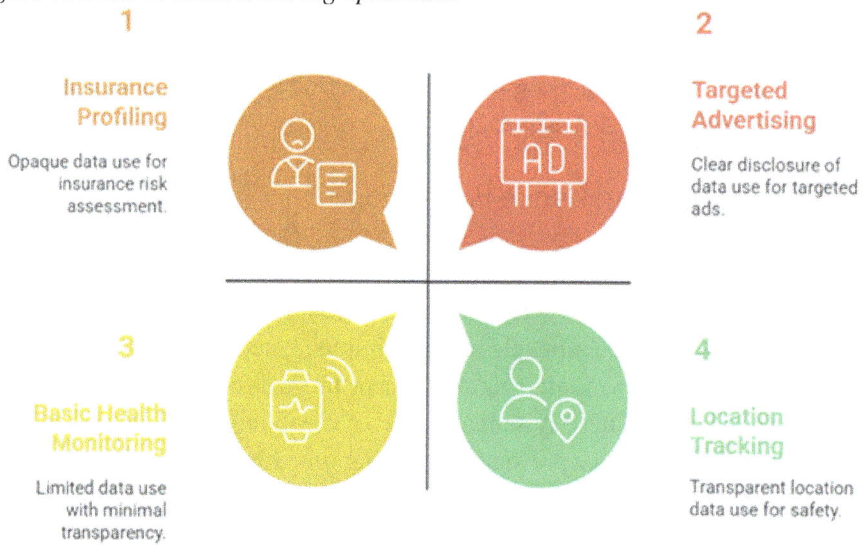

5. LESSONS LEARNED: ETHICAL GUIDELINES FOR FUTURE IMPLEMENTATION

5.1 Governance and Oversight

The epidemic emphasizes the need for a strong management structure for AI and IoT to end in health contexts:

Regulatory Agility: Traditional regulatory structures proved to be inadequate for rapid speed for technical distribution during the health care system. New models for adaptive regulation that can balance the speed with certainty requires (Floridi et al., 2018).
Multi-Interest Management: Effective inspection requires collaboration between technologists, moralists, public health experts and affected communities. The epidemic showed that purely technical solutions without different input often fail to remove moral concerns.
Openness and Responsibility: Algorithm became an essential mechanism of responsibility and openness. This includes documentation of training data, model performance in different populations and continuous monitoring for bias and flow.

5.2 Inclusive Design and Implementation

The epidemic demonstrated the importance of inclusive design methods that consider diverse population from the beginning:

Community Connection: Successful implementation included the affected communities in design and distribution decisions. This participation method helped to identify potential losses and build confidence in technical solutions.
Access and Equity: Future implementation should prioritize access to people with disabilities, limited digital literacy or limited resources. This includes multilingual interfaces, offline options and alternative access methods.

Figure 8. Enhancing project implementations

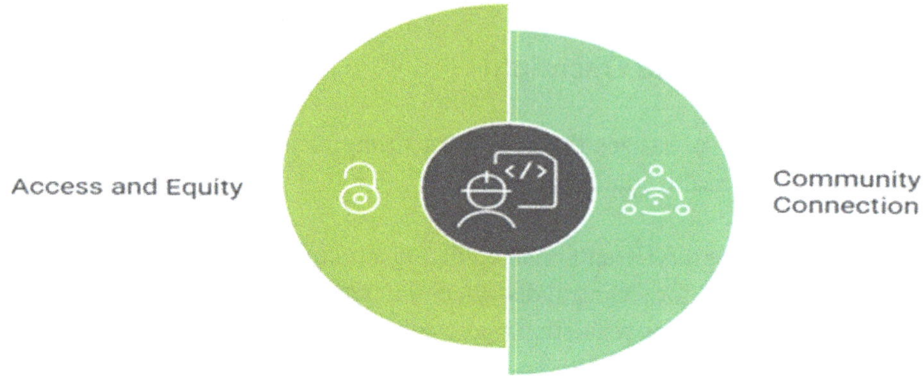

Cultural Sensitivity: AI and IoT systems should be designed with cultural capacity, and identify that health behaviour, privacy expectations and technology -adoption vary in different societies (Sharma et al., 2022).

5.3 Data Protection and Privacy

Epidemic experience gained important lessons to protect personal privacy by enabling public health benefits:

Objective Limit: The data collected for public health goals should be clearly limited to the goals as assignments or strong technical and legal obstacles to secondary use.
Data Minimization: The system should only collect the minimum data required for the declared purpose, and the data should be unknown or collected when possible.
User Control: Individuals should maintain meaningful control of their data, including the ability to reach, correct and remove personal information.

5.4 Algorithmic Fairness and Bias Mitigation

The pandemic exposed the urgent need for systematic approaches to addressing algorithmic partiality:

Various Training Data: The AI system will be trained on different representative data sets that include adequate representation of different demographic groups, geographical areas and clinical presentations. (Pawan Kumar et al., 2020)
Testing and Monitoring: With the on-going monitoring for bias over time, prejudice should be necessary for prejudice in different populations before deploying.
Fairness Matrix: Apparently matrix should be installed and regular evaluation for the righteousness of the algorithm, with transparency on trade ties between different rights norms.

6. FUTURE DIRECTIONS AND RECOMMENDATIONS

6.1 Ethical AI Development

Based on epidemic lessons, future AI development for global health should prioritize:

6.6.1 Federated Learning

Appendix enables the AI model training in many institutes without centralizing sensitive data can address both privacy and diverse training requirements.

Explaining AI: Systems that provide clarification for their decisions are crucial to clinical acceptance and moral responsibility, especially in high-day health decisions.

6.6.2 Participatory Design

Including the latest users including patients and health professionals in AI system design, can help identify potential moral problems and improve system purposes.

The future of artificial intelligence demands a fundamental commitment to ethical development practices that prioritize human welfare, fairness, and transparency. Organizations must establish comprehensive ethical frameworks that guide AI research, development, and deployment from inception through implementation. These frameworks should address bias mitigation, ensuring AI systems do not perpetuate or amplify existing societal inequalities across demographics, socioeconomic groups, or geographic regions. (Upadhyay et al., 2023)

Openness represents another important column, which requires clear AI systems, where decision -making processes can be understood and revised by stakeholders. This is especially important in applications for high dots such as health care, criminal law and financial services. Privacy protection should be built into design, enable innovation and implement strong practical practices for respecting individual rights (Upadhyay et.al, 2022).

Interdisciplinary collaboration is essential, and brings together technologists, moralists, social scientists, decision makers and affected communities to inform development decisions to ensure different approaches. Regular assessment of moral influence should be carried out throughout the AI life cycle, not just as. In addition, the installation of clear responsibility mechanisms and governance structures will help organizations navigate in complex moral dilemmas.

Finally, AI doctors require ongoing education and training to promote the responsible innovation culture, and emphasize the social implications of their work. By entering moral ideas in all phases of AI development, we can create systems that actually serve the good of humanity and reduce potential losses (Saini et al., 2023).

6.2 IoT Ethics and Governance

IoT deployment in global health should be guided by:

6.2.1 Privacy After Design

Privacy protection should be done above the ground in the IoT system, instead of added.

6.2.2 Interoperability Standards

IoT can improve efficiency by maintaining normal standard privacy safety for devices and data sharing.

Life cycle management: Apparently protocols are necessary for long-term privacy safety for data storage, unit deposit and handling of units with end-of-life.

The Internet of Things presents the outstanding moral challenges that require an overview of broader governance to address secrecy, security and social implications. Collecting huge amounts of personal and environmental data with billions of connected devices is crucial to establishing strong data protection protocols. Organizations should apply privacy-for-deciduish principles, so that the user maintains meaningful control of their information by providing clear, understandable consent mechanisms. (Garg et al., 2022)

Safety regimen becomes important because IoT units often act as input points for cyber-attacks, potentially compromise the entire network. Manufacturers should use safe development practices, including regular security updates, encryption standards and vulnerable management protocols in the unit's life cycle. This responsibility extends beyond the first distribution for continuous maintenance and ideas from the end of life.

Openness in data collection and use practice provides necessary trust between users and IoT suppliers. Companies should clearly communicate whether the data is collected, how they are processed, stored and shared, while users provide available options to change or remove the information. Algorithm Responsibility ensures that automated decisions taken by the IoT system can be explained and challenged when needed.

The regulatory structure must develop to solve the IoT-specific challenges and promote innovation. This includes the installation of industry standards, certification processes and enforcement mechanisms. In addition, it ensures digitally distinction similar access to IoT benefits to prevent the construction of surveillance systems that affect the weak population through thoughtful control structures. (Saini et al., 2023)

6.3 Global Health Equity

In order to achieve global equity, there is a need to eliminate systemic obstacles that maintain and maintain health inequalities in the population worldwide. Basic differences in access to health services, quality and consequences have been developed between developed and developing countries, rural and urban communities and various socio -economic groups. Addressing these inequalities requires extensive approaches that only deal with the basic causes instead of treating symptoms.

Resource allocation represents an important challenge, leading to infrastructure, medical supply and lack of trained personnel in many areas. International cooperation and permanent financing mechanisms are crucial to creating strong health systems that are able to provide universal health coverage. This includes investing in primary care facilities, strengthening supply chains and developing local workforce for the health care system through educational and training programs (Garg et al., 2020).

Technology provides powerful tools for bridging equity intervals through telemedicine, mobile health applications and digital health records that can expand the care of the subordinate population. However, digital classification should be addressed to ensure similar access to these innovations. Cultural qualifications and community engagement are equally important, provided that effective health services should match local references, faith and practice (Garg et al., 2019).

Future implementation should prioritize global health equity:

Technology Transfer: Tantra can help address global health inequalities to share AI and IoT innovations with low and medium income countries.

Capacity Building: Investment in digital infrastructure and technical capacity in disqualified areas is necessary for the same access to digital health benefits.

Local Innovation: Supportive adaptation of local innovation and technologies can ensure that solutions are culturally suitable and relevant.

7. CONCLUSION

The COVID-19 epidemic provided a unique opportunity to inspect AI and IoT technologies in global health on a scale. Although these technologies demonstrated remarkable ability to improve health results and epidemic response, they also emphasized significant moral challenges that should be addressed in future implementation.

The lessons from epidemic response highlight the need for the moral structure to be outside traditional biomedical morality to remove the unique challenges of digital health technologies. These frameworks should balance privacy rights with collective public health benefits; algorithm bias and justice should address concerns, and ensure similar access to technological innovations.

In the future, the global health community will be obliged to responsible innovation that prefers moral ideas from the beginning of technology development. This includes strong management mechanisms, inclusive design processes and continuous monitoring for unexpected results. Only through such approaches can we use the full potential of AI and IoT Protect the rights and welfare of technologies, and all individuals and communities (Beniwal et al., 2021).

The epidemic taught us that technology alone is not enough to solve global health challenges. Rather, it is a thoughtful, moral application of technology, implemented with clear principles and used with community engagement, which can actually continue global health capital and welfare.

REFERENCES

Ball, K. (2021). Electronic monitoring and surveillance in the workplace: The effects of digital employee monitoring during COVID-19. *New Technology, Work and Employment, 36*(2), 203–219.

Barocas, S., Hardt, M., & Narayanan, A. (2019). *Fairness and machine learning: Limitations and opportunities*. MIT Press.

Beauchamp, T. L., & Childress, J. F. (2019). *Principles of biomedical ethics* (8th ed.). Oxford University Press.

Beniwal, S., Saini, U., Garg, P., & Joon, R. K. (2021). Improving performance during camera surveillance by integration of edge detection in IoT system. [IJEHMC]. *International Journal of E-Health and Medical Communications, 12*(5), 84–96. DOI: 10.4018/IJEHMC.20210901.oa6

Cavoukian, A. (2009). *Privacy by design: The 7 foundational principles*. Information and Privacy Commissioner of Ontario.

Ferguson, N. M., Laydon, D., Nedjati-Gilani, G., Imai, N., Ainslie, K., Baguelin, M., ...&Ghani, A. C. (2020). Impact of non-pharmaceutical interventions (NPIs) to reduce COVID-19 mortality and healthcare demand. *Imperial College COVID-19 Response Team Report*, 9.

Floridi, L., Cowls, J., Beltrametti, M., Chatila, R., Chazerand, P., Dignum, V., Luetge, C., Madelin, R., Pagallo, U., Rossi, F., Schafer, B., Valcke, P., & Vayena, E. (2018). AI4People—an ethical framework for a good AI society: Opportunities, risks, principles, and recommendations. *Minds and Machines, 28*(4), 689–707. DOI: 10.1007/s11023-018-9482-5 PMID: 30930541

Garg, P., Dixit, A., & Sethi, P. (2019). Wireless sensor networks: An insight review. *International Journal of Advanced Science and Technology, 28*(15), 612–627. http://sersc.org/journals/index.php/IJAST/article/view/1838

Garg, P., Dixit, A., Sethi, P., & Pinheiro, P. R. (2020). Impact of node density on the qos parameters of routing protocols in opportunistic networks for smart spaces. *Mobile Information Systems, 2020*, 1–18. DOI: 10.1155/2020/8868842

Ienca, M., & Vayena, E. (2020). On the responsible use of digital data to tackle the COVID-19 pandemic. *Nature Medicine, 26*(4), 463–464. DOI: 10.1038/s41591-020-0832-5 PMID: 32284619

Jumper, J., Evans, R., Pritzel, A., Green, T., Figurnov, M., Ronneberger, O., Tunyasuvunakool, K., Bates, R., Žídek, A., Potapenko, A., Bridgland, A., Meyer, C., Kohl, S. A. A., Ballard, A. J., Cowie, A., Romera-Paredes, B., Nikolov, S., Jain, R., Adler, J., & Hassabis, D. (2021). Highly accurate protein structure prediction with AlphaFold. *Nature*, *596*(7873), 583–589. DOI: 10.1038/s41586-021-03819-2 PMID: 34265844

Khanna, A., Rani, P., Garg, P., Singh, P. K., & Khamparia, A. (2021). An Enhanced Crow Search Inspired Feature Selection Technique for Intrusion Detection Based Wireless Network System. *Wireless Personal Communications*, 1–18. DOI: 10.1007/s11277-021-09144-1

Mishra, T., Wang, M., Metwally, A. A., Bogu, G. K., Brooks, A. W., Bahmani, A., Alavi, A., Celli, A., Higgs, E., Dagan-Rosenfeld, O., Fay, B., Kirkpatrick, S., Kellogg, R., Gibson, M., Wang, T., Hunting, E. M., Mamic, P., Ganz, A. B., Rolnik, B., & Snyder, M. P. (2020). Pre-symptomatic detection of COVID-19 from smartwatch data. *Nature Biomedical Engineering*, *4*(12), 1208–1220. DOI: 10.1038/s41551-020-00640-6 PMID: 33208926

Morley, J., Cowls, J., Taddeo, M., & Floridi, L. (2020). Ethical guidelines for COVID-19 tracing apps. *Nature*, *582*(7810), 29–31. DOI: 10.1038/d41586-020-01578-0 PMID: 32467596

Nouri, S., Khoong, E. C., Lyles, C. R., & Karliner, L. (2020). Addressing equity in telemedicine for chronic disease management during the COVID-19 pandemic. *NEJM Catalyst Innovations in Care Delivery*, *1*(3).

Park, S., Choi, G. J., & Ko, H. (2020). Information technology–based tracing strategy in response to COVID-19 in South Korea—Privacy controversies. *Journal of the American Medical Association*, *323*(21), 2129–2130. DOI: 10.1001/jama.2020.6602 PMID: 32324202

Prainsack, B., & Buyx, A. (2017). *Solidarity in biomedicine and beyond*. Cambridge University Press. DOI: 10.1017/9781139696593

Pustokhina, I. V., Pustokhin, D. A., Lydia, E. L., Garg, P., Kadian, A., & Shankar, K. (2021). Hyperparameter search based convolution neural network with Bi-LSTM model for intrusion detection system in multimedia big data environment. *Multimedia Tools and Applications*, 1–18. DOI: 10.1007/s11042-021-11271-7

Rajkomar, A., Hardt, M., Howell, M. D., Corrado, G., & Chin, M. H. (2018). Ensuring fairness in machine learning to advance health equity. *Annals of Internal Medicine*, *169*(12), 866–872. DOI: 10.7326/M18-1990 PMID: 30508424

Roberts, M., Driggs, D., Thorpe, M., Gilbey, J., Yeung, M., Ursprung, S., Aviles-Rivero, A. I., Etmann, C., McCague, C., Beer, L., Weir-McCall, J. R., Teng, Z., Gkrania-Klotsas, E., Ruggiero, A., Korhonen, A., Jefferson, E., Ako, E., Langs, G., Gozaliasl, G., & Schönlieb, C. B. (2021). Common pitfalls and recommendations for using machine learning to detect and prognosticate for COVID-19 using chest radiographs and CT scans. *Nature Machine Intelligence*, *3*(3), 199–217. DOI: 10.1038/s42256-021-00307-0

Saini, P., Nagpal, B., Garg, P., & Kumar, S. (2023). CNN-BI-LSTM-CYP: A deep learning approach for sugarcane yield prediction. *Sustainable Energy Technologies and Assessments*, *57*, 103263. DOI: 10.1016/j.seta.2023.103263

Saini, P., Nagpal, B., Garg, P., & Kumar, S. (2023). Evaluation of Remote Sensing and Meteorological parameters for Yield Prediction of Sugarcane (Saccharumofficinarum L.) Crop. *Brazilian Archives of Biology and Technology*, *66*, e23220781. DOI: 10.1590/1678-4324-2023220781

Sharma, N., & Garg, P. (2022). Ant colony based optimization model for QoS-Based task scheduling in cloud computing environment. *Measurement. Sensors*, *100531*, 100531. DOI: 10.1016/j.measen.2022.100531

Sjoding, M. W., Dickson, R. P., Iwashyna, T. J., Gay, S. E., & Valley, T. S. (2020). Racial bias in pulse oximetry measurement. *The New England Journal of Medicine*, *383*(25), 2477–2478. DOI: 10.1056/NEJMc2029240 PMID: 33326721

Troncoso, C., Payer, M., Hubaux, J. P., Salathé, M., Larus, J., Bugnion, E., . . . Lueks, W. (2020). Decentralized privacy-preserving proximity tracing. *arXiv preprint arXiv:2005.12273.*

Upadhyay, D., Garg, P., Aldossary, S. M., Shafi, J., & Kumar, S. (2023). A Linear Quadratic Regression-Based Synchronised Health Monitoring System (SHMS) for IoT Applications. *Electronics (Basel)*, *12*(2), 309. DOI: 10.3390/electronics12020309

Upadhyay, P., & Tomar, P. "Alzheimer's disease Classification using various machine learning approaches: A Review." In 2022 4th International Conference on Advances in Computing, Communication Control and Networking (ICAC3N), pp. 232-237. IEEE, 2022. DOI: 10.1109/ICAC3N56670.2022.10074339

Upadhyay, P., & Tomar, P. "Alzheimer's Disease Classification Using Ensemble Methods." In International Conference on Signal Soni, E., Nagpal, A., Garg, P., &Pinheiro, P. R. (2022). Assessment of Compressed and Decompressed ECG Databases for Telecardiology Applying a Convolution Neural Network. Electronics, 11(17), 2708. DOI: DOI: 10.3390/electronics11172708

Upadhyay, P., Tomar, P., & Yadav, S. P. (2024). Comprehensive Systematic Computation on Alzheimer's Disease Classification. *Archives of Computational Methods in Engineering*, *31*(8), 1–32. DOI: 10.1007/s11831-024-10120-8

Whitelaw, S., Mamas, M. A., Topol, E., & Van Spall, H. G. (2020). Applications of digital technology in COVID-19 pandemic planning and response. *The Lancet. Digital Health*, *2*(8), e435–e440. DOI: 10.1016/S2589-7500(20)30142-4 PMID: 32835201

Compilation of References

Abaee, M. (2023). A systematic review of how cloud computing (CC) and the Internet of Things (IoT) can contribute towards achieving the sustainable development goals (SDGs). *Internet of Things : Engineering Cyber Physical Human Systems, 22*, 100805. DOI: 10.1016/j.iot.2023.100805

Abaee, M., Saeedi, M., & Taherdoost, H. (2024). IoT and cloud computing for sustainable development goals in Industry 4.0: A bibliometric analysis. In Tyagi, A. K., & Rajest, S. S. (Eds.), *Reshaping environmental science through machine learning and IoT* (pp. 98–118). IGI Global., DOI: 10.4018/979-8-3693-2351-9.ch006

Aceto, G., Persico, V., & Pescapé, A. (2018). The role of Information and Communication Technologies in healthcare: Taxonomies, perspectives, and challenges. *Journal of Network and Computer Applications, 107*, 125–154. DOI: 10.1016/j.jnca.2018.02.008

Adhikari, M., & Amgoth, T. (2018). Heuristic-based load-balancing algorithm for IaaS cloud. *Future Generation Computer Systems, 81*, 156–165. DOI: 10.1016/j.future.2017.10.035

Adibi, S., Rajabifard, A., Shojaei, D., & Wickramasinghe, N. (2024). Enhancing healthcare through sensor-enabled digital twins in smart environments: A comprehensive analysis. *Sensors (Basel), 24*(9), 2793. DOI: 10.3390/s24092793 PMID: 38732899

Ahmad, K. A. B., Khujamatov, H., Akhmedov, N., Bajuri, M. Y., Ahmad, M. N., & Ahmadian, A. (2022). Emerging trends and evolutions for smart city healthcare systems. *Sustainable Cities and Society, 80*, 103695. DOI: 10.1016/j.scs.2022.103695

Akbari, M. (2024). Revolutionizing supply chain and circular economy with edge computing: Systematic review, research themes and future directions. *Management Decision, 62*(9), 2875–2899. DOI: 10.1108/MD-03-2023-0412

Al Khatib, I., Shamayleh, A., & Ndiaye, M. (2024, July). Healthcare and the internet of medical things: Applications, trends, key challenges, and proposed resolutions. *Informatics (MDPI), 11*(3), 47. DOI: 10.3390/informatics11030047

Alahi, M. E. E., Sukkuea, A., Tina, F. W., Nag, A., Kurdthongmee, W., Suwannarat, K., & Mukhopadhyay, S. C. (2023). Integration of IoT-enabled technologies and artificial intelligence (AI) for smart city scenario: Recent advancements and future trends. *Sensors (Basel), 23*(11), 5206. DOI: 10.3390/s23115206 PMID: 37299934

Alasiri, A. A., & Mohammed, V. (2022). Healthcare transformation in Saudi Arabia: An overview since the launch of Vision 2030. *Health Services Insights, 15*, 1–7. DOI: 10.1177/11786329221121214 PMID: 36081830

Albreem, M. A., Sheikh, A. M., Alsharif, M. H., Jusoh, M., & Yasin, M. N. M. (2021). Green Internet of Things (GIoT): Applications, practices, awareness, and challenges. *IEEE Access : Practical Innovations, Open Solutions, 9*, 38833–38858. DOI: 10.1109/ACCESS.2021.3061697

Albreem, M. A., Sheikh, A. M., Bashir, M. J. K., & El-Saleh, A. A. (2023). Towards green Internet of Things (IoT) for a sustainable future in Gulf Cooperation Council countries: Current practices, challenges, and future prospects. *Wireless Networks, 29*(2), 539–567. DOI: 10.1007/s11276-022-03133-3

Aledhari, M., Razzak, R., Qolomany, B., Al-Fuqaha, A., & Saeed, F. (2022). Biomedical IoT: Enabling technologies, architectural elements, challenges, and future directions. *IEEE Access : Practical Innovations, Open Solutions, 10*, 31306–31339. DOI: 10.1109/ACCESS.2022.3159235 PMID: 35441062

Alfina, K. N., Ratnayake, R. C., Wibisono, D., Basri, M. H., & Mulyono, N. B. (2022). Analyzing barriers towards implementing circular economy in healthcare supply chains. In *2022 IEEE International Conference on Industrial Engineering and Engineering Management (IEEM)* (pp. 827-831). IEEE. DOI: 10.1109/IEEM55944.2022.9989999

Ali, Z. H., & Ali, H. A. (2021). Towards sustainable smart IoT applications architectural elements and design: Opportunities, challenges, and open directions. *The Journal of Supercomputing, 77*(6), 5668–5725. DOI: 10.1007/s11227-020-03477-7

Allioui, H., & Mourdi, Y. (2023). Exploring the full potentials of IoT for better financial growth and stability: A comprehensive survey. *Sensors (Basel), 23*(19), 8015. DOI: 10.3390/s23198015 PMID: 37836845

Almuaythir, S., Singh, A. K., Alhusban, M., & Daoud, A. O. (2024). Robotics technology: Catalyst for sustainable development—Impact on innovation, healthcare, inequality, and economic growth. *Discover Sustainability*, *5*(1), 486. DOI: 10.1007/s43621-024-00744-y

Almutairi, R., Bergami, G., & Morgan, G. (2024). Advancements and challenges in IoT simulators: A comprehensive review. *Sensors (Basel)*, *24*(5), 1511. DOI: 10.3390/s24051511 PMID: 38475047

Alshammari, H. K. F., Alqoud, H. M. H., Alrimal, M. M. T., Alanazi, H. F. M., Alshammari, M. S., Alharbi, F. S. J., & Almutairi, H. S. M. (2024). Emerging trends in healthcare and medical technology. *Gland Surgery*, *9*(2), 171–179.

Alsharif, M. H., Jahid, A., Kelechi, A. H., & Kannadasan, R. (2023). Green IoT: A review and future research directions. *Symmetry*, *15*(3), 757. DOI: 10.3390/sym15030757

Al-Turjman, F., Nawaz, H., & Ulusar, U. (2019). Intelligence in the Internet of Medical Things era: A systematic review of current and future trends. *Computer Communications*, *150*, 644–660. DOI: 10.1016/j.comcom.2019.12.030

Aminizadeh, S., Heidari, A., Dehghan, M., Toumaj, S., Rezaei, M., Navimipour, N. J., & Unal, M. (2024). Opportunities and challenges of artificial intelligence and distributed systems to improve the quality of healthcare service. *Artificial Intelligence in Medicine*, *149*, 102779. DOI: 10.1016/j.artmed.2024.102779 PMID: 38462281

Amiri, Z., Heidari, A., Navimipour, N. J., & Unal, M. (2023). Resilient and dependability management in distributed environments: A systematic and comprehensive literature review. *Cluster Computing*, *26*(2), 1565–1600. DOI: 10.1007/s10586-022-03738-5

Andrae, A., & Edler, T. (2015). On Global Electricity Usage of Communication Technology: Trends to 2030. *Challenges*, *6*(1), 117–157. DOI: 10.3390/challe6010117

Aragani, V. M., & Maroju, P. K. (2025). Future of blue-green cities emerging trends and innovations in iCloud infrastructure. In *Integrating Blue-Green Infrastructure into Urban Development* (pp. 223-244).

Arega, A., & Sharma, D. P. (2023). Towards smart and green features of cloud computing in healthcare services: A systematic literature review. *Journal of Information Systems Engineering & Business Intelligence*, *9*(2), 161–180. DOI: 10.20473/jisebi.9.2.161-180

Ariffin, N. H. M., Yusoff, F. H., Abu Mansor, N. N., Javed, S., & Nazir, S. N. (2024). Assessment of sustainable healthcare management policies and practices in Oman: A comprehensive analysis using an AI-based literature review. *Multidisciplinary Reviews, 7*(11), 2024257. DOI: 10.31893/multirev.2024257

Arora, A., Rao, A., & Bhutani, M. (2020). A MATLAB simulation model for MAC layer of visible light communication. In *2020 7th International Conference on Signal Processing and Integrated Networks (SPIN)* (pp. 941–945). IEEE. DOI: 10.1109/SPIN48934.2020.9071254

Arthi, R., & Krishnaveni, S. (2024). Optimized Tiny Machine Learning and Explainable AI for Trustable and Energy-Efficient Fog-Enabled Healthcare Decision Support System. *International Journal of Computational Intelligence Systems, 17*(1), 229. DOI: 10.1007/s44196-024-00631-4

Attah, R. U., Garba, B. M. P., Gil-Ozoudeh, I., & Iwuanyanwu, O. (2024). Strategic partnerships for urban sustainability: Developing a conceptual framework for integrating technology in community-focused initiatives. *GSC Advances in Research and Reviews, 21*(2), 409–418. DOI: 10.30574/gscarr.2024.21.2.0454

Awan, U., Sroufe, R., & Shahbaz, M. (2021). Industry 4.0 and the circular economy: A literature review and recommendations for future research. *Business Strategy and the Environment, 30*(4), 2038–2060. DOI: 10.1002/bse.2731

Bai, C. A., Cordeiro, J., & Sarkis, J. (2020). Blockchain technology: Business, strategy, the environment, and sustainability. *Business Strategy and the Environment, 29*(1), 321–322. DOI: 10.1002/bse.2431

Ball, K. (2021). Electronic monitoring and surveillance in the workplace: The effects of digital employee monitoring during COVID-19. *New Technology, Work and Employment, 36*(2), 203–219.

Bamakan, S. M. H., Malekinejad, P., & Ziaeian, M. (2022). Towards blockchain-based hospital waste management systems: Applications and future trends. *Journal of Cleaner Production, 349*, 131440. DOI: 10.1016/j.jclepro.2022.131440

Barocas, S., Hardt, M., & Narayanan, A. (2019). *Fairness and machine learning: Limitations and opportunities*. MIT Press.

Barwal, R. K., & Rohilla, N. (2024). Internet of Things (IoT): A comprehensive review, emerging trends, and future directions. *International Journal of Engineering Science and Humanities, 14*(Special Issue 1), 164–169.

Bates, D. W., Saria, S., Ohno-Machado, L., Shah, A., & Escobar, G. (2014). Big Data In Health Care: Using Analytics To Identify And Manage High-Risk And High-Cost Patients. *Health Affairs, 33*(7), 1123–1131. DOI: 10.1377/hlthaff.2014.0041 PMID: 25006137

Bauer, W., & Dorn, J. (2016). Requirements for Product-service Description at e-marketplaces in the Manufacturing Domain. *Procedia CIRP, 47*, 406–411. DOI: 10.1016/j.procir.2016.03.040

Beauchamp, T. L., & Childress, J. F. (2019). *Principles of biomedical ethics* (8th ed.). Oxford University Press.

Beniwal, S., Saini, U., Garg, P., & Joon, R. K. (2021). Improving performance during camera surveillance by integration of edge detection in IoT system. [IJEHMC]. *International Journal of E-Health and Medical Communications, 12*(5), 84–96. DOI: 10.4018/IJEHMC.20210901.oa6

Bennett, L. (2024). The Role of Blockchain Technology in Enhancing Supply Chain Transparency in Italy.

Bevere, D., & Faccilongo, N. (2024). Shaping the future of healthcare: Integrating ecology and digital innovation. *Sustainability (Basel), 16*(9), 3835. DOI: 10.3390/su16093835

Bhushan, B., Kumar, A., Agarwal, A. K., Kumar, A., Bhattacharya, P., & Kumar, A. (2023). Towards a secure and sustainable internet of medical things (IoMT): Requirements, design challenges, security techniques, and future trends. *Sustainability (Basel), 15*(7), 6177. DOI: 10.3390/su15076177

Bhutani, M., Lall, B., & Agrawal, M. (2023). An efficient and adaptive superframe structure for IEEE 802.15.7-based real-time sensor networks.

Bhutani, M., Lall, B., & Agrawal, M. (2023). An efficient and adaptive superframe structure for IEEE 802.15.7-based real-time sensor networks. [No journal information available]

Bhutani, M., Gupta, M., Jain, T., Sharma, T. P., Solanki, A., & Malhotra, A. (2023). Data-driven robotics: A performance analysis of wearable sensor-based wound assessment system for sensor data collection. In *Bio-Inspired Data-Driven Distributed Energy in Robotics and Enabling Technologies* (pp. 249–266). CRC Press.

Bhutani, M., Lall, B., & Agrawal, M. (2022). Optical wireless communications: Research challenges for MAC layer. *IEEE Access : Practical Innovations, Open Solutions, 10*, 126969–126989. DOI: 10.1109/ACCESS.2022.3225913

Bhutani, M., Lall, B., & Agrawal, M. (2023). A novel energy-efficient adaptive superframe structure for OWC-based real-time bio-sensor networks. *Peer-to-Peer Networking and Applications, 16*(6), 3021–3031. DOI: 10.1007/s12083-023-01544-3

Bhutani, M., Lall, B., & Dixit, A. (2021). MAC layer performance modelling for IEEE 802.15.7 based on discrete-time Markov chain. *IET Communications, 15*(14), 1883–1896. DOI: 10.1049/cmu2.12199

Bianco, G. M. (2024). *Firmware Development and Certification for IoT Devices* (Doctoral dissertation, Politecnico di Torino).

Birtel, M., David, A., Hermann, J., Mohr, F., & Ruskowski, M. (2019). FutureFit: A strategy for getting a production asset to an industry 4.0 component – a human-centered approach. *Procedia Manufacturing, 38*, 1000–1007. DOI: 10.1016/j.promfg.2020.01.185

Bressanelli, G., Adrodegari, F., Perona, M., & Saccani, N. (2018). Exploring How Usage-Focused Business Models Enable Circular Economy through Digital Technologies. *Sustainability (Basel), 10*(3), 639. DOI: 10.3390/su10030639

Cappellini, I., & Schirru, E. (2024). Bridging anesthesia and sustainability: A special article for a path towards eco-conscious practice. *Anesthesia Research, 1*(3), 168–179. DOI: 10.3390/anesthres1030016

Cassee, F. R., Bleeker, E. A. J., Durand, C., Exner, T., Falk, A., Friedrichs, S., Heunisch, E., & Wijnhoven, S. W. P. (2024). Roadmap towards safe and sustainable advanced and innovative materials. *Computational and Structural Biotechnology Journal, 25*, 105–126. DOI: 10.1016/j.csbj.2024.05.018 PMID: 38974014

Cavalieri, A., Reis, J., & Amorim, M. (2021). Circular economy and Internet of Things: Mapping science of case studies in manufacturing industry. *Sustainability (Basel), 13*(6), 3299. DOI: 10.3390/su13063299

Cavoukian, A. (2009). *Privacy by design: The 7 foundational principles*. Information and Privacy Commissioner of Ontario.

Chataut, R., Phoummalayvane, A., & Akl, R. (2023). Unleashing the power of IoT: A comprehensive review of IoT applications and future prospects in healthcare, agriculture, smart homes, smart cities, and industry 4.0. *Sensors (Basel), 23*(16), 7194. DOI: 10.3390/s23167194 PMID: 37631731

Chaudhary, V., Gaur, P., & Rustagi, S. (2024). Sensors, society, and sustainability. *Sustainable Materials and Technologies*, Article e00952.

Chauhan, D., Bahad, P., & Jain, J. K. (2024). Sustainable AI: Environmental implications, challenges, and opportunities. In Explainable AI (XAI) for Sustainable Development (pp. 1-15). Springer.

Chauhan, S., Singh, M., & Garg, P. (2021). Rapid forecasting of pandemic outbreak using machine learning. *Enabling Healthcare 4.0 for Pandemics: A Roadmap Using AI, Machine Learning, IoT and Cognitive Technologies*, 59-73.

Choudhary, S., Vijitha, S., Bhavani, D. D., Bhuvaneswari, N., Tiwari, M., & Subburam, S. (2025). Edge AI Deploying Artificial Intelligence Models on Edge Devices for Real-Time Analytics. In *ITM Web of Conferences* (Vol. 76, p. 01009). EDP Sciences. DOI: 10.1051/itmconf/20257601009

Corigliano, O., Morrone, P., & Algieri, A. (2025). Navigating the challenges of sustainability in the food processing chain: Insights into energy interventions to reduce footprint. *Energies*, *18*(2), 296. DOI: 10.3390/en18020296

Costa, E., Soares, A. L., & de Sousa, J. P. (2020). Industrial business associations improving the internationalisation of SMEs with digital platforms: A design science research approach. *International Journal of Information Management*, *53*, 102070. DOI: 10.1016/j.ijinfomgt.2020.102070

D'Alessandro, C., Szopik-Depczyńska, K., Tarczyńska-Łuniewska, M., Silvestri, C., & Ioppolo, G. (2024). Exploring circular economy practices in the healthcare sector: A systematic review and bibliometric analysis. *Sustainability (Basel)*, *16*(1), 401. DOI: 10.3390/su16010401

Das, S. K., Chinnappan, A., Jayathilaka, W. A. D. M., Gosh, R., Baskar, C., & Ramakrishna, S. (2021). Challenges and potential solutions for 100% recycling of medical textiles. *Materials Circular Economy, 3*(1), Article 13.

Dautov, R., Song, H., & Ferry, N. (2021, September). Towards a sustainable IoT with last-mile software deployment. In *2021 IEEE Symposium on Computers and Communications (ISCC)* (pp. 1-6). IEEE. DOI: 10.1109/ISCC53001.2021.9631250

De Pretis, F., van Gils, M., & Forsberg, M. M. (2022). A smart hospital-driven approach to precision pharmacovigilance. *Trends in Pharmacological Sciences*, *43*(6), 473–481. DOI: 10.1016/j.tips.2022.03.009 PMID: 35490032

Dhikale, P., Shrivastava, S., & Srinivasan, S. (2020). Perspectives about professionalism among undergraduate students in a medical college in India: A qualitative study. *Indian Journal of Community Medicine : Official Publication of Indian Association of Preventive & Social Medicine*, *45*(2), 230. DOI: 10.4103/ijcm.IJCM_238_19 PMID: 32905183

Dietrich, F., & Krenkel, N. (2019). Preface. *Procedia CIRP*, *86*, 1–2. DOI: 10.1016/j.procir.2020.01.001

Dimitrov, D. (2016). Medical Internet of Things and Big Data in Healthcare. *Healthcare Informatics Research*, *22*(3), 156. DOI: 10.4258/hir.2016.22.3.156 PMID: 27525156

Dodig, I., Cafuta, D., Kramberger, T., & Cesar, I. (2021). A novel software architecture solution with a focus on long-term IoT device security support. *Applied Sciences (Basel, Switzerland)*, *11*(11), 4955. DOI: 10.3390/app11114955

Eckelman, M. J., & Sherman, J. (2016). Environmental Impacts of the U.S. Health Care System and Effects on Public Health. *PLoS One*, *11*(6), e0157014. DOI: 10.1371/journal.pone.0157014 PMID: 27280706

Elendu, C., Elendu, T. C., & Elendu, I. D. (2024). 5G-enabled smart hospitals: Innovations in patient care and facility management. *Medicine*, *103*(20), e38239. DOI: 10.1097/MD.0000000000038239 PMID: 38758872

Fahal, A., Smith, D. J., Nyaoke, B., Asiedu, K., Falves, F., Warusavithana, S., Argaw, D., & Hay, R. (2024). Towards enhanced control of mycetoma: A roadmap to achieve the UN's sustainable development goals by 2030. *Transactions of the Royal Society of Tropical Medicine and Hygiene*, *118*(10), 646–651. DOI: 10.1093/trstmh/trae016 PMID: 38530874

Ferguson, N. M., Laydon, D., Nedjati-Gilani, G., Imai, N., Ainslie, K., Baguelin, M., ...&Ghani, A. C. (2020). Impact of non-pharmaceutical interventions (NPIs) to reduce COVID-19 mortality and healthcare demand. *Imperial College COVID-19 Response Team Report*, 9.

Fernandes, L., Rosado da Cruz, A. M., Cruz, E. F., & Lopes, S. I. (2023). A review on adopting blockchain and IoT technologies for fostering the circular economy in the electrical and electronic equipment value chain. *Sustainability (Basel)*, *15*(5), 4574. DOI: 10.3390/su15054574

Floridi, L., Cowls, J., Beltrametti, M., Chatila, R., Chazerand, P., Dignum, V., Luetge, C., Madelin, R., Pagallo, U., Rossi, F., Schafer, B., Valcke, P., & Vayena, E. (2018). AI4People—an ethical framework for a good AI society: Opportunities, risks, principles, and recommendations. *Minds and Machines*, *28*(4), 689–707. DOI: 10.1007/s11023-018-9482-5 PMID: 30930541

Fu, L., Pan, R., Gu, T., Cheng, X., & Zhang, H. (2024). IoT-Enabled Supply Chain Management From A Customer Perspective: Challenges and Opportunities. *IEEE Internet of Things Journal*, *PP*, *1*. DOI: 10.1109/JIOT.2024.3505412

Gagneja, A., Lall, B., & Bhutani, M. (2024). Unveiling CM-Det: Leveraging ConvMixer architecture for advanced object detection. *International Journal of Information Technology : an Official Journal of Bharati Vidyapeeth's Institute of Computer Applications and Management, 16*(7), 4273–4278. DOI: 10.1007/s41870-024-02069-5

Gallo, M., Viola, A., & Romagnoli, G. (2024). The contribution of RFID to circular economy deployment: A literature review. *International Journal of RF Technologies*, Article 17545730251331373.

Garg, P., Pranav, S., & Prerna, A. (2021). Green internet of things (G-IoT): A solution for sustainable technological development. In *Green Internet of Things for Smart Cities* (pp. 23-46). CRC Press. Garg, P., Sharma, N., &Sonal, B. S. INTELLIGENT SYSTEMS AND APPLICATIONS IN ENGINEERING.

Garg, P. (2024). Prediction of female pregnancy complication using artificial intelligence. In *Artificial Intelligence and Machine Learning for Women's Health Issues* (pp. 17–35). Academic Press.

Garg, P. (2025). Explainable AI & Model Interpretability in Healthcare: Challenges & Future Directions. *EKSPLORIUM-BULETIN PUSAT TEKNOLOGI BAHAN GALIAN NUKLIR, 46*(1), 104–133.

Garg, P., Dixit, A., & Sethi, P. (2019). Wireless sensor networks: An insight review. *International Journal of Advanced Science and Technology, 28*(15), 612–627. http://sersc.org/journals/index.php/IJAST/article/view/1838

Garg, P., Dixit, A., Sethi, P., & Pinheiro, P. R. (2020). Impact of node density on the qos parameters of routing protocols in opportunistic networks for smart spaces. *Mobile Information Systems, 2020*, 1–18. DOI: 10.1155/2020/8868842

Garg, P., Dixit, A., Sethi, P., & Pruthi, J. (2023, December). Strengthening Smart City with Opportunistic Networks: An Insight. In *2023 International Conference on Advanced Computing & Communication Technologies (ICACCTech)* (pp. 700-707). IEEE. DOI: 10.1109/ICACCTech61146.2023.00117

Garg, P., Srivastava, A. K., Anas, A., Gupta, B., & Mishra, C. (2023). Pneumonia Detection Through X-Ray Images Using Convolution Neural Network. In *Advancements in Bio-Medical Image Processing and Authentication in Telemedicine* (pp. 201–218). IGI Global. DOI: 10.4018/978-1-6684-6957-6.ch011

Gautam, V. K., Gupta, S., & Garg, P. (2024, March). Automatic Irrigation System using IoT. In *2024 International Conference on Automation and Computation (AUTOCOM)* (pp. 100-103). IEEE.

Geissdoerfer, M., Savaget, P., Bocken, N. M. P., & Hultink, E. J. (2017). The Circular Economy – A new sustainability paradigm? *Journal of Cleaner Production, 143*, 757–768. DOI: 10.1016/j.jclepro.2016.12.048

Ghisellini, P., Cialani, C., & Ulgiati, S. (2016). A review on circular economy: The expected transition to a balanced interplay of environmental and economic systems. *Journal of Cleaner Production, 114*, 11–32. DOI: 10.1016/j.jclepro.2015.09.007

Ghulam, S. T., & Abushammala, H. (2023). Challenges and opportunities in the management of electronic waste and its impact on human health and environment. *Sustainability (Basel), 15*(3), 1837. DOI: 10.3390/su15031837

Goyal, R., Sharma, A., Jain, A., Elawadhi, O., & Bhutani, M. (2023). Adaptive density-based traffic management system utilizing ultrasonic units and RODEM algorithm. In *2023 10th International Conference on Signal Processing and Integrated Networks (SPIN)* (pp. 571-577). IEEE. DOI: 10.1109/SPIN57001.2023.10116660

Goyal, R., Elawadhi, O., Sharma, A., Bhutani, M., & Jain, A. (2024). Cloud-connected central unit for traffic control: Interfacing sensing units and centralized control for efficient traffic management. *International Journal of Information Technology : an Official Journal of Bharati Vidyapeeth's Institute of Computer Applications and Management, 16*(2), 841–851. DOI: 10.1007/s41870-023-01527-w

Gupta, A., Bhutani, M., Dixit, A., & Lall, B. (2021). FD-OMAC: A novel full-duplex optical media access control for IEEE 802.15.7. *IEEE Access : Practical Innovations, Open Solutions, 9*, 148894–148910. DOI: 10.1109/ACCESS.2021.3124975

Gupta, A., & Singh, A. (2023). Healthcare 4.0: Recent advancements and futuristic research directions. *Wireless Personal Communications, 129*(2), 933–952. DOI: 10.1007/s11277-022-10164-8 PMID: 36594000

Hameed, K., Naha, R., & Hameed, F. (2024). Digital transformation for sustainable health and well-being: A review and future research directions. *Discover Sustainability, 5*(1), 104. DOI: 10.1007/s43621-024-00273-8

Haque, N., Hughes, A., Lim, S., & Vernon, C. (2014). Rare Earth Elements: Overview of Mining, Mineralogy, Uses, Sustainability and Environmental Impact. *Resources, 3*(4), 614–635. DOI: 10.3390/resources3040614

Hernandez, D., Pasha, L., Yusuf, D. A., Nurfaizi, R., & Julianingsih, D. (2024). The role of artificial intelligence in sustainable agriculture and waste management: Towards a green future. *International Transactions on Artificial Intelligence, 2*(2), 150–157. DOI: 10.33050/italic.v2i2.552

Hira, F. A., Khalid, H., & Moshiul, A. M. (2024). Blockchain-based electronic health records: Revolutionizing healthcare in Malaysia. *Asia Pacific Journal of Health Management, 19*(1), 226–233. DOI: 10.24083/apjhm.v19i1.2621

Ho, H.-J., Iizuka, A., Shibata, E., Tomita, H., Takano, K., & Endo, T. (2021). Utilization of CO2 in direct aqueous carbonation of concrete fines generated from aggregate recycling: Influences of the solid–liquid ratio and CO2 concentration. *Journal of Cleaner Production, 312*, 127832. DOI: 10.1016/j.jclepro.2021.127832

Hollander, J. E., & Carr, B. G. (2020). Virtually Perfect? Telemedicine for Covid-19. *The New England Journal of Medicine, 382*(18), 1679–1681. DOI: 10.1056/NEJMp2003539 PMID: 32160451

Holmner, Å., Ebi, K. L., Lazuardi, L., & Nilsson, M. (2014). Carbon Footprint of Telemedicine Solutions - Unexplored Opportunity for Reducing Carbon Emissions in the Health Sector. *PLoS One, 9*(9), e105040. DOI: 10.1371/journal.pone.0105040 PMID: 25188322

Holt, S. G., Koornneef, E., Al Obeidli, A. A. K., Hubbert, L., & Nicholson, L. (2024). Decarbonisation of kidney care in the United Arab Emirates: A roadmap to an environmentally sustainable care. *International Journal of Nephrology and Renovascular Disease, 17*, 241–253. DOI: 10.2147/IJNRD.S481121 PMID: 39430398

Huang, B., Xing, K., Ness, D., Liao, L., Huang, K., Xie, P., & Huang, J. (2022). Rethinking carbon–neutral built environment: Urban dynamics and scenario analysis. *Energy and Building, 255*, 111672. DOI: 10.1016/j.enbuild.2021.111672

Hu, J. J., Bonnichsen, C. R., Dearani, J. A., Miranda, W. R., Johnson, J. N., Cetta, F., Stephens, E. H., Aganga, D. O., & van Dorn, C. S. (2021). Adults With Tetralogy of Fallot: Early Postoperative Outcomes and Risk Factors for Complications. *Mayo Clinic Proceedings, 96*(9), 2398–2406. DOI: 10.1016/j.mayocp.2021.01.032 PMID: 34412856

Ienca, M., & Vayena, E. (2020). On the responsible use of digital data to tackle the COVID-19 pandemic. *Nature Medicine, 26*(4), 463–464. DOI: 10.1038/s41591-020-0832-5 PMID: 32284619

Islam, M. M., & Bhuiyan, Z. A. (2023). An integrated scalable framework for cloud and IoT based green healthcare system. *IEEE Access : Practical Innovations, Open Solutions, 11*, 22266–22282. DOI: 10.1109/ACCESS.2023.3250849

Islam, M. M., Nooruddin, S., Karray, F., & Muhammad, G. (2022). Internet of things: Device capabilities, architectures, protocols, and smart applications in healthcare domain. *IEEE Internet of Things Journal, 10*(4), 3611–3641. DOI: 10.1109/JIOT.2022.3228795

Izhar, M., Naqvi, S. A. A., Ahmed, A., Abdullah, S., Alturki, N., & Jamel, L. (2023). Enhancing healthcare efficacy through IoT-edge fusion: A novel approach for smart health monitoring and diagnosis. *IEEE Access : Practical Innovations, Open Solutions, 11*, 136456–136467. DOI: 10.1109/ACCESS.2023.3337092

Jamoussi, B., Abu-Rizaiza, A., & Al-Haij, A. (2022). Sustainable building standards, codes, and certification systems: The status quo and future directions in Saudi Arabia. *Sustainability (Basel), 14*(16), 10314. DOI: 10.3390/su141610314

Jauhar, S., Pratap, S., Lakshay, , Paul, S., & Gunasekaran, A. (2023). Internet of Things-based innovative solutions and emerging research clusters in circular economy. *Operations Management Research : Advancing Practice Through Research, 16*(4), 1968–1988. DOI: 10.1007/s12063-023-00421-9

Jayaraman, P. P., Forkan, A. R. M., Morshed, A., Haghighi, P. D., & Kang, Y. (2020). Healthcare 4.0: A review of frontiers in digital health. *Wiley Interdisciplinary Reviews. Data Mining and Knowledge Discovery, 10*(2), e1350. DOI: 10.1002/widm.1350

Jumaan, I. A., Hashim, N. H., & Al-Ghazali, B. M. (2020). The role of cognitive absorption in predicting mobile internet users' continuance intention: An extension of the expectation-confirmation model. *Technology in Society, 63*, 101355. DOI: 10.1016/j.techsoc.2020.101355

Jumper, J., Evans, R., Pritzel, A., Green, T., Figurnov, M., Ronneberger, O., Tunyasuvunakool, K., Bates, R., Žídek, A., Potapenko, A., Bridgland, A., Meyer, C., Kohl, S. A. A., Ballard, A. J., Cowie, A., Romera-Paredes, B., Nikolov, S., Jain, R., Adler, J., & Hassabis, D. (2021). Highly accurate protein structure prediction with AlphaFold. *Nature, 596*(7873), 583–589. DOI: 10.1038/s41586-021-03819-2 PMID: 34265844

Karrouk, Y., Debasa, F., & Sanchez, L. M. F. (2025). The digital transformation of smart hospitals: Challenges and opportunities. *Utilizing AI of Medical Things for Healthcare Security and Sustainability*, 1-54.

Kasinathan, P., Pugazhendhi, R., Elavarasan, R. M., Ramachandaramurthy, V. K., Ramanathan, V., Subramanian, S., Kumar, S., Nandhagopal, K., Raghavan, R. R. V., Rangasamy, S., Devendiran, R., & Alsharif, M. H. (2022). Realization of sustainable development goals with disruptive technologies by integrating Industry 5.0, Society 5.0, smart cities and villages. *Sustainability (Basel)*, *14*(22), 15258. DOI: 10.3390/su142215258

Kaur, G., Gupta, V., Settia, N., Duggal, S., & Bhutani, M. (n.d.). Reilluminating wireless horizons: Exploring complete information transmission through visible light. [No journal or source information available]

Kaur, M., Singh, D., Kumar, V., Gupta, B. B., & Abd El-Latif, A. A. (2021). Secure and energy efficient-based E-health care framework for green internet of things. *IEEE Transactions on Green Communications and Networking*, *5*(3), 1223–1231. DOI: 10.1109/TGCN.2021.3081616

Kelly, J. T., Campbell, K. L., Gong, E., & Scuffham, P. (2020). The Internet of Things: Impact and Implications for Health Care Delivery. *Journal of Medical Internet Research*, *22*(11), e20135. DOI: 10.2196/20135 PMID: 33170132

Khan, N., Hossain, S., Khadka, U., & Sarkar, S. (2024). Blockchain in Supply Chain Management: Enhancing Transparency, Efficiency, and Trust. AIJMR-Advanced International Journal of Multidisciplinary Research, 2(5).

Khanfar, A. A., Iranmanesh, M., Ghobakhloo, M., Senali, M. G., & Fathi, M. (2021). Applications of blockchain technology in sustainable manufacturing and supply chain management: A systematic review. *Sustainability (Basel)*, *13*(14), 7870. DOI: 10.3390/su13147870

Khanna, A., Rani, P., Garg, P., Singh, P. K., & Khamparia, A. (2021). An Enhanced Crow Search Inspired Feature Selection Technique for Intrusion Detection Based Wireless Network System. *Wireless Personal Communications*, 1–18. DOI: 10.1007/s11277-021-09144-1

Khan, R. A., Sharma, B., & Thakare, N. M. (2024). A systematic review of protecting cloud healthcare data using blockchain: Target-oriented approach. In. AIP Conference Proceedings: Vol. 3217. *No. 1* (pp. 1–7). AIP Publishing. DOI: 10.1063/5.0234398

Kirchherr, J., Reike, D., & Hekkert, M. (2017). Conceptualizing the circular economy: An analysis of 114 definitions. *Resources, Conservation and Recycling*, *127*, 221–232. DOI: 10.1016/j.resconrec.2017.09.005

Krishnamoorthy, S., Dua, A., & Gupta, S. (2023). Role of emerging technologies in future IoT-driven Healthcare 4.0 technologies: A survey, current challenges and future directions. *Journal of Ambient Intelligence and Humanized Computing*, *14*(1), 361–407. DOI: 10.1007/s12652-021-03302-w

Kulkarni, M. H., & Inamdar, P. R. (2025). Healthcare unbound: Navigating emerging trends and future applications in IoT-based innovations for daily well-being. In *Evolution of Machine Learning and Internet of Things Applications in Biomedical Engineering* (pp. 238–250). Springer.

Kumar, B., Garg, V., Ahmed, K., Garg, P., Choudhary, S., & Baniya, P. (2025, March). Enhancing Healthcare with Blockchain: Innovations in Data Privacy, Security, and Interoperability. In *2025 3rd International Conference on Disruptive Technologies (ICDT)* (pp. 932-938). IEEE.

Kumar, M., Choudhary, C., & ElBarachi, M. (2024, December). Sustainability of Energy Saving in Healthcare IoT Devices. In *2024 IEEE International Conference on Big Data (BigData)* (pp. 6494-6501). IEEE. DOI: 10.1109/BigData62323.2024.10825806

Kumar, M., Kumar, A., Verma, S., Bhattacharya, P., Ghimire, D., Kim, S. H., & Hosen, A. S. (2023). Healthcare Internet of Things (H-IoT): Current trends, future prospects, applications, challenges, and security issues. *Electronics (Basel)*, *12*(9), 2050. DOI: 10.3390/electronics12092050

Kumar, R. (2025). A comprehensive review of MCDM methods, applications, and emerging trends. *Decision Making Advances*, *3*(1), 185–199. DOI: 10.31181/dma31202569

Kumar, R., Gupta, S. K., Wang, H. C., Kumari, C. S., & Korlam, S. S. V. P. (2023). From efficiency to sustainability: Exploring the potential of 6G for a greener future. *Sustainability (Basel)*, *15*(23), 16387. DOI: 10.3390/su152316387

Kumar, R., Singh, A., Ahmed Kassar, A. S., Humaida, M. I., Joshi, S., & Sharma, M. (2025). Leveraging artificial intelligence to achieve sustainable public healthcare services in Saudi Arabia: A systematic literature review of critical success factors. *Computer Modeling in Engineering & Sciences*, *142*(2), 1191–1228. DOI: 10.32604/cmes.2025.059152

Kutybayeva, K., Razaque, A., & Rai, H. M. (2025). Enhancing Pharmaceutical Supply Chain Transparency and Security with Blockchain and Big Data Integration. *Procedia Computer Science*, *259*, 1511–1522. DOI: 10.1016/j.procs.2025.04.106

Lacwik, P., & Kupczyk, M. (2021). Asthma Phenotype Overlap: More May Be Less. *The Journal of Allergy and Clinical Immunology. In Practice*, *9*(1), 363–364. DOI: 10.1016/j.jaip.2020.10.032 PMID: 33429708

Lawan, I., Zhou, W., Garba, Z. N., Zhang, M., Yuan, Z., & Chen, L. (2019). Critical insights into the effects of bio-based additives on biodiesels properties. *Renewable & Sustainable Energy Reviews*, *102*, 83–95. DOI: 10.1016/j.rser.2018.12.008

Lenzen, M., Malik, A., Li, M., Fry, J., Weisz, H., Pichler, P.-P., Chaves, L. S. M., Capon, A., & Pencheon, D. (2020). The environmental footprint of health care: A global assessment. *The Lancet. Planetary Health*, *4*(7), e271–e279. DOI: 10.1016/S2542-5196(20)30121-2 PMID: 32681898

Li, L., & Zhang, S. (2021). Techno-economic and environmental assessment of multiple distributed energy systems coordination under centralized and decentralized framework. *Sustainable Cities and Society*, *72*, 103076. DOI: 10.1016/j.scs.2021.103076

Li, M., Wang, L., Liu, R., Li, J., Zhang, Q., Shi, G., Li, Y., Hou, C., & Wang, H. (2021). A highly integrated sensing paper for wearable electrochemical sweat analysis. *Biosensors & Bioelectronics*, *174*, 112828. DOI: 10.1016/j.bios.2020.112828 PMID: 33250335

Liu, Q., & Zhang, Z. (2020). The analysis of coverage probability, ASE and EE in heterogeneous ultra-dense networks with power control. *Digital Communications and Networks*, *6*(4), 524–533. DOI: 10.1016/j.dcan.2020.02.002

Mallo, S. F., Abdulqader, D. M., Abdullah, R. M., Ismael, H. R., Rashid, Z. N., & Sami, T. M. G. (2024). A review on feasibility of web technology and cloud computing for sustainable ES: Leveraging AI, IoT, and security for green operations. *Journal of Information Technology and Informatics*, *3*(2).

Mallo, S. F., Abdulqader, D. M., Abdullah, R. M., Ismael, H. R., Rashid, Z. N., & Sami, T. M. G. (2024). A Review on Feasibility of Web Technology and Cloud Computing for Sustainable ES: Leveraging AI, IoT, and Security for Green Operations. *Journal of Information Technology and Informatics*, *3*(2).

Mani, Z. A., & Goniewicz, K. (2024). Transforming healthcare in Saudi Arabia: A comprehensive evaluation of Vision 2030's impact. *Sustainability (Basel)*, *16*(8), 3277. DOI: 10.3390/su16083277

Manjula Devi, C., Sivakarthi, G., Srinivasan, A., Gobinath, A., & Rajeswari, P. (2025). 13 Edge AI in LoRa-Based Health Monitoring. *Edge AI for Industry 5.0 and Healthcare 5.0 Applications*, 185.

Mann, S., Potdar, V., Gajavilli, R. S., & Chandan, A. (2018, December). Blockchain technology for supply chain traceability, transparency and data provenance. In *Proceedings of the 2018 international conference on blockchain technology and application* (pp. 22-26). DOI: 10.1145/3301403.3301408

Martínez-Ramón, N., Calvo-Rodríguez, F., Iribarren, D., & Dufour, J. (2024). Frameworks for the application of machine learning in life cycle assessment for process modeling. *Cleaner Environmental Systems, 14*, 100221. DOI: 10.1016/j.cesys.2024.100221

Mashallahi, A., Ardalan, A., Nejati, A., & Ostadtaghizadeh, A. (2022). Climate adaptive hospital: A systematic review of determinants and actions. *Journal of Environmental Health Science & Engineering, 20*(2), 983–1013. DOI: 10.1007/s40201-022-00810-5 PMID: 36406601

Matin, A., Islam, M. R., Wang, X., Huo, H., & Xu, G. (2023). AIoT for sustainable manufacturing: Overview, challenges, and opportunities. *Internet of Things : Engineering Cyber Physical Human Systems, 24*, 100901. DOI: 10.1016/j.iot.2023.100901

Ma, X., Sheikholeslami, M., Jafaryar, M., Shafee, A., Nguyen-Thoi, T., & Li, Z. (2020). Solidification inside a clean energy storage unit utilizing phase change material with copper oxide nanoparticles. *Journal of Cleaner Production, 245*, 118888. DOI: 10.1016/j.jclepro.2019.118888

Mazumder, S., & Garg, S. (2021). Decoding digital transformational outsourcing: The role of service providers' capabilities. *International Journal of Information Management, 58*, 102295. DOI: 10.1016/j.ijinfomgt.2020.102295

Mishra, P., & Singh, G. (2023). Internet of medical things healthcare for sustainable smart cities: Current status and future prospects. *Applied Sciences (Basel, Switzerland), 13*(15), 8869. DOI: 10.3390/app13158869

Mishra, T., Wang, M., Metwally, A. A., Bogu, G. K., Brooks, A. W., Bahmani, A., Alavi, A., Celli, A., Higgs, E., Dagan-Rosenfeld, O., Fay, B., Kirkpatrick, S., Kellogg, R., Gibson, M., Wang, T., Hunting, E. M., Mamic, P., Ganz, A. B., Rolnik, B., & Snyder, M. P. (2020). Pre-symptomatic detection of COVID-19 from smartwatch data. *Nature Biomedical Engineering, 4*(12), 1208–1220. DOI: 10.1038/s41551-020-00640-6 PMID: 33208926

Missbauer, H., & Uzsoy, R. (2022). Order release in production planning and control systems: Challenges and opportunities. *International Journal of Production Research, 60*(1), 256–276. DOI: 10.1080/00207543.2021.1994165

Mondejar, M. E., Avtar, R., Baños Diaz, H. L., Dubey, R. K., Esteban, J., Gómez-Morales, A., & Hallam, B. (2021). Digitalization to achieve sustainable development goals: Steps towards a smart green planet. *The Science of the Total Environment, 794*, 148539. DOI: 10.1016/j.scitotenv.2021.148539 PMID: 34323742

Morley, J., Cowls, J., Taddeo, M., & Floridi, L. (2020). Ethical guidelines for COVID-19 tracing apps. *Nature, 582*(7810), 29–31. DOI: 10.1038/d41586-020-01578-0 PMID: 32467596

Moshkbid, E., Cree, D. E., Bradford, L., & Zhang, W. (2024). Biodegradable alternatives to plastic in medical equipment: Current state, challenges, and the future. *Journal of Composites Science, 8*(9), 342. DOI: 10.3390/jcs8090342

Moutaoukil, O., Mahfoud, H., & Bouhaddou, I. (2024). A novel conceptual framework for predictive maintenance of medical devices in a circular healthcare supply chain. In *2024 International Conference on Control, Automation and Diagnosis (ICCAD)* (pp. 1-6). IEEE. DOI: 10.1109/ICCAD60883.2024.10553792

Muhoza, A. C., Bergeret, E., Brdys, C., & Gary, F. (2023). Power consumption reduction for IoT devices thanks to Edge-AI: Application to human activity recognition. *Internet of Things : Engineering Cyber Physical Human Systems, 24*, 100930. DOI: 10.1016/j.iot.2023.100930

Musa, A. A., Hussaini, A., Qian, C., Guo, Y., & Yu, W. (2024). Open radio access networks for smart IoT systems: State of art and future. *International Journal of Semantic Computing, 11*(3), 310–327.

Nanda, S. K., Panda, S. K., & Dash, M. (2023). Medical supply chain integrated with blockchain and IoT to track the logistics of medical products. *Multimedia Tools and Applications, 82*(21), 32917–32939. DOI: 10.1007/s11042-023-14846-8 PMID: 37362711

Nanwal, J., Garg, P., Sethi, P., & Dixit, A. (2021). Green IoT and Big Data: Succeeding towards Building Smart Cities. In *Green Internet of Things for Smart Cities* (pp. 83-98). CRC Press.

Nijhawan, R., & Bhatia, S. (2023). Edge-AI tools and techniques for healthcare. In *Edge-AI in Healthcare* (pp. 17–30). CRC Press. DOI: 10.1201/9781003244592-2

Nižetić, S., Šolić, P., López-de-Ipiña González-de-Artaza, D., & Patrono, L. (2020). Internet of Things (IoT): Opportunities, issues and challenges towards a smart and sustainable future. *Journal of Cleaner Production, 274*, 122877. DOI: 10.1016/j.jclepro.2020.122877 PMID: 32834567

Nouri, S., Khoong, E. C., Lyles, C. R., & Karliner, L. (2020). Addressing equity in telemedicine for chronic disease management during the COVID-19 pandemic. *NEJM Catalyst Innovations in Care Delivery*, *1*(3).

Nti, I. K., Adekoya, A. F., Weyori, B. A., & Keyeremeh, F. (2023). A bibliometric analysis of technology in sustainable healthcare: Emerging trends and future directions. *Decision Analytics Journal*, *8*, 100292. DOI: 10.1016/j.dajour.2023.100292

Nwobodo-Anyadiegwu, E. N., Ditend, M. N., & Lumbwe, A. K. (2022). The benefits and challenges of implementing smart hospital projects: A systematic review. In *2022 IEEE 28th International Conference on Engineering, Technology and Innovation (ICE/ITMC) & 31st International Association for Management of Technology (IAMOT) Joint Conference* (pp. 1-7). IEEE. DOI: 10.1109/ICE/ITMC-IAMOT55089.2022.10033238

Olabi, A. G., Shehata, N., Obaideen, K., Sayed, E. T., Mahmoud, M., AlMallahi, M. N., & Abdelkareem, M. A. (2024). COVID-19: Medical waste management, impact on sustainable development goals, and bibliometric analysis. *Chemical Engineering & Technology*, *47*(1), 4–19. DOI: 10.1002/ceat.202300046

Ongaro, A. E., Ndlovu, Z., Sollier, E., Otieno, C., Ondoa, P., Street, A., & Kersaudy-Kerhoas, M. (2022). Engineering a sustainable future for point-of-care diagnostics and single-use microfluidic devices. *Lab on a Chip*, *22*(17), 3122–3137. DOI: 10.1039/D2LC00380E PMID: 35899603

Osama, M., Ateya, A. A., Sayed, M. S., Hammad, M., Pławiak, P., Abd El-Latif, A. A., & Elsayed, R. A. (2023). Internet of medical things and healthcare 4.0: Trends, requirements, challenges, and research directions. *Sensors (Basel)*, *23*(17), 7435. DOI: 10.3390/s23177435 PMID: 37687891

Oyeniran, O. C., Modupe, O. T., Otitoola, A. A., Abiona, O. O., Adewusi, A. O., & Oladapo, O. J.Oyekunle Claudius OyeniranOluwole Temidayo ModupeAanuoluwapo Ayodeji OtitoolaOluwatosin Oluwatimileyin AbionaAdebunmi Okechukwu AdewusiOluwatayo Jacob Oladapo. (2024). A comprehensive review of leveraging cloud-native technologies for scalability and resilience in software development. *International Journal of Science and Research Archive*, *11*(2), 330–337. DOI: 10.30574/ijsra.2024.11.2.0432

Panchal, S. (2022). A road map to achieve sustainable goals in the healthcare system: A perspective of a pharmacist. *Nirma University Journal of Pharmaceutical Sciences*, *9*(1–2), 61–65.

Panda, S. K., & Satapathy, S. C. (2024). Drug traceability and transparency in medical supply chain using blockchain for easing the process and creating trust between stakeholders and consumers. *Personal and Ubiquitous Computing*, *28*(1), 75–91. DOI: 10.1007/s00779-021-01588-3

Pandey, A. (2025). Energy-efficient communication in edge computing iot networks. *Metaversalize*, *2*(1), 21–30.

Pandiyan, P., Saravanan, S., Kannadasan, R., Krishnaveni, S., Alsharif, M. H., & Kim, M. K. (2024). A comprehensive review of advancements in green IoT for smart grids: Paving the path to sustainability. *Energy Reports*, *11*, 5504–5531. DOI: 10.1016/j.egyr.2024.05.021

Park, S., Choi, G. J., & Ko, H. (2020). Information technology–based tracing strategy in response to COVID-19 in South Korea—Privacy controversies. *Journal of the American Medical Association*, *323*(21), 2129–2130. DOI: 10.1001/jama.2020.6602 PMID: 32324202

Pelekoudas-Oikonomou, F., Zachos, G., Papaioannou, M., De Ree, M., Ribeiro, J. C., Mantas, G., & Rodriguez, J. (2022). Blockchain-based security mechanisms for IoMT Edge networks in IoMT-based healthcare monitoring systems. *Sensors (Basel)*, *22*(7), 2449. DOI: 10.3390/s22072449 PMID: 35408064

Perez, M., Mahaffey, K. W., Hedlin, H., Rumsfeld, J. S., Garcia, A., Ferris, T., Balasubramanian, V., Russo, A. M., Rajmane, A., Cheung, L., Hung, G., Lee, J., Kowey, P., Talati, N., Nag, D., Gummidipundi, S. E., Beatty, A., Hills, M. T., Desai, S., & Turakhia, M. P. (2019). Large-Scale Assessment of a Smartwatch to Identify Atrial Fibrillation. *The New England Journal of Medicine*, *381*(20), 1909–1917. DOI: 10.1056/NEJMoa1901183 PMID: 31722151

Pournader, M., Shi, Y., Seuring, S., & Koh, S. L. (2020). Blockchain applications in supply chains, transport and logistics: A systematic review of the literature. *International Journal of Production Research*, *58*(7), 2063–2081. DOI: 10.1080/00207543.2019.1650976

Pradere, B., Mallet, R., de La Taille, A., Bladou, F., Prunet, D., Beurrier, S., Bardet, F., Game, X., Fournier, G., Lechevallier, E., Meria, P., Matillon, X., Polguer, T., Abid, N., De Graeve, B., Kassab, D., Mejean, A., Misrai, V., & Pinar, U. (2023). Climate-smart actions in the operating theatre for improving sustainability practices: A systematic review. *European Urology*, *83*(4), 331–342. DOI: 10.1016/j.eururo.2022.01.027 PMID: 35151515

Prainsack, B., & Buyx, A. (2017). *Solidarity in biomedicine and beyond*. Cambridge University Press. DOI: 10.1017/9781139696593

Prakash, A., Avasthi, S., Kumari, P., Rawat, M., & Garg, P. (2023). 18 Modern healthcare system: unveiling the possibility of quantum computing in medical and biomedical zones.

Psillaki, M., Apostolopoulos, N., Makris, I., Liargovas, P., Apostolopoulos, S., Dimitrakopoulos, P., & Sklias, G. (2023). Hospitals' energy efficiency in the perspective of saving resources and providing quality services through technological options: A systematic literature review. *Energies, 16*(2), 755. DOI: 10.3390/en16020755

Pustokhina, I. V., Pustokhin, D. A., Lydia, E. L., Garg, P., Kadian, A., & Shankar, K. (2021). Hyperparameter search based convolution neural network with Bi-LSTM model for intrusion detection system in multimedia big data environment. *Multimedia Tools and Applications*, 1–18. DOI: 10.1007/s11042-021-11271-7

Putra, K. T., Arrayyan, A. Z., Hayati, N., Damarjati, C., Bakar, A., & Chen, H. C. (2024). A review on the application of internet of medical things in wearable personal health monitoring: A cloud-edge artificial intelligence approach. *IEEE Access : Practical Innovations, Open Solutions, 12*, 21437–21452. DOI: 10.1109/ACCESS.2024.3358827

Qadri, Y. A., Nauman, A., Zikria, Y., Vasilakos, A. V., & Kim, S. W. (2020). The Future of Healthcare Internet of Things: A Survey of Emerging Technologies. *IEEE Communications Surveys and Tutorials, 22*(2), 1121–1167. DOI: 10.1109/COMST.2020.2973314

Qamar, N., Siddiqui, N., Ehatisham-ul-Haq, M., Awais Azam, M., & Naeem, U. (2020). An Approach towards Position-Independent Human Activity Recognition Model based on Wearable Accelerometer Sensor. *Procedia Computer Science, 177*, 196–203. DOI: 10.1016/j.procs.2020.10.028

Quan, J., Miyake, Y., & Nozawa, T. (2021). Incorporating Interpersonal Synchronization Features for Automatic Emotion Recognition from Visual and Audio Data during Communication. *Sensors (Basel), 21*(16), 5317. DOI: 10.3390/s21165317 PMID: 34450759

Quinto, S., Law, N., Fletcher, C., Le, J., Antony Jose, S., & Menezes, P. L. (2025). Exploring the E-waste crisis: Strategies for sustainable recycling and circular economy integration. *Recycling, 10*(2), Article 72.

Qwaider, S., Al-Ramadan, B., Shafiullah, M., Islam, A., & Worku, M. Y. (2023). GIS-based progress monitoring of SDGs towards achieving Saudi Vision 2030. *Remote Sensing (Basel), 15*(24), 5770. DOI: 10.3390/rs15245770

Rajagopalan, S., McAlister, S., Jay, J., Pham, R. D., Brook, R. D., Nasir, K., & Nieuwenhuijsen, M. J. (2024). Environmental sustainability in cardiovascular practice: Current challenges and future directions. *Nature Reviews. Cardiology*, 1–14. PMID: 39455886

Raj, G., Verma, A., Dalal, P., Shukla, A. K., & Garg, P. (2023). Performance comparison of several LPWAN technologies for energy constrained IOT network. *International Journal of Intelligent Systems and Applications in Engineering*, *11*(1s), 150–158.

Rajkomar, A., Hardt, M., Howell, M. D., Corrado, G., & Chin, M. H. (2018). Ensuring fairness in machine learning to advance health equity. *Annals of Internal Medicine*, *169*(12), 866–872. DOI: 10.7326/M18-1990 PMID: 30508424

Rautela, R., Arya, S., Vishwakarma, S., Lee, J., Kim, K.-H., & Kumar, S. (2021). E-waste management and its effects on the environment and human health. *The Science of the Total Environment*, *773*, 145623. DOI: 10.1016/j.scitotenv.2021.145623 PMID: 33592459

Regona, M., Yigitcanlar, T., Hon, C., & Teo, M. (2024). Artificial intelligence and sustainable development goals: Systematic literature review of the construction industry. *Sustainable Cities and Society*, *108*, 105499. DOI: 10.1016/j.scs.2024.105499

Rejeb, A., Rejeb, K., Appolloni, A., Treiblmaier, H., & Iranmanesh, M. (2023). Circular economy research in the COVID-19 era: A review and the road ahead. *Circular Economy and Sustainability*, *3*(4), 2127–2157. DOI: 10.1007/s43615-023-00265-2

Rejeb, A., Rejeb, K., Keogh, J. G., & Zailani, S. (2022). Barriers to blockchain adoption in the circular economy: A fuzzy Delphi and best-worst approach. *Sustainability (Basel)*, *14*(6), 3611. DOI: 10.3390/su14063611

Rejeb, A., Suhaiza, Z., Rejeb, K., Seuring, S., & Treiblmaier, H. (2022). The Internet of Things and the circular economy: A systematic literature review and research agenda. *Journal of Cleaner Production*, *350*, 131439. DOI: 10.1016/j.jclepro.2022.131439

Roberts, M., Driggs, D., Thorpe, M., Gilbey, J., Yeung, M., Ursprung, S., Aviles-Rivero, A. I., Etmann, C., McCague, C., Beer, L., Weir-McCall, J. R., Teng, Z., Gkrania-Klotsas, E., Ruggiero, A., Korhonen, A., Jefferson, E., Ako, E., Langs, G., Gozaliasl, G., & Schönlieb, C. B. (2021). Common pitfalls and recommendations for using machine learning to detect and prognosticate for COVID-19 using chest radiographs and CT scans. *Nature Machine Intelligence*, *3*(3), 199–217. DOI: 10.1038/s42256-021-00307-0

Rønn, C., Wieland, A., Lehrer, C., Márton, A., LaRoche, J., Specker, A., Leroy, P., & Fürstenau, D. (2023). Circular business model for digital health solutions: Protocol for a scoping review. *JMIR Research Protocols*, *12*(1), e47874. DOI: 10.2196/47874 PMID: 37999949

Rosen, J. M., Adams, L. V., Geiling, J., Curtis, K. M., Mosher, R. E., Ball, P. A., Grigg, E. B., & Grigg, E. B. (2021). Telehealth's new horizon: Providing smart hospital-level care in the home. *Telemedicine Journal and e-Health*, *27*(11), 1215–1224. DOI: 10.1089/tmj.2020.0448 PMID: 33656918

Ru, X., Wang, S., & Yan, S. (2018). Exploring the effects of normative factors and perceived behavioral control on individual's energy-saving intention: An empirical study in eastern China. *Resources, Conservation and Recycling*, *134*, 91–99. DOI: 10.1016/j.resconrec.2018.03.001

Saberi, S., Kouhizadeh, M., Sarkis, J., & Shen, L. (2019). Blockchain technology and its relationships to sustainable supply chain management. *International Journal of Production Research*, *57*(7), 2117–2135. DOI: 10.1080/00207543.2018.1533261

Sabet, M. (2025). Sustainable halogen-free polymer composites for next-generation flexible electronics. *Journal of Nanoparticle Research*, *27*(2), 33. DOI: 10.1007/s11051-025-06234-0

Saini, P., Nagpal, B., Garg, P., & Kumar, S. (2023). CNN-BI-LSTM-CYP: A deep learning approach for sugarcane yield prediction. *Sustainable Energy Technologies and Assessments*, *57*, 103263. DOI: 10.1016/j.seta.2023.103263

Saini, P., Nagpal, B., Garg, P., & Kumar, S. (2023). Evaluation of Remote Sensing and Meteorological parameters for Yield Prediction of Sugarcane (Saccharumofficinarum L.) Crop. *Brazilian Archives of Biology and Technology*, *66*, e23220781. DOI: 10.1590/1678-4324-2023220781

Sallam, K., Mohamed, M., & Mohamed, A. W. (2023). Internet of Things (IoT) in supply chain management: Challenges, opportunities, and best practices. *Sustainable Machine Intelligence Journal*, *2*, 3–1. DOI: 10.61185/SMIJ.2023.22103

Santoso, A., & Surya, Y. (2024). Maximizing Decision Efficiency with Edge-Based AI Systems: Advanced Strategies for Real-Time Processing, Scalability, and Autonomous Intelligence in Distributed Environments. *Quarterly Journal of Emerging Technologies and Innovations*, *9*(2), 104–132.

Sathiya, A., Angel, D., Iswarya, M., Poonkodi, R., & Angelo, K. M. (2025, January). IoT Enabled Healthcare Framework Using Edge AI and Advanced Wearable Sensors for Real Time Health Monitoring. In *2025 International Conference on Multi-Agent Systems for Collaborative Intelligence (ICMSCI)* (pp. 384-392). IEEE. DOI: 10.1109/ICMSCI62561.2025.10894492

Saunders, C., Brown, S., Bygstad, B., Dennis, A., Ferran, C., Galletta, D., Liang, T.-P., Lowry, P., Recker, J., & Sarker, S. (2017). Goals, Values, and Expectations of the AIS Family of Journals. *Journal of the Association for Information Systems*, *18*(9), 633–647. DOI: 10.17705/1jais.00468

Selvaraj, S., Prasadh, S., Fuloria, S., Subramaniyan, V., Sekar, M., Ahmed, A. M., Bouallegue, B., Hari Kumar, D., Sharma, V. K., Maziz, M. N. H., Sathasivam, K. V., Meenakshi, D. U., & Fuloria, N. K. (2022). COVID-19 biomedical plastics wastes—Challenges and strategies for curbing the environmental disaster. *Sustainability (Basel)*, *14*(11), 6466. DOI: 10.3390/su14116466

Senarathna, W. G. C. M., Sulaksha, L. G. T., Weerarathna, D. M., Jayathma, W. M. V. A., Gamage, D. G. M. L., Thennakoon, T. M. T. N., Hewage, H. T. M., Panagoda, L. P. S. S., Sandunika, D. M. I., & Perera, M. D. V. (2023). Paper recycling for a sustainable future: Global trends. *Journal of Research in Technology and Engineering*, *4*(3), 169–186.

Settia, N., Bhutani, M., & Saini, V. (2025). Exploring the impact of artificial intelligence on healthcare: A comprehensive review. In *Digitalization and the Transformation of the Healthcare Sector* (pp. 115-138).

Settia, N., Bhutani, M., & Saini, V. (2025). Exploring the impact of artificial intelligence on healthcare: A comprehensive review. In *Digitalization and the Transformation of the Healthcare Sector* (pp. 115–138). Springer.

Settia, N., Bhutani, M., & Saini, V. (2025). Exploring the impact of artificial intelligence on healthcare: A comprehensive review. In Tyagi, A. K., & Rajest, S. S. (Eds.), *Digitalization and the transformation of the healthcare sector* (pp. 115–138). IGI Global., DOI: 10.4018/979-8-3693-5577-0.ch006

Shaikh, F. K., Zeadally, S., & Exposito, E. (2017). Enabling Technologies for Green Internet of Things. *IEEE Systems Journal*, *11*(2), 983–994. DOI: 10.1109/JSYST.2015.2415194

Sharma, H. B., Vanapalli, K. R., Samal, B., Cheela, V. S., Dubey, B. K., & Bhattacharya, J. (2021). Circular economy approach in solid waste management system to achieve UN-SDGs: Solutions for post-COVID recovery. *The Science of the Total Environment*, *800*, 149605. DOI: 10.1016/j.scitotenv.2021.149605 PMID: 34426367

Sharma, N., & Garg, P. (2022). Ant colony based optimization model for QoS-Based task scheduling in cloud computing environment. *Measurement. Sensors, 100531*, 100531. DOI: 10.1016/j.measen.2022.100531

Sharma, T. P., Solanki, A., Jain, T., Malhotra, A., & Bhutani, M. (2024). Wi-Fi based quadcopter drone with battery monitoring and optimization using crazyflie platform. *International Journal of Information Technology : an Official Journal of Bharati Vidyapeeth's Institute of Computer Applications and Management, 16*(3), 1887–1898. DOI: 10.1007/s41870-023-01639-3

Shi, X., Wang, L., Zhang, F., Zheng, K., Mühlhäuser, M., & Liu, Z. (2019). PABO: Mitigating congestion via packet bounce in data center networks. *Computer Communications, 140–141*, 1–14. DOI: 10.1016/j.comcom.2019.04.002

Shoomal, A., Jahanbakht, M., Componation, P. J., & Ozay, D. (2024). Enhancing supply chain resilience and efficiency through internet of things integration: Challenges and opportunities. *Internet of Things : Engineering Cyber Physical Human Systems, 27*, 101324. DOI: 10.1016/j.iot.2024.101324

Shumba, A. T., Montanaro, T., Sergi, I., Fachechi, L., De Vittorio, M., & Patrono, L. (2022). Leveraging IoT-aware technologies and AI techniques for real-time critical healthcare applications. *Sensors (Basel), 22*(19), 7675. DOI: 10.3390/s22197675 PMID: 36236773

Singh, N., Ogunseitan, O. A., & Tang, Y. (2022). Medical waste: Current challenges and future opportunities for sustainable management. *Critical Reviews in Environmental Science and Technology, 52*(11), 2000–2022. DOI: 10.1080/10643389.2021.1885325

Singh, Y., Kumar, R., Kabdal, S., & Upadhyay, P. (2023). Youtube video summarizer using nlp: A review. *International Journal of Performability Engineering, 19*(12), 817.

Sjoding, M. W., Dickson, R. P., Iwashyna, T. J., Gay, S. E., & Valley, T. S. (2020). Racial bias in pulse oximetry measurement. *The New England Journal of Medicine, 383*(25), 2477–2478. DOI: 10.1056/NEJMc2029240 PMID: 33326721

Smith, A. C., Thomas, E., Snoswell, C. L., Haydon, H., Mehrotra, A., Clemensen, J., & Caffery, L. J. (2020). Telehealth for global emergencies: Implications for coronavirus disease 2019 (COVID-19). *Journal of Telemedicine and Telecare, 26*(5), 309–313. DOI: 10.1177/1357633X20916567 PMID: 32196391

Stahel, W. R. (2016). The circular economy. *Nature, 531*(7595), 435–438. DOI: 10.1038/531435a PMID: 27008952

Stahel, W. R., & MacArthur, E. (2019). *The Circular Economy*. Routledge., DOI: 10.4324/9780429259203

Steinhubl, S. R., Waalen, J., Edwards, A. M., Ariniello, L. M., Mehta, R. R., Ebner, G. S., Carter, C., Baca-Motes, K., Felicione, E., Sarich, T., & Topol, E. J. (2018). Effect of a Home-Based Wearable Continuous ECG Monitoring Patch on Detection of Undiagnosed Atrial Fibrillation. *Journal of the American Medical Association, 320*(2), 146. DOI: 10.1001/jama.2018.8102 PMID: 29998336

Tariq, U., Ahmed, I., Khan, M. A., & Bashir, A. K. (2025). Bridging biosciences and deep learning for revolutionary discoveries: A comprehensive review. IAES International Journal of Artificial Intelligence (IJ-AI), 14(2), 867-883.

Thakur, A., Mukhopadhyay, T., & Ahirwar, A. K. (2024). Approaching sustainability in laboratory medicine. [CCLM]. *Clinical Chemistry and Laboratory Medicine, 62*(9), 1787–1794. DOI: 10.1515/cclm-2023-0973 PMID: 38557335

Troncoso, C., Payer, M., Hubaux, J. P., Salathé, M., Larus, J., Bugnion, E., . . . Lueks, W. (2020). Decentralized privacy-preserving proximity tracing. *arXiv preprint arXiv:2005.12273.*

Tunkelang, D., Capra, R., Golovchinsky, G., Kules, B., Smith, C., & White, R. (2013). Symposium on Human–Computer Information Retrieval. *Big Data, 1*(1), 65–70. DOI: 10.1089/big.2012.0003 PMID: 27447040

Turskis, Z., & Šniokienė, V. (2024). IoT-driven transformation of circular economy efficiency: An overview. *Mathematical & Computational Applications, 29*(4), 49. DOI: 10.3390/mca29040049

Upadhyay, P., & Tomar, P. (2022, August). Alzheimer's Disease Classification Using Ensemble Methods. In International Conference on Signal Processing and Integrated Networks (pp. 3-15). Singapore: Springer Nature Singapore.

Upadhyay, P., & Tomar, P. (2022, December). Alzheimer's disease Classification using various machine learning approaches: A Review. In 2022 4th International Conference on Advances in Computing, Communication Control and Networking (ICAC3N) (pp. 232-237). IEEE. DOI: 10.1109/ICAC3N56670.2022.10074339

Upadhyay, P., & Tomar, P. "Alzheimer's Disease Classification Using Ensemble Methods." In International Conference on Signal Soni, E., Nagpal, A., Garg, P., &Pinheiro, P. R. (2022). Assessment of Compressed and Decompressed ECG Databases for Telecardiology Applying a Convolution Neural Network. Electronics, 11(17), 2708. DOI: DOI: 10.3390/electronics11172708

Upadhyay, D., Garg, P., Aldossary, S. M., Shafi, J., & Kumar, S. (2023). A linear quadratic regression-based synchronised health monitoring system (SHMS) for IoT applications. *Electronics (Basel), 12*(2), 309. DOI: 10.3390/electronics12020309

Upadhyay, P., Agarwal, R., Dhiman, S., Sarkar, A., & Chaturvedi, S. (2024). A comprehensive survey on answer generation methods using NLP. *Natural Language Processing Journal*, *8*, 100088. DOI: 10.1016/j.nlp.2024.100088

Upadhyay, P., & Tomar, P. “Alzheimer’s Disease Classification Using Ensemble Methods.” In *International Conference on Signal*

Upadhyay, P., Tomar, P., & Yadav, S. P. (2024). Advancements in Alzheimer’s disease classification using deep learning frameworks for multimodal neuroimaging: A comprehensive review. *Computers & Electrical Engineering*, *120*, 109796. DOI: 10.1016/j.compeleceng.2024.109796

Upadhyay, P., Tomar, P., & Yadav, S. P. (2024). Comprehensive Systematic Computation on Alzheimer’s Disease Classification. *Archives of Computational Methods in Engineering*, *31*(8), 4773–4804. DOI: 10.1007/s11831-024-10120-8

Upadhyay, P., Tomar, P., & Yadav, S. P. (2025). ADNet: Deep learning based model for Alzheimer’s disease classification. *Iran Journal of Computer Science*, *8*(2), 1–13. DOI: 10.1007/s42044-025-00243-x

Velmurugan, J. S. A. (2024). A review on green human resource management in hospitals: Paving the way for sustainable development goals.

Venkatesh, V. G., Kang, K., Wang, B., Zhong, R. Y., & Zhang, A. (2020). System architecture for blockchain based transparency of supply chain social sustainability. *Robotics and Computer-integrated Manufacturing*, *63*, 101896. DOI: 10.1016/j.rcim.2019.101896

Venkateswaran, M. R., Khosravi, A., Zarepour, A., Iravani, S., & Zarrabi, A. (2024). Self-healing materials in biomedicine and the circular economy. *Environmental Science. Nano*, *11*(7), 2771–2802. DOI: 10.1039/D4EN00235K

Verdejo Espinosa, Á., López, J. L., Mata Mata, F., & Espinilla Estevez, M. (2021). Application of IoT in healthcare: Keys to implementation of the sustainable development goals. *Sensors (Basel)*, *21*(7), 2330. DOI: 10.3390/s21072330 PMID: 33810606

Vergara, R., Théate, I., Boor, P., Gordon, I. O., West, J., Abdelmoula, S., Tilmant, C., Wiseman Pinto, R. G., Gaillot-Durand, L., Scott, S., Trecourt, A., & Rullier, A. (2025). Surgical pathology and sustainable development: International landscape and prospects. *Journal of Clinical Pathology*, *78*(4), 233–239. DOI: 10.1136/jcp-2024-209555 PMID: 39638426

Vishwakarma, L. P., Singh, R. K., Mishra, R., & Kumari, A. (2025). Application of artificial intelligence for resilient and sustainable healthcare system: Systematic literature review and future research directions. *International Journal of Production Research, 63*(2), 822–844. DOI: 10.1080/00207543.2023.2188101

Wang, B., Lin, Z., Wang, M., Wang, F., Xiangli, P., & Li, Z. (2023). Applying blockchain technology to ensure compliance with sustainability standards in the PPE multi-tier supply chain. *International Journal of Production Research, 61*(14), 4934–4950. DOI: 10.1080/00207543.2022.2025944

Wan, P. K., Huang, L., & Holtskog, H. (2020). Blockchain-enabled information sharing within a supply chain: A systematic literature review. *IEEE Access : Practical Innovations, Open Solutions, 8*, 49645–49656. DOI: 10.1109/ACCESS.2020.2980142

Webster, D. E., Tummalacherla, M., Higgins, M., Wing, D., Ashley, E., Kelly, V. E., McConnell, M., Muse, E. D., Olgin, J. E., Mangravite, L. M., Godino, J., Kellen, M. R., & Omberg, L. (2021). Smartphone-Based VO2max Measurement With Heart Snapshot in Clinical and Real-world Settings With a Diverse Population: Validation Study. *JMIR mHealth and uHealth, 9*(6), e26006. DOI: 10.2196/26006 PMID: 34085945

Whitelaw, S., Mamas, M. A., Topol, E., & Van Spall, H. G. (2020). Applications of digital technology in COVID-19 pandemic planning and response. *The Lancet. Digital Health, 2*(8), e435–e440. DOI: 10.1016/S2589-7500(20)30142-4 PMID: 32835201

Wicklum, S. C., Nuique, K., Kelly, M. A., Nesbitt, C. C., Zhang, J. J., & Svrcek, C. P. (2023). Greening family medicine clinic operations and clinical care: Where do we start? A scoping review of toolkits and aids. *Family Practice, 40*(3), 473–485. DOI: 10.1093/fampra/cmad006 PMID: 36730055

Woll, A., & Tørresen, J. (2022). What is a smart hospital? A review of the literature. *Human-Automation Interaction: Manufacturing, Services and User Experience*, 145-165.

Xames, M. D., & Topcu, T. G. (2025). How can digital twins support the economic, environmental, and social sustainability of healthcare systems: A systematic review focused on the triple-bottom-line. *IEEE Access : Practical Innovations, Open Solutions, 13*, 1594–1613. DOI: 10.1109/ACCESS.2025.3559502

Yallamelli, A. R. G., & Devarajan, M. V. (2023). Hybrid edge-AI and cloudlet-driven IoT framework for real-time healthcare. *International Journal of Computer Science and Engineering Technology, 7*(1).

Yang, L., Nguyen, H., Bui, X.-N., Nguyen-Thoi, T., Zhou, J., & Huang, J. (2021). Prediction of gas yield generated by energy recovery from municipal solid waste using deep neural network and moth-flame optimization algorithm. *Journal of Cleaner Production, 311*, 127672. DOI: 10.1016/j.jclepro.2021.127672

Zeng, J., Zhang, T., Mishra, D., Yuan, J., & Seneviratne, A. (2025). A survey on green designs for energy harvesting backscatter communications to enable sustainable IoT. *Energies, 18*(4), 840. DOI: 10.3390/en18040840

Zhang, R., & Chan, W. K. V. (2020, July). Evaluation of energy consumption in block-chains with proof of work and proof of stake. *Journal of Physics: Conference Series, 1584*(1), 012023. DOI: 10.1088/1742-6596/1584/1/012023

Zhang, Y., & Feng, J. (2024). *Towards a smart and sustainable future with edge computing-powered internet of things: Fundamentals, applications, challenges, and future research directions. Journal of The Institution of Engineers*. Series B.

Zhu, R., Wong, M. S., Kwan, M.-P., Chen, M., Santi, P., & Ratti, C. (2022). An economically feasible optimization of photovoltaic provision using real electricity demand: A case study in New York city. *Sustainable Cities and Society, 78*, 103614. DOI: 10.1016/j.scs.2021.103614

Zou, H., Qin, J., Yang, P., & Dai, B. (2018). A Coordinated Revenue-Sharing Model for a Sustainable Closed-Loop Supply Chain. *Sustainability (Basel), 10*(9), 3198. DOI: 10.3390/su10093198

About the Contributors

Chich-Jen has been a professor at Chang Jung Christian University, Taiwan for more than 12 years. Since 2015, He has been appointed as a Distinguished Professor at Huaqiao University, China. He is also the Executive Director of International Association of Organization Innovation (IAOI).

Prashant Upadhyay is an Assistant Professor at, the Department of CSE, School of Engineering and Technology, Sharda University Greater Noida, Uttar Pradesh, India. He completed his Doctorate of Philosophy in Computer Science and Engineering, Gautam Buddha University. He has more than 07 years of teaching and research experience. He published several research papers in internationally reputed journals indexed in Web of Science, SCOPUS & SCIE and top-rank conferences as author or co-author. His research findings are published in Archives of Computational Methods in Engineering Springer, Computers and Electrical Engineering Elsevier, IGI Global USA, IEEE Xplore, Taylor and Francis. He served as a TPC and reviewer in Various International Conferences. He has edited 11 books with various publishers, including IGI Global, Bentham Science, and CRC Press, among others. He served as session chair in several international conferences- NGISE-2025, ICDT-2024, ICDETGT-2023, and ICDT-2023. He has authored a book for Wiley Publication- Introduction to Python, ISBN: 978-93-5746-219-8 and currently he is working on the book titled "Data Science in Business Analytics". He has published 8 Indian patents and holds 2 grants.

Monica Bhutani is a distinguished academician and researcher in the field of Electronics and Communications Engineering and Computer Science. She is currently serving as an Associate Professor in the Department of Electronics and Communication Engineering at Bharati Vidyapeeth's College of Engineering (BVCOE), New Delhi. She is also holding a position of Postdoctoral Researcher and Research Associate at Lincoln University College, Malaysia, where she is further expanding her expertise in next-generation communication technologies and interdisciplinary applications. Dr. Bhutani holds the esteemed position of Editor-in-Chief of the Scienxt Journal of Electrical & Electronics Communication, and also serves as an Associate Editorial Board Member for Wireless and Communication Letters published by Bentham Science. Dr. Bhutani earned her Ph.D. from the Indian Institute of Technology (IIT) Delhi, where her research focused on MAC layer issues in Optical Wireless Communication. With over 18 years of academic experience, her core research interests include Optical Wireless Communications, Visible Light Communication (VLC), Li-Fi, Wireless Sensor Networks, Zigbee and IEEE 802.15.x standards, Computer Networks, and Machine Learning applications in IoT. She plays an active role in IEEE, serving as the Head of the IEEE BVCOE Student Chapter under Delhi Section R10. Dr. Bhutani has authored and co-authored over 75 research papers in reputed national and international journals and conferences, and has chaired sessions and contributed to technical committees in more than 50 international conferences. A highly sought-after speaker, she frequently delivers keynote addresses and conducts training sessions in STTPs and FDPs on research, emerging technologies, and AI tools for academia. Her contributions include two patents, editorial roles in multiple academic books on wireless networks, smart systems, and blockchain, and peer-review responsibilities for renowned journals such as IEEE Access, OSA-JOCN, and IEEE Sensors Journal. Recipient of several Best Researcher Awards, Dr. Bhutani is also deeply committed to promoting STEM education, especially for young women, through outreach initiatives like WIEHACK 4.0 and WIESTAR. Her innovative research and academic leadership continue to shape the future of intelligent communication systems and inspire the next generation of engineers and technologists.

Vamsi Krishna Reddy Bandaru is currently working for Cisco. With a strong foundation in technology and innovation, he has successfully led multiple projects that enhanced system efficiencies and drove digital transformation. Vamsi's work has been recognized for its impact on improving network security and scalability across diverse industries. Known for his problem-solving skills and strategic thinking, he has contributed to key initiatives that foster growth and operational excellence. Beyond his technical expertise, Vamsi is committed to continuous learning and knowledge sharing, making him a respected figure in his field. His dedication and achievements reflect his passion for pushing the boundaries of technology to create meaningful solutions.

Pranav Dass is currently working as an Assistant Professor in the Department of Computer Science, Shyam Lal College, University of Delhi. He is also currently Visiting Professor at Shinawatra University, Bangkok, Thailand. He has served as Professor and Head of the Department of Computer Science and Engineering and also in-charge of Higher Study Cell in Bharati Vidyapeeth's College of Engineering, New Delhi and has bagged several prestigious awards in his illustrious career which includes multiple Young Researcher National Science Foundation (NSF) Awards worth US $500 at several International conferences organized by University of North Carolina at Greensboro, USA. Besides these, he has been a pioneer research team member of Network for Non-Profit and Social Impact (NNSI) project titled Collaborative Research: Mind mapping consumers and activists' response to NGO-corporate partnerships and Brand Associative Network Diagrams (BrANDi) fully funded by NSF Grant from October 2014 till July 2015. Additionally, Dr. Dass' research paper Difference Equations and RaoHamming Bound had been awarded the Best paper award in the pre-doctoral student category at an International conference held at University of Southern Maine, Portland, USA in 2003; he also received a certificate, plaque and a cheque worth US $100 for the same. Dr. Dass had his college education from North Dakota State University, North Dakota, USA from where he earned his Ph.D. degree in Computer Science in 2017. Before this, he earned a Master's degree in Computer Science from the same University in 2009. Talking about his academic accolades, he has published several international conference and journal articles. He has provided research guidance to students at both UG and PG levels including 5 Master's students and 8 Bachelor's students. Dr. Dass has delivered honor speech and served as session chair at international conference in South Korea. He is currently the Associate Editor of 2 SCOPUS journals and lifetime member of Computer Society of India (CSI).

***

Neetu Anand received her BSc., MIT, and PhD degrees in Computer Science. She is working as an Associate Professor in the Department of Computer Applications(M), Maharaja Surajmal Institute, Affiliated to GGSIPU. She has 25 years of teaching experience. She has published more than 55 research papers in reputed International and National Journals and conferences including SCI, ESCI, Scopus. Her main area of research includes Web Mining, Big Data Analysis, and Statistical Data Mining. She has made an impact on the lives of the students and contributed to the educational field to make the surroundings a better place to learn by adopting the latest innovations or best global practices. She has received many national and international awards, including the FoXclues India Prime Icon Award 2022, in the category of Education and Social Leadership Award, and Awarded the most prestigious International Eminent Researcher Award 2021, instituted by the International Institute of Organized Research (I2OR) which is a registered MSME with the Ministry of Micro, Small and Medium Enterprises, Government of India.

Rohan Chaudhary is an Assistant Professor in the Department of Computer Science and Engineering. He is passionate about teaching and guiding students in various areas of computer science. With a strong academic background and practical knowledge, he aims to help students build a solid foundation in the subject.

Pawan Kr. Goel, is a accomplished academician and researcher with 19.5 years of experience, is an Associate Professor at Raj Kumar Goel Institute of Technology, Ghaziabad. He holds a Ph.D. in Computer Science Engineering and is UGC NET qualified. His expertise spans various domains, including wireless sensor networks, cloud computing, and artificial intelligence. Dr. Goel has published extensively in prestigious journals and conferences, with notable papers on topics such as cybersecurity, IoT, and machine learning. He is an active member of numerous professional bodies, including the IEEE, Computer Society of India and the International Association of Engineers. Recognized for his contributions, he has received several awards and certificates of appreciation. Dr. Goel is dedicated to fostering industry-academia collaborations and has organized numerous workshops and seminars. He is also involved in various training programs, MOOCs, and NPTEL certifications, contributing significantly to the advancement of education and research in his field.

Dedeepya Sai Gondi (Datta) is a seasoned AI/ML researcher and engineer with over 11 years of experience in designing and developing cutting-edge artificial intelligence solutions across diverse industries. Currently serving as Lead Software Development Engineer (AI/ML) at Ascendion, he previously held key machine learning roles at Amazon Web Services, where he architected scalable ML systems and led teams in developing innovative AI-powered solutions. Datta holds a Master of Science in Information Technology and Management from The University of Texas at Dallas and a Bachelor of Science in Computer Science & Engineering from the prestigious Indian Institute of Technology (IIT), Jodhpur. His research expertise spans generative AI, natural language processing, computer vision, and healthcare AI applications. As a published author and thought leader in AI, Datta has contributed significantly to the intersection of artificial intelligence and healthcare through his research work. He has extensive experience in building resilient, cloud-based infrastructure using technologies like Terraform, Ansible, and Kubernetes, with systems capable of handling over 500,000 requests per second while maintaining 99.99% uptime. His expertise in distributed computing frameworks including Apache Spark and Hadoop, combined with his work on real-time data processing pipelines, positions him uniquely to address the scalability challenges inherent in IoT healthcare systems. His technical expertise encompasses the full spectrum of modern AI technologies, from traditional machine learning frameworks like

TensorFlow and PyTorch to cutting-edge generative AI models. Datta's research focuses on developing sustainable, efficient systems that can process large-scale healthcare data while optimizing resource utilization—a critical consideration in the development of eco-conscious IoT medical devices.

Seema Gulati is a skilled academician and researcher specializing in Computer Science and Engineering. She earned her Ph.D. from Chitkara University, Punjab, and has a robust teaching career spanning over nine years. She is actively involved in research, focusing on deep learning, federated learning, and their applications in healthcare. She has contributed to several peer-reviewed journal articles and conference papers, with work published in high-impact journals like IEEE Access and SN Computer Science.

B. Mamatha, currently serving as an Assistant Professor in the Department of Computer Science and Engineering at Tirumala Engineering College, Narasaraopet, Andhra Pradesh, is a dedicated academician and researcher with a strong focus on Artificial Intelligence, Deep Learning, and Internet of Things (IoT). She holds an M.Tech in Computer Science and Engineering from JNTUA and has completed her Pre-Ph.D. in 2025 at Sri Padmavati Mahila Visvavidyalayam (SPMVV), Tirupati, where she is presently pursuing her Ph.D. Her professional journey includes academic positions and a stint as a Consultant at Capgemini Technology Solutions, Bangalore, enriching her profile with a blend of teaching and industry insights. Ms. Mamatha has contributed to several national and international conferences and journals, with notable work in intelligent healthcare systems, IoT-driven disease detection, and AI-powered optimization models. She received the Best Paper Award at the 2024 International Conference on Innovations in Advanced AI and Augmented Reality. Actively engaged in workshops, FDPs, and technical symposia, she continuously upgrades her knowledge and teaching methodologies. Her participation in this edited volume showcases her commitment to research excellence and her ability to translate complex computational ideas into practical solutions for emerging challenges in smart and secure systems.

Y.V. Narayana has obtained his B.E. in Electronics and communication Engineering from JNTU Ananthapuram, P.G and Ph.D. in Antennas from Andhra University, Visakhapatnam. He is currently working as a Principal in Tirumala engineering College, Narsaraopet, Guntur (Dt). He has 28 years of Experience and he has executed various responsibilities as an academician, researcher and best administrator. Apart from this he has 4 years of Industrial experience as a senior engineer in instrumentation at Rain Calcining Ltd. Visakhapatnam. Despite his hectic administrative responsibilities, he continued his everlasting interest on research in the form of publications. He is also head of Engineering Research Groups in various Research sectors and Industry Incubation Centre, Tirumala Engineering College, Narsaraopet, Guntur(Dt.). India. He has published several papers in National and International journals like Springer, Elsevier,AMSE France, Inderscience and in good impact factor Journals. He attended several number of workshops and Faculty development programs across the Country and he has made good contributions to the respective workshops as a speaker. He is an invited author for an International Book Chapters. He has received Centre of Excellence award for skill development centre from Dr.K. Satya Paul Sing,Minister of MHRD and received "Distinguished Professor Award" from the Honourable Governor, Andhra Pradesh. His areas of interest includes Antennas, Cloud Computing, Wireless sensors networks and Applied Electromagnetics. He is a fellow member of IETE and member of various professional bodies like CSI,IEEE,IE,ISTE.

Kondragunta Rama Krishnaiah (M.Sc., Ph.D. & M.Tech., Ph.D.) MIE., LMCSI., LMISTE & FIETE, is a highly qualified person, an efficient and eminent academician. He is an outstanding administrator; a prolific researcher published 95+ research papers in various National and International Journals and a forward-looking educationist. He has guided several students for their M.Tech and Ph.D Degrees and also authored 5 text books. He worked in prestigious K L (Deemed to be) University for 11.5 Years and has contributed his service for NBA accreditation in May 2004, Aug 2007 with 'record rating', ISO 9001:2000 in 2004, Autonomous status in 2006, NAAC accreditation of UGC in 2008 and University status in 2009. Later on he worked as Principal at Nova College of Engineering and Technology, Vijayawada for a period of 3.5 Yrs and NVR College of Engineering & Technology, Tenali for a period of 1.5 Yrs. He took charge as the Principal, RKCE, Vijayawada in December 2015. Under his able leadership, the college is rising towards a bright future with high academic values and the college got accredited by NAAC with "A" Grade and obtained UGC "AUTONOMOUS STATUS" from the Academic Year 2024 – 25. He has Received "Appreciation Award" for the Senior Principals from JNTUK, Kakinada in the Year 2024.

Kuppani Sathish received bachelor's degree and Master of Technology degree in Computer Science Engineering and Completed Ph. D in computer science and engineering from sri Venkateswara University, College of engineering, Tirupati. . He is working as an Professor in Tirumala Engineering college, Narasaraopet, which is affiliated to Jawaharlal Nehru Technological University, Kakinada. He is having 22 years of experience in teaching and research and taught bachelor and master degree courses. His area of interest includes Grid and cloud Computing, Bio inspired Networks, Software Architecture and Communication Systems. He made significant contributions in Networking in Rayalaseema region to provide internet access to common man. He attended many conferences and workshops and he has published 15 Sci papers and 10 Scopus papers to reputed National and International journals. He has granted many patents and published. He has received many awards from well-regarded organizations

Narinderjit Singh Sawaran Singh is an Associate Professor at INTI International University, Malaysia. He graduated from Universiti Teknologi PETRONAS (UTP) in 2016 with Ph.D in Electrical & Electronic Engineering specialized in Probabilistic Methods for Fault Tolerant Computing. Currently, he is appointed as the research cluster head for Computational Mathematics and Technology focuses on the areas like Expert Systems and Artificial Intelligence, Exploratory Data Analysis, Optimization Algorithms, Data Modelling and Computational Intelligence - Swarm Intelligence. He is one of the appointed Editorial Board Members for Journal of Data Science (JoDS) - an international journal indexed by Google Scholar and MyJurnal. He is a very dedicated and committed academician with 20+ years of experience in teaching and more than 10 years of research experience. He is a very resourceful researcher with wide-ranging research interests - Computational, Modeling & Simulation, Mathematical & Statistical Modeling, Fault-tolerant Computing (reliability - VLSI circuit design), Data Modelling & Computational Intelligence and Healthcare Data Analysis are on the top of the list.

Siva Rama Krishna T. is an Assistant Professor in the Department of Computer Science and Engineering at University College of Engineering Kakinada, JNTUK, Kakinada. He holds a Ph.D. in Computer Science and Engineering and several industry certifications in cyber security, cloud computing and blockchain domains. His core areas of expertise include Artificial Intelligence, Cybersecurity, Digital Forensics, and Blockchain Technologies. With over 15 years of academic experience, he has guided numerous undergraduate, postgraduate and doctoral theses, and actively contributed to curriculum development and faculty training initiatives. Dr. Siva Rama Krishna has authored and edited books, published extensively in reputed journals, and holds multiple patents in AI-driven security systems. He regularly delivers invited lectures and workshops on emerging technologies and serves on technical committees of national and international conferences. His ongoing work reflects a strong commitment to interdisciplinary research, ethical technology adoption, and academic leadership in advanced computing domains.

Himanshu Tyagi is currently an Assistant Professor in the Department of Computer Science & Engineering at Quantum University, Roorkee. With a strong academic and research background, Dr. Tyagi has been actively contributing to the field of Machine Learning,Computer Vision.

Index

A

B

C

D

E

G

H

I

L

M

R

S

T

W

www.ingramcontent.com/pod-product-compliance
Lightning Source LLC
LaVergne TN
LVHW062326300926
849161LV00007B/1592